Climate Change and Human Rights

Do anthropogenic greenhouse gas emissions affect human rights? Should fundamental rights shape climate policies? Scientific evidence demonstrates that anthropogenic greenhouse gas emissions contribute to increasing atmospheric temperatures, soon passing the compromising threshold of 2°C. Consequences such as Typhoon Haiyan prove that climate alteration has the potential to significantly impair critical needs. Although the United Nations Framework Convention on Climate Change and human rights regulatory regimes have so far proceeded separately, awareness is arising about their reciprocal implications. Based on tripartite fundamental obligations, this volume explores the relationship between climate change and interdependent human rights, through the lens of an international and comparative perspective. Along the lines of the metaphor of the 'wall', the research ultimately investigates the possibility of overcoming the divide between universal rights and climate change, and underlying barriers.

This book aims to be a useful resource not only for practitioners, policymakers, academics, and students in international, comparative, environmental law and politics and human rights, but also for the wider public.

Ottavio Quirico is a Senior Lecturer in the School of Law at the University of New England, Australia.

Mouloud Boumghar is Professor in the Faculty of Law at Université de Picardie, France.

Routledge Research in International Environmental Law

Available titles in this series:

International Environmental Law and the Conservation of Coral Reefs
Edward J Goodwin

Environmental Governance in Europe and Asia
A Comparative Study of Institutional and Legislative Frameworks
Jona Razzaque

Climate Change, Forests and REDD
Lessons for Institutional Design
Joyeeta Gupta, Nicolien van der Grijp and Onno Kuik

Common Pools of Genetic Resources
Equity and Innovation in International Biodiversity Law
Evanson Chege Kamau and Gerd Winter

International Environmental Law and Distributive Justice
The Equitable Distribution of CDM Projects under the Kyoto Protocol
Tomilola Akanle Eni-Ibukun

The Precautionary Principle in Marine Environmental Law
With Special Reference to High Risk Vessels
Benedicte Sage-Fuller

International Climate Change Law and Policy
Cultural Legitimacy in Adaptation and Mitigation
Thoko Kaime

Law and Practice on Public Participation in Environmental Matters
The Nigerian Example in Transnational Comparative Perspective
Uzuazo Etemire

Climate Change and Human Rights
An International and Comparative Law Perspective
Ottavio Quirico and Mouloud Boumghar

Forthcoming titles in this series:

International Liability Regime for Biodiversity Damage
The Nagoya-Kuala Lumpur Supplementary Protocol
Akiho Shibata

Climate Change and Forest Governance
Lessons from Indonesia
Simon Butt, Rosemary Lyster and Tim Stephens

Ecological Restoration in International Environmental Law
Afshin Akhtarkhavari, An Cliquet and Anastasia Telesetsky

Human Rights Approaches to Climate Change
Challenges and Opportunities
Sumudu Atapattu

Natural Resources Law, Investment and Sustainability
Shawkat Alam, Jahid Hossain Bhuiyan, Carmen Gonzalez and Jona Razzaque

Sustainable Development Principles in the Decisions of International Courts and Tribunals
1992–2012
Marie-Claire Cordonier Segger and Yolanda Saito

Climate Change and Human Rights

An international and comparative law perspective

Edited by
Ottavio Quirico and Mouloud Boumghar

Routledge
Taylor & Francis Group
LONDON AND NEW YORK

First published 2016
by Routledge
4 Park Square, Milton Park, Abingdon, Oxon OX14 4RN

and by Routledge
605 Third Avenue, New York, NY 10017

First issued in paperback 2017

Routledge is an imprint of the Taylor & Francis Group, an informa business

British Library Cataloguing in Publication Data
A catalogue record for this book is available from the British Library

Library of Congress Cataloging-in-Publication Data
A catalogue record for this book has been requested

ISBN 13: 978-0-8153-5534-2 (pbk)
ISBN 13: 978-1-138-78321-8 (hbk)

Typeset in Baskerville
by Florence Production Ltd, Stoodleigh, Devon, UK

‘Is there anybody out there?’

The Wall, Live in Berlin, 1990

Contents

Foreword

Climate change and human rights: this volume systemically sheds light on an increasingly vital subject, tackling critical issues, such as the centrality of the basic claim to (a sustainable) environment for resolving problems of causation and imputation. Whereas these questions have been mostly approached politically, the book interestingly discloses above all both a theoretical and practical legal perspective. The contributors thus comprehensively explore the reciprocal regulatory implications of human rights and global warming, whose scope and consequences have so far been largely underestimated.

From a realistic and cosmopolitan standpoint, actual damages and future threats posed by climate change to fundamental rights bolster the idea of 'risk society', first posited by Ulrich Beck.[1] Global warming has a universal impact, so much so that it affects Australia and Goudier Island, Alaska, Paris, Beijing, Rio de Janeiro or New Delhi alike: its effects concern human beings as such, not as nationals of different countries. Everyone is concerned: together with the right to environment, which is progressively developing in domestic and supranational case law and regulation, the claim of humanity to survival is eventually at stake. Individual and collective perspectives therefore merge with one another: threatened by climate change, intertwined particular and collective rights may allow overcoming national and cultural barriers. Along these lines, as postulated by Jürgen Habermas,[2] globalised human relationships at best facilitate the conception of a universal community and 'global citizenship'.[3] Based on the Kantian idea of a cosmopolitan society centred on individuals and peoples,[4] international and domestic law, or even 'global' law, can thus play a key role in shaping the relationship between climate change and human rights. Within this context, the book develops a robust practical analysis of current and prospective positive regulation, outlining its potential, loopholes and limits, against a consistent theoretical background, sharply inspired by tripartite fundamental obligations.

1 Ulrich Beck, *Risikogesellschaft, Auf dem Weg in eine andere Moderne* (Suhrkamp, 1986); Anthony Giddens, 'Risk and Responsibility' (1999) 62(1) *Modern Law Review* 1.

2 Jürgen Habermas, *Die postnationale Konstellation: Politische Essays* (Suhrkamp, 1998).

3 Louis Lourme, *Le nouvel âge de la citoyenneté mondiale* (PUF, 2014) 221.

4 Immanuel Kant, *Zum ewigen Frieden. Ein philosophischer Entwurf* (1795).

Owing to the rapid melting of the Arctic ice cap, the Inuit are currently in the frontline and face the most tangible and immediate effects of climate change. They are the guardians of the planet and remind us how crucial global warming is to human and peoples' rights, including the ultimate question of survival. Temperatures keep increasing and call for action, involving physicists as well as lawyers, in order to prompt governments to review their outdated perception of the passing of time. In light of this, the volume provides first and foremost a particularly rigorous scientific investigation of climate change and human rights governance, but can also be interpreted as a programme of action in favour of the rights of the earth and its inhabitants.

Pierre-Marie Dupuy
Geneva
8 June 2015

Introduction

Scientific evidence shows that anthropogenic GHG emissions have an impact on climate change. They contribute to rising atmospheric temperatures, and values higher than 2° Celsius will have serious consequences in terms of phenomena such as ice melting, sea-level rise, droughts, floods and hurricanes.[1] These can further cause, inter alia, loss of life, territories, biodiversity and health problems, and thus significantly affect human rights.[2] In addition, mitigation and adaptation policies may impinge upon fundamental claims; for instance, REDD programmes have the potential to affect the life of indigenous peoples.[3] However, so far climate change and human rights have basically been kept legally separate, inter alia, because the consequences of hybridisation are largely unexplored and might reshape established political balances.[4]

The question has nevertheless been raised in a set of relatively recent instruments. In 2005, the Inuit submitted to the IAComHR a petition denouncing human rights breaches caused by US GHG emissions.[5] A conference of small island States convened by the Maldives in 2007 adopted the Malé Declaration on the Human Dimension of Global Climate Change, meaningfully calling for the UNFCCC COP to cooperate with the OHCHR and the HRC in assessing relevant connections.[6] Subsequently, the HRC recognised the relevance of the issue via Resolutions 7/23 of 23 March 2008, 10/4 of 25 March 2009 and 18/22 of 30 September 2011. This prompted the OHCHR to elaborate an exploratory report in 2009.[7] Meanwhile, the Cancun Agreements acknowledged the importance of human

1 IPCC, *Climate Change: the Physical Science Basis* (2013).

2 IPCC, *Climate Change: Impacts, Adaptation and Vulnerability* (2014).

3 See Ole W Pedersen, 'The Janus-Head of Human Rights and Climate Change: Adaptation and Mitigation' (2011) 80 *Nordic Journal of International Law* 403.

4 See Paul Wapner, 'The Challenges of Planetary Bandwagoning' (2011) 11(3) *Global Environmental Politics* 137.

5 Inuit, *Petition to the Inter-American Commission on Human Rights, Seeking Relief from Violations Resulting from Global Warming Caused by Acts and Omissions of the United States* (7 December 2005).

6 *Malé Declaration on the Human Dimension of Global Climate Change* (14 November 2007).

7 OHCHR, *Report on the Relationship between Climate Change and Human Rights*, UN Doc A/HRC/10/61 (15 January 2009).

rights obligations for the States parties to the UNFCCC.[8] In 2013, the Arctic Athabaskan Council filed a second petition with the IAComHR, pointing out human rights breaches resulting from Arctic warming caused by Canada's black carbon emissions.[9] Latterly, proposals have been put forward to include fundamental rights in a future climate agreement, to be negotiated in Paris in December 2015.[10] Such initiatives are nonetheless still in their infancy and have not yet resulted in binding regulation.

This volume aims to contribute to exploring the relationship between climate change and human rights, by investigating positive legal issues from an international and comparative perspective. Systemically, the research is underpinned by tripartite human rights obligations, that is, the duties to respect, protect and fulfil first, second and third generation human rights. Against the background of the universality, indivisibility and interdependence of fundamental rights, the scope of the analysis includes both substantive and procedural obligations *de lege lata* and *ferenda*. Along these lines, the first part of the volume outlines a general framework addressing the main issues raised by the impact of climate change on human rights. The second part deals with the relationship between climate change and specific fundamental rights, while the third part explores their interaction within key regulatory regimes. Finally, the fourth part considers how international, regional and domestic institutions approach the reciprocal implications of climate change and human rights.

Part I focuses on climate change constraints imposed on States and private corporations by fundamental rights. Ottavio Quirico, Jürgen Bröhmer and Marcel Szabó consider State climate policies in light of the obligations to respect, protect and fulfil human rights. Particularly based on the limited number of States composing the international society and the no-harm rule, possibly interpreted as a fundamental claim, the authors explore prospective avenues to overcome issues of causation, imputation, policy discretion and extraterritoriality. According to Bridget Lewis, further balancing different human rights constraints leads to the conclusion that a case-by-case assessment is necessary to minimise the negative impact of State climate policies on fundamental rights. As to the responsibility of private corporations, Anna Riddell argues that, whereas it is difficult to find these innumerable entities in breach of human rights for GHG emissions, the State should act as a 'catalyst' for responsibility under the obligation to protect fundamental rights.

8 UNFCCC COP, Decision 1/CP.16, *The Cancun Agreements: Outcome of the Work of the Ad Hoc Working Group on Long-term Cooperative Action under the Convention*, UN Doc FCCC/CP/2010/7/Add.1 (15 March 2011).

9 Arctic Athabaskan Council on Behalf of All Arctic Athabaskan Peoples of the Arctic Regions of Canada and the United States, *Petition to the Inter-American Commission on Human Rights, Seeking Relief from Violations of the Rights of Arctic Athabaskan Peoples Resulting from Rapid Arctic Warming and Melting Caused by Emissions of Black Carbon by Canada* (23 April 2013).

10 UN Special Procedures Mandate Holders, *Statements on Climate Change and Human Rights* (10 December 2014 and 5 June 2015).

Contributions in Part II provide a critical analysis of the relationship between climate change and individual and collective human rights, moving from first to second and third generation claims. Christine Bakker shows that only in combination with collective fundamental claims, specifically the prospective right to a healthy environment, may the intrinsically individual right to life become a suitable tool for defining widespread climate policies. According to Alessandra Franca, the rights to food and water, underpinned by the right to health, particularly compel adaptation and should ideally foster a systemically integrated approach to climate policies by governmental and non-governmental organisations. Cameron Moore contends that the right to self-determination should prompt a progressively extensive interpretation of the law of the sea, so as to preserve existing claims over natural resources of peoples inhabiting disappearing low-lying island States, possibly beyond Statehood. In the views of Same Varayudej, despite its uncertain legal status, the two-pronged right to (sustainable) development, including an economic and environmental component, should at least play a crucial role in facilitating participation of developing countries in decision-making processes as to climate change. Francesco Francioni and Ottavio Quirico further assess the status of the human right to (a sustainable) environment and argue that, although it is not yet a customary rule of international law, this claim could be considered a general principle of law based on domestic constitutions. However, such a right can only be useful in shaping climate policies to the extent that it is clearly defined as to both its substantive and procedural elements, which is not the case for the time being.

Part III contextualises the relationship between climate change and human rights within specific regulatory frameworks. According to Federico Lenzerini and Erika Piergentili, climate policies are crucial to preserve basic claims grounded in biodiversity, particularly the rights of indigenous peoples and cultural rights. They must nonetheless be respectful of fundamental procedural claims and adequately balance different human rights obligations. Benoît Mayer and Christel Cournil question the notion of 'climate refugees' and raise the issue as to whether it is more appropriate to regulate this phenomenon based on broader categories, such as that of 'survival migrants', rather than providing specific-group protection. Valentina Vadi shows how fundamental rights can be positively or negatively affected by measures governing foreign investment, either fostering or repealing climate-friendly investment regulation. In the views of Olivier De Schutter, although current trade liberalisation is inconsistent with climate change mitigation and adaptation, human rights can provide guidance to coordinate these diverging trends, notably in favour of developing countries.

Contributions in Part IV take an institutional focus on the relationship between climate change and human rights. Spyridon Aktypis, Emmanuel Decaux and Bronwen Leroy show how the UNFCCC regime and UN human rights protection systems have so far proceeded separately, and argue that the recognition of a fundamental right to (a sustainable) environment is crucial to make them converge. Particularly in light of the problem of human displacement, Ben Boer demonstrates that the current regulatory framework in the Asia-Pacific, which is mainly based on first and second generation human rights, is insufficient to tackle climate change. He therefore advocates the establishment of a comprehensive human rights protection system, including third generation fundamental claims. Conversely, Faustin Ntoubandi and Roland Adjovi contend that African regulation encompasses advanced third generation tools, specifically the right to a satisfactory environment favourable to development, which are adequate to address climate change threats. Their practical implementation is nonetheless still insufficient and hampered by endemic continental problems. Although it is not theoretically as advanced as the African regime, according to Mouloud Boumghar the Inter-American human rights protection system includes several means to work out the threats posed by climate change to human rights. However, the IAComHR is not particularly reactive and is thus missing important opportunities to clarify these issues. Against the background of the Aarhus Convention, Ottavio Quirico assesses the relationship between climate change and human rights in the EU and Council of Europe systems. He argues that accession of the former to the latter might generate a paradoxical imbalance, thus prompting the recognition of a substantive human right to (a sustainable) environment as a precious tool to define climate policies. Finally, although the case law is still insufficient to make conclusive inferences, according to Tineke Lambooy and Hanneke Palm domestic jurisdictions provide room for the view that issues of causation, imputation and discretion can be overcome in establishing a link between climate change and human rights violations.

Overall, along the lines of the universal nature of human rights, this script explores in a positivist manner the possibility of tearing down old 'walls' inherited from the 20th century hampering the new Millennium, particularly the theoretical divide between climate change and human rights, the practical distinction between developed and developing countries, state frontiers preventing a global approach to climate change, intra-systemic barriers, such as separate human rights protection in Europe, the gap between regulation and its implementation, and ultimately the greenhouse effect itself.

Hello? Is there anybody in there?

Ottavio Quirico and
Mouloud Boumghar

Part I

General framework

1 States, climate change and tripartite human rights

The missing link

Ottavio Quirico, Jürgen Bröhmer and Marcel Szabó

Introduction

Relevant scientific evidence demonstrates that anthropogenic GHG emissions almost certainly have an impact on climate change (general causation). Currently, GHGs are estimated at 430 ppm CO_2e,[1] approximately doubling pre-industrial concentrations. In order to contain temperature increase within the sustainable threshold of no more than 2° Celsius, GHGs must be stabilised between 450 and 550 ppm, halving current emissions by 2050.[2] Higher atmospheric temperatures would have negative environmental consequences, increasing phenomena such as rising sea level, droughts, floods, land degradation, and cyclones.[3]

As recognised by the UN HRC in Resolutions 7/23 (2008) and 10/4 (2009), this chain of facts impacts on legal obligations (specific causation), including human rights, such as the rights to life, adequate food, housing, health, water and self-determination.[4]

Briefly, it is possible to represent the outlined causation pattern as follows:

> (1) anthropogenic GHG emissions → (2) rising atmospheric temperatures (climate change) → (3) further environmental changes (general causation) → (4) legal effects, including breaches of (human) rights of (present and future) generations (specific causation).[5]

1 Parts per million of CO_2 equivalent.

2 Stern Review, *The Economics of Climate Change, Summary of Conclusions* (2007) vi–vii.

3 IPCC, *Climate Change: Synthesis Report* (2014) 2 ff (2007) 26.

4 HRC, Resolution 10/4, UN Doc A/HRC/10/29 (25 March 2009); Resolution 7/23, UN Doc A/HRC/7/78 (29 March 2008); OHCHR, *Report on the Relationship between Climate Change and Human Rights*, UN Doc A/HRC/10/61 (15 January 2009) [16] ff and Annex. See also Timo Koivurova, Sébastien Duyck and Leena Heinämäki, 'Climate Change and Human Rights', in Erkki J Hollo, Kati Kulovesi and Michael Mehling (eds), *Climate Change and the Law* (Springer, 2013) 287, 292–304.

5 See Derek Bell, 'Climate Change and Human Rights' (2013) 4 *WIREs Climate Change* 159, 166 and 168.

Within the context of such a complex chain, the Cancun Agreements provide that the 'Parties [to the UNFCCC] should, in all climate change-related actions, fully respect human rights'.[6] Climate change-related action encompasses mitigation, that is 'anthropogenic intervention to reduce the sources of GHGs or enhance their sinks',[7] and adaptation, that is 'adjustment in natural or human systems in response to actual or expected climatic stimuli or their effects, which moderates harm or exploits beneficial opportunities'.[8] According to the OHCHR, such conduct is informed by the State duties to respect, protect and fulfil human rights.[9]

This chapter considers the impact of tripartite human rights obligations on anthropogenic GHG emissions. More specifically, the analysis addresses the question as to whether or not, and to what extent, the obligations to respect, protect and fulfil human rights compel States to take mitigation and adaptation measures against climate change. The issue is considered by taking a backward approach to the GHG–human rights chain of causation, exploring its objective and subjective elements. Furthermore, the chapter begins by tackling existing first and second generation human rights and is concluded by taking into account evolving third generation human rights, in particular, the right to (a sustainable) environment.

1. GHG emissions and tripartite human right obligations

Who has to do what to avoid cumulatively causing climate change and adapt to its effects? Around 30 per cent of current GHG emissions come from the electricity sector, 28 per cent from transportation, 20 per cent from other industrial sectors, 10 per cent from the agricultural industry, and 10 per cent from residential and commercial activities.[10] Within this context, the picture of State responsibility is not excessively complex. Out of the currently existing 200 States of the international community, almost a half of total GHG emissions are produced by the US and China, more than a half by these two States plus India and Russia, and more than two-thirds by all these countries plus European States.[11]

With regard to tripartite human rights obligations, since States' GHG emissions are the result of aggregate emissions by public and private corporations and natural

6 UNFCCC COP, Decision 1/CP.16, *The Cancun Agreements: Outcome of the Work of the Ad Hoc Working Group on Long-term Cooperative Action under the Convention*, UN Doc FCCC/CP/2010/7/Add.1 (15 March 2011) 4, [8] (emphasis added).

7 IPCC, *Climate Change Synthesis Report* (2001), Annex B, Glossary of Terms.

8 Ibid, 365.

9 OHCHR (2009) 27 ff, [80] ff, in particular, 28, [86]. On the distinction between the obligations to respect, protect and fulfil human rights, see Asbjorn Eide, *The Right to Adequate Food as a Human Right*, E/CN.4/Sub.2/1987/23 (1987); Olivier De Schutter, *International Human Rights Law* (CUP, 2014) 280.

10 US EPA, *Sources of GHG Emissions* (2014) <www.epa.gov/climatechange/ghgemissions/sources.html>. See also Hari M Osofsky, 'Is Climate Change "International"? Litigation's Diagonal Regulatory Role' (2009) 49 *Va J Int'l L* 585, 592 ff; Christina Voigt, 'State Responsibility for Climate Change Damage' (2008) 77 *Nordic J Int'l L* 9.

11 World Bank Group, *State and Trends of Carbon Pricing* (2014) 19.

persons,[12] States might be held overall liable under the obligation to respect human rights, which entails a duty not to take action interfering with fundamental claims and is therefore negative.[13] Part of the Inuit petition to the IAComHR is based on this approach. In this case, the Inuit filed their complaint with the Commission against the US for emitting unsustainable GHGs, causing the melting of the Arctic ice-cap and thus breaching several human rights.[14] Grounded in the ADRDM, ACHR, ICCPR, ICESCR, UNFCCC and the Kyoto Protocol, the petition claims violations of fundamental rights such as the rights to property, health, life, physical integrity, residence, movement and inviolability of home, inter alia, because of positive 'action', that is, aggregate GHG emissions, by the US.[15] More analytically, when GHGs are emitted by public agents and corporations, State responsibility might be established for positive action, in breach of the negative obligation to respect human rights.[16] This responsibility may be fostered by conduct aimed at increasing GHG emissions, such as 'denying, distorting and suppressing scientific evidence' on climate change, which the Inuit petition imputes to the US government.[17]

With respect to non-State natural and legal persons, in particular private energy corporations, a more correct human rights framework to approach State mitigation measures is provided by the obligation to protect human rights. This obligation implies for States a duty to prevent and punish human rights breaches caused by third parties, and is thus positive.[18] There is therefore room for the view that States are compelled to implement regulatory measures that exclude or limit GHGs emitted by third persons, in order to achieve sustainable emissions. This is the stance taken by the Arctic Athabaskan Council in a petition filed with the IAComHR against Canada for insufficient regulation of black carbon emissions.[19]

12 See, for instance, the determination of GHG reduction targets under Annex A to the Kyoto Protocol.

13 De Schutter (2014) 280.

14 Inuit, *Petition to the Inter-American Commission on Human Rights, Seeking Relief from Violations Resulting from Global Warming Caused by Acts and Omissions of the United States* (7 December 2005) 103–4, <www.ciel.org/Publications/ICC_Petition_7Dec05.pdf>.

15 *Inuit Petition* (2005) 103:

> By its acts . . . the United States violates the human rights of the Inuit . . . the United States is, by any measure, the world's largest contributor to global warming and its damaging effect on the Inuit. As the world's largest consumer of energy, both historically and at present, it emits the most fossil fuels and is responsible for the largest amount of cumulative emissions of any nation on Earth. It follows that the United States has contributed more than any other nation to the rise in global temperature.

16 John Knox, 'Climate Change and Human Rights Law' (2009) 50 *Va J Int'l L* 163, 171.

17 *Inuit Petition* (2005) 109.

18 De Schutter (2014) 427. A merely negative lecture of human rights affected by climate change is thus insufficient to understand their reciprocal relationship (see Bell (2013) 165).

19 Arctic Athabaskan Council on Behalf of All Arctic Athabaskan Peoples of the Arctic Regions of Canada and the United States, *Petition to the Inter-American Commission on Human Rights Seeking Relief from Violations of the Rights of Arctic Athabaskan Peoples Resulting from Rapid Arctic Warming and Melting Caused by Emissions of Black Carbon by Canada* (23 April 2013) 49 ff, <http://earthjustice.org/sites/default/files/AAC_PETITION_13-04-23a.pdf>: 'Canada's failure to sufficiently regulate black carbon emissions is violating Arctic Athabaskan peoples' human rights'.

In turn, third entities, such as energy corporations, might be held responsible under the obligation to respect human rights.[20]

The obligation to fulfil human rights, which is positive and compels States to facilitate, provide and promote the enjoyment of fundamental claims,[21] is also relevant to mitigation measures, to the extent that the latter can be interpreted as a means of facilitating satisfactory fundamental rights. However, this duty is particularly important for adaptation measures.[22] In fact, if rising average temperatures affect the enjoyment of specific human rights, for instance, the right to water owing to desertification, it is sensible to assume that the State has an obligation to provide water and thus facilitate the enjoyment of that specific fundamental right.

Finally, States may be held responsible for actively breaching human rights when they take positive action aiming to mitigate GHG emissions and deploy adaptation measures.[23] Not taking action should initially generate State responsibility under the obligations to respect, protect and fulfil human rights, but the negative obligation to respect human rights and the positive obligation to fulfil them are also particularly relevant when States take action.[24]

2. Respecting human rights

Along the lines of the *Trail Smelter* arbitration, transboundary pollution can give rise to State responsibility,[25] and that is also true of anthropogenic GHG emissions.[26]

Non-State natural and legal persons, in particular energy corporations, may be held responsible for exceeding GHG emissions limits. Although it is problematic internationally, because of the limited personality of non-State entities, this is domestically possible when emission caps are established under administrative law.[27]

20 See the contribution by Anna Riddell in this volume. See also Siobhán McInerney-Lankford, Mac Darrow and Lavanya Rajamani, *Human Rights and Climate Change: a Review of the International Legal Dimensions* (World Bank, 2011) 61.

21 De Schutter (2014) 527.

22 On the notions of proactive and reactive adaptation, see Margaux J Hall and David C Weiss, 'Avoiding Adaptation Apartheid: Climate Change Adaptation and Human Rights Law' (2012) 39 *Yale J Int'l L* 309, 364; Alyssa Johl and Sébastien Duyck, 'Promoting Human Rights in the Future Climate Change Policy' (2012) 15 *Ethics Policy and Environment* 298, 299; Knox (2009) 197.

23 See Naomi Roht-Arriaza, '"First, Do No Harm": Human Rights and Efforts to Combat Climate Change' (2010) 38 *Georgia J Int'l & Comp L* 593 ff; Dinah Shelton, *Human Rights and Climate Change* (Northwestern University, 2009) 19.

24 UNFCCC COP (2011) [8]. See also the contribution by Bridget Lewis in this volume.

25 *Trail Smelter (USA v Canada)*, Award of 16 April 1938 and 11 March 1941 (2006) III *RIAA* 1905.

26 Voigt (2008) 15; Daniel A Farber, 'Apportioning Climate Change Costs' (2008) 26 *UCLA J Envtl L and Pol'y* 21, 39–40.

27 See, for instance, *Directive 2003/87/EC of the European Parliament and of the Council Establishing a Scheme for Greenhouse Gas Emission Allowance Trading within the Community and Amending Council Directive 96/61/EC* (13 October 2013).

As for States, in addition to the hypothesis of aggregate emissions, responsibility might be engaged both domestically and internationally for emissions by public agencies breaching the obligation to respect human rights, provided a causal link can be demonstrated between unsustainable GHG emissions and specific first and second generation human rights.[28] Such a logical pattern underpins the request of the Federal State of Micronesia to the Czech Minister for Environment to conduct an environmental impact assessment (EIA) of the project of modernisation of Prunerov II power plant under Article 1(b)(11) of Czech EIA Law 100/2001. Prunerov II is operated by CEZ, a State-owned monopoly. It is the largest source of CO_2 in the Czech Republic, and its modernisation would lead to the emission of 0.021 per cent of global CO_2 in the next 25 years.[29]

Similarly, with regard to first and second generation human rights and environmental protection, in *Fadeyeva v Russia* the ECtHR held Russia responsible for failing to regulate a large iron smelter located in Cherepovets. The plant was run by the Ministry of Black Metallurgy of the Russian Soviet Federative Socialist Republic and released high levels of climate pollutants beyond the nationally set threshold. More specifically, according to the Court, Russian authorities failed to design and apply effective measures to reduce the industrial pollution to acceptable levels.[30] Along these lines, in *Öneryıldız v Turkey* the ECtHR adjudicated upon the case of an applicant's dwelling built without authorisation on land surrounding a municipal rubbish tip under the authority of a local mayor answerable to the District Council.[31] In April 1993, a methane explosion occurred at the tip and the refuse erupting from the pile of waste engulfed more than ten surrounding houses, including that of the applicant, who lost nine relatives. The Court found that the Turkish government violated Article 2 ECHR, because it had not provided the slum inhabitants with information about the risks of living in a dangerous area; and even if it had done so, the government was to be considered responsible for not taking measures necessary to prevent the explosion, in spite of an alarming expert report. The State regulatory framework had proved defective, since the tip had been allowed to open and operate without a coherent supervisory system. Also the town-planning policy was considered inadequate and part of the sequence of events leading to the accident.[32] In our view, in these cases the public nature of the legal persons operating the defective plants triggers State responsibility under the obligation to respect, rather than protect, human rights.

Conversely, as to emissions by private persons, the international and domestic responsibility of the State should be primarily engaged under the obligation to protect human rights, which compels prevention and punishment of human rights

28 OHCHR (2009) [18].

29 Greenpeace, *Legal Steps Taken by the Federated States of Micronesia against the Prunéřov II Coal-fired Power Plant, Czech Republic* (Background Information, 2010).

30 *Fadeyeva v Russia*, Appl No 55723/00 (ECtHR, 9 June 2005).

31 *Öneryıldız v Turkey*, Appl No 48939/99 (ECtHR, 30 November 2004) especially [97] ff.

32 Ibid, Findings.

violations committed by private entities.[33] However, at least under international law, it is also possible to think of directly attributing the conduct of non-State agents to the State, thus triggering the responsibility of the latter for breaching the obligation to respect human rights. The attribution process should take place under Article 8 of the ILC's 2001 Draft Articles on State Responsibility, which provides that the conduct of a 'person or a group of persons' can be imputed to the State 'if the person or group of persons is in fact acting on the instructions of, or under the direction or control of, that State in carrying out the conduct'.[34] This mechanism raises the question as to whether GHGs emitted by private legal persons can be attributed to the State. A positive answer would permit the consideration of direct State responsibility for breaching the obligation to respect human rights. Based on the case law of the ICJ and the ICTY, Article 8 of the ILC's Draft Articles on State Responsibility can be interpreted either narrowly or extensively. Initially, in *Military and Paramilitary Activities in and against Nicaragua* the ICJ applied the restrictive 'effective control' test.[35] The threshold established for attribution of responsibility of private legal persons to the State was therefore based on 'instructions for the commission of specific acts'. Subsequently, in *Tadić* the ICTY adopted the broader 'overall control' test, based on the notion of 'general planning'.[36] However, this interpretation has been latterly rejected by the ICJ in the *Bosnian Genocide* case. In this dispute, adjudicating upon the same facts as the ICTY in *Tadić*, the ICJ applied the very narrow criterion of 'complete dependence', based on a factual evaluation of the position of the agent as a 'mere instrument' of the State.[37] The ICJ thus explicitly rejected the ICTY extensive interpretation and narrowed the initial view provided in *Nicaragua*.

With specific regard to climate change, it is possible that a State takes policy measures entailing the emission of GHGs beyond sustainable limits, which is actually claimed in the petitions filed with the IAComHR by the Inuit and Athabaskan people respectively against the US and Canada.[38] A State might, for instance, not commit to significant reduction targets, or more simply decide to provide financial support to the oil extraction and car production industry, thus also fostering oil consumption and increasing the release of GHGs into the atmosphere. In this case, it can certainly be maintained that the State has 'overall' control over oil extraction and consumption, but it is difficult to assume that the State has 'effective control' or 'complete control' over each act of oil extraction and consumption. Furthermore, in order to establish whether or not oil production

33 See De Schutter (2014) 527.

34 *ILC's Draft Articles on the Responsibility of States for Internationally Wrongful Acts*, UN Doc A/56/10 (2001) 53 UN GAOR Suppl (No 10) 43.

35 *Military and Paramilitary Activities in and against Nicaragua (Nicaragua v United States)* [1986] ICJ Rep 1, 64–5, [115].

36 *Tadić* (ICTY, IT-94-1-A, 15 July 1999) 1, 49, [120].

37 *Application of the Convention on the Prevention and Punishment of the Crime of Genocide (Bosnia and Herzegovina v Serbia and Montenegro)* [2007] ICJ Rep 43, 205, [392].

38 *Inuit Petition* (2005) 105; *Arctic Athabaskan Council Petition* (2013) 49 ff.

and consumption are in breach of sustainable GHG emission limits, it is necessary to balance measures financing oil production and consumption and other policies, for instance, State measures aiming at financing renewable energy. In light of this, it can be concluded that the State has 'overall control' over GHGs emitted by private corporations, and even by natural persons, but it is difficult to demonstrate 'effective' or 'complete' control. Given the privileged position of the ICJ as to the interpretation of international law, a narrow approach to Article 8 of the ILC's Draft Articles on State Responsibility is to be preferred. The international responsibility of the State for breaching the obligation to respect human rights based on GHGs emitted by private corporations should therefore be attentively assessed on a case-by-case basis, in light of the degree of control exercised by the State on private persons owing to public policy measures, and preferably excluded. The State should thus rather be held co-responsible with private legal persons for GHG emissions under the obligation to respect human rights. Otherwise, the State could be considered liable under the duty to protect human rights. In this respect, a relevant case in the human rights and environment field is *Ogoniland*. In this dispute, the AComHPR held that oil extraction in the Ogoniland region by the State of Nigeria and partner private corporations, such as Shell Oil, had an impact on the (right to a satisfactory) environment. In its findings, the Commission focused on a breach by Nigeria of the obligation to protect several human rights, including the rights to life, health and food.[39] In fact, the Commission ultimately held the Federal Republic of Nigeria 'in violation of Articles 2, 4, 14, 16, 18(1), 21 and 24 of the African [Banjul] Charter' and recommended that it 'ensure that appropriate environmental and social impact assessments are prepared for any future oil development and that the safe operation of any further oil development is guaranteed through effective and independent oversight bodies for the petroleum industry'.[40]

3. Protecting human rights

Under the obligation to protect human rights, States have an international and domestic duty to intervene in order to prevent acts of third parties threatening human rights and punish actual violations.[41] This obligation might thus be interpreted as compelling States to adopt pre-emptive measures in order to ensure that violations of human rights by third natural and legal persons' GHG emissions do not materialise and, if these violations materialise, punish them.[42] Such a mechanism is particularly relevant to GHGs emitted by private corporations, specifically, energy-providing companies, which are some of the most significant GHG emitters.

39 *Social and Economic Rights Action Center and the Center for Economic and Social Rights v Nigeria*, Comm No 155/96 (27 October 2001) [44]–[47], [56]–[57], [64]–[70].

40 Ibid, Findings. See also Ntoubandi and Adjovi in this volume.

41 De Schutter (2014) 427.

42 Shelton (2009) 41.

Such logic underpins the Athabaskan petition to the IAComHR, claiming that Canada is responsible for poorly regulating black carbon emissions that accelerate Arctic global warming and thus breach indigenous peoples' rights.[43] This approach also underpins the part of the Inuit petition to the IAComHR that puts a spotlight on the failure of the US to act against climate change, according to the obligation to protect human rights, rather than on the obligation to respect them. Namely, the petition targets the 2002 US climate policy, specifically the Global Climate Change Initiative, which allegedly does not set out binding reduction targets for business corporations. In particular, with regard to 'indirect regulation' the petition argues that the US has not set out sufficient reduction standards for GHG emissions from vehicles and power plants. Allegedly, some US federated States have established low or non-existent GHG reduction targets, thus providing safe havens for private corporations.[44]

Along these lines, in the domestic case of *Massachusetts v Environmental Protection Agency (EPA)*, where the Inuit also introduced a brief on the Arctic impact of climate change,[45] the American Supreme Court reversed a decision of the Appeals Court for the District Court of Columbia and held the EPA liable for failing to adequately regulate GHG emissions from cars as 'air pollutants' under the Environmental Protection Act 1994, thus contributing to global warming and affecting sea level at the coast of Massachusetts.[46] Although the EPA was aware of the potential danger resulting from GHG emissions and had the authority to restrain them, it failed to take appropriate measures, by way of which at least part of the damage could have been avoided. The Court thus requested the EPA to review its contention that it has discretion in regulating GHG emissions, articulating reasonable grounds to avoid regulating the matter.[47]

With regard to human rights case law on environmental protection and health, in *Lopez Ostra v Spain* the ECtHR held that the treatment of liquid and solid waste by a tannery belonging to the limited company SACURSA built close to the applicant's home was in breach of Article 8 ECHR. The Court determined that prolonged noise and emissions from the tannery required the relevant Spanish authorities to take appropriate steps to protect the right to private and family life

43 *Arctic Athabaskan Council Petition* (2013) 49 ff, particularly 52: 'By failing to regulate black carbon emissions sufficiently, Canada has violated its international responsibility to prevent activities within its jurisdiction from damaging the environment outside its borders.'

44 *Inuit Petition* (2005) 107–8. See also Stephanie Stern, 'State Action as Political Voice in Climate Change Policy: a Case Study of the Minnesota Environmental Cost Valuation Regulation', in William GC Burns and Hari M Osofsky (eds), *Adjudicating Climate Change: State, National, and International Approaches* (CUP, 2009) 31 ff.

45 Brief of Amici Curiae Alaska Inter-Tribal Council *et al* in Support of Petitioners, *Massachusetts v EPA*, 127 SCt 1438, Case No 05–1120.

46 *Massachusetts v EPA*, Case No 05–1120 (US Supreme Court, 2 April 2007).

47 This logic is particularly relevant to GHG reduction mechanisms provided for in the UNFCCC regime, such as Clean Development Mechanisms (CDMs – Article 12 of the Kyoto Protocol), which require the State to promote the employment and transfer of clean technology, often involving interaction between the State and private industries.

of the applicants, for example by relocating or closing down the plant.[48] In *Guerra v Italy*, the ECtHR adjudicated upon the case of applicants living in Manfredonia, Italy, around a kilometre away from a chemical factory producing fertilisers, owned by Enichem. Following accidents due to the plant's malfunctioning, on one occasion 150 people were admitted to hospital with acute poisoning because of the release of tonnes of substances containing toxic arsenic in the environment. The applicants complained that the lack of State measures to reduce pollution levels and major-accident hazards was in breach of the right to respect for their lives and physical integrity. The Court finally held that Italy violated its obligation to secure the applicants' right to respect for private and family life, in breach of Article 8 ECHR.[49] Similarly, in *Community of San Mateo de Huanchor and Its Members v Peru* the IAComHR adjudicated upon the alleged responsibility of the Republic of Peru for failure to prevent the effects of the environmental pollution emitted by a field of toxic waste sludge belonging to Lizandro Proaño SA mining company. The Commission held the petition admissible based on a possible violation of several human rights under the ACHR.[50]

On a final note, the State obligation to protect human rights from breaches by third parties may also operate internationally on a merely inter-State basis. Indeed, States may be held responsible for failing to cooperate in international initiatives aiming to reduce GHGs, especially within the framework of the UNFCCC. As pointed out in the Inuit petition to the IAComHR, this can be interpreted as a means for hindering GHG-mitigating action by third States, and thus as a breach of the positive duty to act and prevent breaches of fundamental human rights obligations by third States.[51] The same stand cannot, instead, be easily assumed with regard to non-State third parties operating abroad, because of the uncertain extraterritorial application of human rights duties.[52]

As to the obligation to punish third persons possibly breaching human rights obligations by means of GHG emissions, it is noteworthy that in *Ogoniland* the AComHPR finally invited the federal State of Nigeria to 'conduct an investigation into the human rights violations' committed, inter alia, by oil extraction corporations.[53]

4. Fulfilling human rights

The obligation to fulfil human rights requires States to take appropriate measures in order to fully realise fundamental claims. In other words, States are requested

48 *Lopez Ostra v Spain*, Appl No 16798/90 (ECtHR, 9 December 1994) [44] ff.

49 *Guerra and Others v Italy*, Appl No 116/1996/735/932 (ECtHR, 19 February 1998) [35], [60].

50 *Community of San Mateo de Huanchor and Its Members v Peru*, Admissibility, Petition 504/03 (IAComHR, 15 October 2004) [66].

51 *Inuit Petition* (2005) 110.

52 See section 6.2 below.

53 *Social and Economic Rights Action Center* (2001) Findings.

to adopt policies that make it possible for individuals to enjoy human rights. This obligation can be further disaggregated into the duties to facilitate, promote and provide human rights.[54]

The obligation to facilitate human rights compels the taking of positive measures to assist individuals in enjoying fundamental rights. The duty to provide human rights compels States to ensure the material enjoyment of fundamental rights whenever an individual or a group is unable to enjoy them for reasons beyond their control. Finally, the obligation to promote human rights requires States to ensure appropriate awareness about fundamental rights.[55]

Unlike the obligation to respect human rights, which entails State inaction, and the obligation to protect human rights, which entails positive State action to ensure third parties' inaction, the obligation to fulfil human rights basically entails positive State conduct to ensure that human rights can be enjoyed. This matches mitigation measures, but is also particularly relevant to adaptation policies.

4.1. Complementary mitigation

GHG mitigation measures can be envisaged not only from the perspective of the obligation to respect human rights, by avoiding emissions, and from that of the duty to protect human rights from third parties' emissions, but also from the standpoint of the obligation to fulfil human rights. If the obligations to respect and protect human rights are directly and indirectly relevant to negative conduct, that is, non-emission of GHGs, the obligation to fulfil human rights is important to the extent that it can potentially compel States to take measures, that is, positive conduct, mitigating GHG emissions, thus facilitating, providing and promoting fundamental rights. This duty can therefore be seen as complementary to the obligations to respect and protect human rights as to GHG emissions. In this regard, it has been held that, while the responsibility to respect and protect human rights constrains policy-making, the responsibility to fulfil human rights guides and supports policy-making.[56]

In practice, the Inuit petition to the IAComHR also encompasses this stance. Indeed, the petition alleges that the US should foster the fundamental rights of indigenous people in the Arctic by safeguarding their land and environment, including the adoption of special measures to give proper effect to such rights.[57] The petition mentions, in particular, the necessity that States 'take positive measures' to protect the integrity of minority rights under Article 27 ICESCR and to provide 'special safeguard' for the rights of tribal peoples in independent

54 De Schutter (2014) 427–31.

55 CESCR, General Comment No 15, *The Right to Water (Arts 11 and 12 of the Covenant)*, E/C.12/2002/11 (20 January 2003) [25]–[29].

56 Sébastien Jodoin and Katherine Loft, *Economic, Social and Cultural Rights: a Reference Guide* (CSD, 2013) 6.

57 *Inuit Petition* (2005) 70–1.

countries under ILO Convention 169. Specific positive action is thus invoked based on the need to 'preserve' the cultural identity of the Inuit, which is grounded in the ice-covered Arctic environment. This obligation is regarded as part of the more general negative State duty 'not to degrade the environment to an extent that threatens indigenous peoples' culture, health, life, property, or ecological security'.[58]

Along the same lines, particular aspects of the general principles regulating environmental protection and, specifically, climate change, can be interpreted in light of the obligation to fulfil human rights. For instance, the State obligation to cooperate in reducing GHG emissions can be regarded as a tool by means of which States ultimately fulfil human rights affected by climate change. On a broader scale, the obligation to fulfil fundamental rights is relevant to mitigation measures within the context of the evolving human right to development, which is a third generation fundamental claim naturally entangled with several other human rights, such as food, water and health. Given that the low level of development in developing countries hampers their capacity to fight climate change and thus to fulfil human rights, the obligation to fulfil the right to development might be read as imposing on developed countries a duty to assist developing countries by means such as technology transfer and financial assistance.

4.2. Spotlight on adaptation

Adaptation measures aim at reacting to the impacts of climate change by implementing national strategies, by means such as flood control and raising the levels of dykes, developing drought-tolerant crops, using scarce water resources more efficiently, choosing tree species and forestry practices less vulnerable to storms and fires, adapting building codes to future climate conditions and extreme weather events, and creating land corridors to help the migration of animals. This implies positive national, regional and international measures.[59] Systemically, such positive initiatives of States can be read in light of the obligation to fulfil human rights. In fact, the question arises as to whether GHG adaptation measures are necessary and adequate to make human rights actually enjoyable. For instance, desertification caused by climate change impinges upon the enjoyment of fundamental claims such as the rights to food, water and health, in which the obligation to fulfil human rights is originally rooted.

Under Article 11(1) ICESCR States have to take measures necessary to ensure adequate living conditions, including sufficient food. Along these lines, Article 11(2) ICESCR provides:

> The States Parties to the present Covenant, recognising the fundamental right of everyone to be free from hunger, shall take, individually and through

58 Ibid, 72; See also *Arctic Athabaskan Council Petition* (2013) 57.

59 See, for instance, the *EU Strategy on Adaptation to Climate Change*, COM (2013) 0216.

international co-operation, the measures, including specific programmes, which are needed:

(a) To improve methods of production, conservation and distribution of food by making full use of technical and scientific knowledge, by disseminating knowledge of the principles of nutrition and by developing or reforming agrarian systems in such a way as to achieve the most efficient development and utilisation of natural resources;

(b) Taking into account the problems of both food-importing and food-exporting countries, to ensure an equitable distribution of world food supplies in relation to need.

Interpreting this rule, the CESCR has held that 'the obligation to fulfil (facilitate) means the State must pro-actively engage in activities intended to progressively realise people's access to and utilisation of resources and means to ensure their livelihood, including food security'.[60] Furthermore, whenever 'an individual or group is unable, for reasons beyond their control, to enjoy the right to adequate food by the means at their disposal, States have the obligation to fulfil (provide) that right directly'.[61] Moreover, Article 11(2)(a) ICESCR explicitly embeds an obligation to promote the right to food, since States are requested to 'disseminate knowledge of the principles of nutrition'. The CESCR also relied upon Article 11 to assume that 'the obligation to facilitate requires the State to take positive measures to assist individuals and communities to enjoy the right [to water].'[62]

Similar considerations have been developed with regard to Article 12 ICESCR, which interlinks health and environment and provides:

1. The States Parties to the present Covenant recognise the right of everyone to the enjoyment of the highest attainable standard of physical and mental health.
2. The steps to be taken by the States Parties to the present Covenant to achieve the full realisation of this right shall include those necessary for:
   (a) The provision for the reduction of the stillbirth-rate and of infant mortality and for the healthy development of the child;
   (b) The improvement of all aspects of environmental and industrial hygiene;

60 CESCR, General Comment No 12, *The Right to Adequate Food (Art 11)*, E/C.12/1999/5 (12 May 1999) [15].

61 Ibid.

62 CESCR (2003) [25]. The Committee maintains that 'States parties are obliged to fulfil (provide) the right [to water] when individuals or a group are unable, for reasons beyond their control, to realise that right themselves by the means at their disposal'. Furthermore, according to the Committee the 'obligation to promote [human rights] obliges the State party to take steps to ensure that there is appropriate education concerning the hygienic use of water, protection of water sources and methods to minimize water wastage' (ibid).

(c) The prevention, treatment and control of epidemic, endemic, occupational and other diseases;
(d) The creation of conditions which would assure to all medical service and medical attention in the event of sickness.

According to the CESCR, the duty to fulfil the right to health 'contains obligations to facilitate, provide and promote' this claim and 'requires States to adopt appropriate legislative, administrative, budgetary, judicial, promotional and other measures towards [its] full realisation'.[63]

Based on these premises, since it is demonstrated that climate change has a negative impact on food, water, and health conditions,[64] it follows that States have an obligation to take positive adaptation measures to fulfil human rights, such as food, water and health, as a response to climate change-induced environmental effects. For instance, in light of the obligation to fulfil the human rights to food, water and health, relocation of people or the construction of aqueducts may be deemed necessary in progressively desertified areas.[65]

This logic is highlighted in the request for relief in the Inuit petition, which requires the IAComHR to 'establish and implement, in coordination with Petitioner and the affected Inuit, a plan to protect Inuit culture and resources, including, inter alia, the land, water, snow, ice, and plant and animal species . . . and mitigate any harm to these resources caused by US greenhouse gas emissions'.[66] Even more explicitly, the petition requests the Commission to recommend that the US government 'establish and implement, in coordination with Petitioner and the affected Inuit communities, a plan to provide assistance necessary for Inuit to adapt to the impacts of climate change that cannot be avoided'.[67] Similar, although less explicit, language is adopted in the Athabaskan petition, which requests the Canadian government to 'establish and implement, in coordination with Petitioners and affected Arctic Athabaskan peoples, a plan to protect Athabaskan culture and resources from the effects of accelerated Arctic warming and melting'.[68]

The obligation to fulfil human rights might also assume particular importance as to adaptation within the context of the right to development. Indeed, Article 4(4) and (9) UNFCCC compels industrialised States to make assistance available to developing countries for addressing their adaptation needs, especially through technology transfer.[69] Within this context, initiatives such as the Green Climate

63 CESCR, General Comment No 14, *The Right to the Highest Attainable Standard of Health (Art 12)*, E/C.12/2000/4 (11 August 2000) [33].
64 OHCHR (2009) [25]–[27], [28]–[30], and [31]–[34].
65 See the contribution by Alessandra Franca in this volume.
66 *Inuit Petition* (2005) 118.
67 Ibid.
68 *Arctic Athabaskan Council Petition* (2013) 87, Request for Relief, 3(c).
69 OHCHR (2009) 28, [87].

Fund, the proposal for an international air passengers adaptation levy,[70] and adaptation funding under the International Maritime Emission Reduction Scheme (IMERS),[71] can be regarded as tools by means of which developed countries comply with their obligation to fulfil the right to development.

5. Redressing human rights breaches

The obligations to protect, respect and fulfil human rights encompass both primary and secondary rules,[72] and thus entail for States a duty to redress human rights breaches.[73] In other words, if a State does not take measures to prevent human rights breaches caused by GHG emissions or measures already taken fail, the State might also be considered bound by an obligation to provide effective remedies, specifically via restitution and compensation. Restitution can encompass both mitigation and adaptation measures, and thus the obligations to respect, protect and fulfil human rights might be interpreted as a means to prompt the State to act with respect to all the range of pre-emptive and reactive climate measures.

With regard to environmental protection and first and second generation human rights,[74] the operation of the duty to provide a remedy has been illustrated by the AComHPR with respect to the obligation to protect human rights in *Ogoniland*. Indeed, in this case the AComHPR finally recommended that the State of Nigeria 'ensure adequate compensation to victims of the human rights violations' committed, inter alia, by private corporations in the oil extraction sector. The Commission included within the concept of compensation the notions of 'relief and resettlement assistance to victims' and 'comprehensive cleanup of lands and rivers damaged by oil operations'.[75]

Specifically as to GHG emissions, the Inuit petition to the IAComHR concludes by claiming that the US must remedy its breaches of human rights obligations. Indeed, the petition states that '[by] failing to act to reduce greenhouse gas emissions, the United States has allowed domestic emitters to impose environmental costs', inter alia, on the Inuit population.[76] As a consequence, the petition claims that the US is obliged to 'take responsibility for its contributions to global climate change both by limiting emissions and by paying reparations to those that it has harmed and continues to harm . . . to provide appropriate remedy and

70 *A Proposal by the Maldives on behalf of the Group of Least Developed Countries (LDCs) within the Framework of the Bali Action Plan* (12 December 2008).

71 IMERS, <www.imers.org>.

72 Based on the Kelsenian categorisation ('If A, B ought to be': Hans Kelsen, *Pure Theory of Law* (Peter Smith, 1989) 77).

73 CESCR (1999) [32]: 'All victims of such violations [of the right to food] are entitled to adequate reparation, which may take the form of restitution, compensation, satisfaction or guarantees of non-repetition'.

74 De Schutter (2014) 428.

75 *Social and Economic Rights Action Center* (2001) Findings.

76 *Inuit Petition* (2005) 101.

redress to the Inuit.'[77] Besides GHG mitigation measures, the Inuit petition asks the US to implement 'a plan to protect Inuit culture and resources' and 'a plan to provide assistance necessary for Inuit to adapt to the impacts of climate change that cannot be avoided.'[78] Similarly, the Athabaskan petition claims that Canada 'has a duty to provide appropriate remedy and redress', including 'environmental protection measures' in favour of the Arctic Athabaskan peoples.'[79]

6. Obstacles to implementation

Although theoretically tripartite human rights obligations allow the assumption that States must reduce GHG emissions and adapt to them, some caveats must be taken into account. These concern different subjective and objective aspects of the relationship between States, GHG emissions and human rights breaches, in particular: (1) the margin of discretion enjoyed by the State in exercising power in light of its triadic separation; (2) the spatial scope of application of human rights obligations; (3) the causal link; and (4) imputation of responsibility.

6.1. State discretion as to mitigation and adaptation policies

States enjoy a margin of discretion in implementing their policies, and thus also GHG mitigation and adaptation measures. With specific reference to positive obligations, this has been pointed out by the the ECtHR in *Hämäläinen v Finland*:

> In implementing their positive obligation under Article 8 the States enjoy a certain margin of appreciation. A number of factors must be taken into account when determining the breadth of that margin . . . Where, however, there is no consensus within the member States of the Council of Europe, either as to the relative importance of the interest at stake or as to the best means of protecting it, particularly where the case raises sensitive moral or ethical issues, the margin will be wider . . . There will also usually be a wide margin if the State is required to strike a balance between competing private and public interests or Convention [ECHR] rights.[80]

Of course, the weaker the consensus on the means to protect a right, the greater the difficulty in striking a fair balance with conflicting rights or even conflicting positions within the same right, and the more discretion States have. As a consequence, the more judicial organs have to step back from decisions of the legislative and executive branches.

77 Ibid, 103.
78 Ibid, 118.
79 *Arctic Athabaskan Council Petition* (2013) 54.
80 *Hämäläinen v Finland*, Appl No 37359/09 (ECtHR, 16 July 2014) [67]. The (unsuccessful) case concerned alleged violations of Article 8 ECHR resulting from linking the full recognition of the applicant's new gender to the transformation of her marriage into a registered partnership.

With respect to environmental law and human rights, the issue has been illustrated by the German Federal Supreme Court (BGH) and Constitutional Court in the wake of the acid rain debate in the 1980s, dealing with the case of private owners of forests who had sustained damages by acid rain intake.[81] The plaintiffs sued the Federal Republic of Germany and the State of Baden-Württemberg on the basis of property protection provided for in the fundamental rights catalogue of the German Constitution (Article 14 Basic Law). Claims concerning the authorisation and failure to prevent emissions that combine in the atmosphere and acidify rainwater and the omission of just compensation were rejected in all instances. In particular, based on the doctrine of the separation of powers, the Constitutional Court denied that German legislation on air pollution breached the duty to protect the right to property and provide compensation on discretionary grounds:

> It is necessary to exercise restraint in assessing constitutionality in these instances, because the determination of how the State is to realise a duty to protect which is inferred by way of interpretation of a fundamental right is a highly complex matter. Under the doctrine of the separation of powers and the principle of democracy, such decisions fall to the legislature as the directly legitimised representative of the people [. . .][82]

This judicial restraint should apply *a fortiori* in the context of climate change and human rights. Whereas we know what needs to be done to mitigate climate change, there is little consensus as to the adoption of regulatory measures imposing significant reduction of GHG emissions. Indeed, the US has not ratified the Kyoto Protocol, Canada withdrew from it in 2012, and Japan and Russia have indicated that they do not intend to assume GHG reduction targets under the second commitment period.[83] This is all the more true with respect to positive human rights duties, which are obligations of means rather than result, and thus do not exactly match binding GHG reduction targets.[84] From the standpoint of international law, uncertainty may be fostered by the assumption that climate change law embeds soft duties.[85] This reasoning also applies to adaptation measures, but only to a limited extent, in light of their more localised nature.

81 Constitutional Court, BVerfG, 1 BvR 180/88 (26 May 1998); Federal Supreme Court (BGH), III ZR 220/86 (10 December 1987) BGHZ 102, 350–68.

82 BVerfG, 1 BvR 180/88 (26 May 1998) [23].

83 UNFCCC, *Outcome of the Work of the Ad Hoc Working Group on Further Commitments for Annex I Parties under the Kyoto Protocol at Its Sixteenth Session*, Draft decision -/CMP.7 (2011) 6.

84 De Schutter (2014) 479 and 531.

85 See Antto Vihma, 'Analysing Soft Law and Hard Law in Climate Change', in Erkki J Hollo, Kati Kulovesi and Michael Mehling (eds), *Climate Change and the Law* (Springer, 2013) 143, 162–3. Vihma points out that:

> the UN climate negotiations can be framed as efforts to operationalise the soft law of the Framework Convention [UNFCCC] into decisions or legal instruments, with a greater degree

The issue emerged in *Kivalina v ExxonMobil Corporation and Others*. In this case, based on State and federal common law on nuisance, the plaintiffs claimed compensation from oil, energy and utility companies, for breaching fundamental rights, including the right to use and enjoy public and private property, because of GHG emissions. In support of a State and federal common law nuisance claim, the community of Kivalina alleged that massive amounts of GHGs contribute to global warming, severely eroding the land where the city of Kivalina sits and threatening it with imminent destruction. In this respect, the US Court for the Northern District of California held that, besides a lack of standing, the claims were barred by the political question doctrine, which is 'a species of the separation of powers doctrine and provides that certain questions are political as opposed to legal, and thus must be resolved by the political branches rather than by the judiciary'.[86] Subsequently, the Court of Appeals for the Ninth Circuit determined that the Congress had taken sufficient action as to stationary sources of GHG emissions by means of the Clean Air Act (CAA), displacing nuisance claims.[87]

6.2. Transboundary pollution and scope of application of human rights obligations

With regard to first generation human rights, the question has been raised as to whether their scope of application makes them a valid tool in framing policy measures against climate change. In fact, Article 2(1) ICCPR provides that each State party 'undertakes to *respect and to ensure to all individuals within its territory and subject to its jurisdiction* the rights recognised in the present Covenant'.[88]

The issue is less problematic with respect to second generation human rights, in which the obligations to protect and fulfil human rights are grounded. In fact, Article 2(1) ICESCR simply provides that each State party '*undertakes to take steps*, individually and *through international assistance and co-operation*, especially economic

of obligation, precision, and delegation. From this legalisation perspective there indeed seems to be a notable drive towards soft law within the climate change regime. This is not only due to the form of the agreement that is relying on COP-decisions that "take note" of parties' actions and not a ratifiable protocol, but to the broader tendency towards less obligation, precision and delegation for developed country parties in mitigation.

86 *Native Village of Kivalina, City of Kivalina v ExxonMobil Corporation and Others*, Order Granting Defendants' Motions to Dismiss for Lack of Subject Matter Jurisdiction, Case No C 08–1138 SBA (United States District Court for the Northern District of California, Oakland Division, 30 September 2009) III.A and Conclusion.

87 *Native Village of Kivalina, City of Kivalina v ExxonMobil Corporation and Others*, Case No 09–17490 (US Court of Appeals, Ninth Circuit, 21 September 2012) II.B: 'The test for whether congressional legislation excludes the declaration of federal common law is simply whether the statute speak[s] directly to [the] question at issue . . . Put more plainly, "how much congressional action is enough?" . . . The majority opinion holds that the Clean Air Act ("CAA") and the Environmental Protection Agency ("EPA") action the Act authorises displace Kivalina's claims.' See also the contribution by Tineke Lambooy and Hanneke Palm in this volume.

88 Emphasis added.

and technical, to the maximum of its available resources' aiming at '*achieving progressively the full realisation of the rights* recognised in the present Covenant by all appropriate means, including particularly the *adoption of legislative measures.*'[89]

Although the distinction between the obligations to respect, protect and fulfil human rights is grounded in the difference between first and second human rights generations, this categorisation must not be rigidly interpreted. Therefore, it must be assumed that States currently have an obligation to respect, protect and fulfil first and second generation human rights, be they embedded in the ICCPR or in the ICESCR.[90] The question therefore is: is it possible to apply the obligations to respect, protect and fulfil human rights, which originally have territorial application, to anthropogenic GHG emissions, which, by their very nature, have a transboundary effect, and are thus extraterritorial? This issue is relevant to the decision-making process leading to the adoption of general mitigation measures, whereas adaptation measures are more localised and jurisdiction-based. However, within the context of international cooperation and, in particular, the right to development, adaptation measures may also assume an extraterritorial dimension. The question can thus be re-formulated as follows: should States shape general GHG mitigation and adaptation measures so as to respect, protect and fulfil human rights extraterritorially?

In the end, with regard to mitigation and internal adaptation measures, the issue is not so crucial. In fact, in taking mitigation and national adaptation policies a State must respect, protect and fulfil human rights it owes to its own citizens. A purely internal obligation can thus be envisaged compelling the State to act in order to mitigate its own GHG emissions and take local adaptation measures. With regard to external State action, a narrow approach compels an extraterritorial extension of jurisdiction in order to hold a State responsible for human rights breaches. In this respect, the case law of the ECtHR has evolved from the initial objective requirement of a 'jurisdictional link' between the State and the foreign territory where a breach takes place,[91] specifically based on 'authority and control',[92] to the subjective criterion of State responsibility for extraterritorial conduct of 'all persons under [State] actual authority and responsibility'.[93] These criteria make it particularly complex to attribute to States GHGs emitted by private corporations located abroad, especially in light of the difficulty of establishing an 'effective' or 'complete' control of States over private legal persons.[94]

89 Emphasis added.

90 HRCte, General Comment No 31, *The Nature of the General Legal Obligation Imposed on States Parties to the Covenant*, CCPR/C/21Rev.1/Add.13 (26 May 2004) [6]; CESCR, General Comment No 13, *The Right to Education (Art 13)*, E/C.12/1999/10 (8 December 1999) [46]. See also Henry Shue, *Basic Rights: Subsistence, Affluence and US Foreign Policy* (Princeton University Press, 1980) 52.

91 *Banković and Others v Belgium and 16 Other Contracting States*, Admissibility, Appl No 52207/99 (ECtHR, 12 December 2001) [61].

92 *Issa and Others v Turkey*, Appl No 31821/96 (ECtHR, 16 November 2004) [71].

93 *Cyprus v Turkey*, Appl Nos 6780/74 and 6950/75 (ECtHR, 26 May 1975) 136, [8].

94 See section 2 above. But see CESCR (2003) [33].

From a broader standpoint, under international human rights treaties domestic obligations between a State and non-State legal persons are complemented by inter-State obligations.[95] On this basis, inter-State action is possible if human rights obligations are not respected, even as a result of climate change-related conduct, within regional or universal procedural mechanisms, such as the ECtHR, the HRC and the ICJ.[96]

Within this framework, the extraterritorial application of a State's (A) human rights obligations is possible by thinking of an inter-State obligation not to interfere with the capacity of other States (C, D, E, etc) to comply with their obligations to respect, protect and fulfil human rights. This is the minimum threshold implied by the duty to cooperate.[97] Article 55 UN Charter, which requires 'universal *respect for*, and *observance of*, human rights and fundamental freedoms for *all*', provides the basis for a possible *erga omnes* interpretation of this duty.[98] More broadly, this (meta-)obligation could be regarded as an indirect compulsion on a State (A) to respect, protect and fulfil fundamental rights of peoples in other States (C, D, E, etc). Thus, for instance, since sea-level rise caused by unsustainable GHG emissions in China threatens the existence of low-lying islands, it hampers the capacity of States such as the Maldives to respect, protect and fulfil the fundamental rights of their citizens. Therefore, it can be claimed that unsustainable GHG emitters such as China have a fundamental obligation to respect, protect and fulfil the right of the Maldives to implement an effective human rights policy. China would thus have an indirect extraterritorial obligation to reduce GHG emissions in order to respect, protect and fulfil the fundamental rights of people in the Maldives.

It is doubtful, however, whether or not international human rights treaties create direct obligations between States and extraterritorial non-State entities. This question is particularly relevant to extraterritorial climate change adaptation measures. Furthermore, a direct extraterritorial application of the obligations to respect, protect and fulfil human rights would be more constraining on States as to mitigation policies and would facilitate international cooperation,[99] within the limits of the best efforts standard.[100]

95 Dinah Shelton, *Litigating a Rights-Based Approach to Climate Change* (GWU Legal Studies Research Paper No 2013–57, 2009) 236 ff.

96 For instance, the ICESCR does not provide for a binding dispute resolution mechanism, but any States making a special declaration on the recognition of the ICJ jurisdiction under Article 36(2) of its Statute could be summoned before the Court for breaching obligations under the Covenant (on this issue see, more extensively, Andrew Strauss, 'Climate Change Litigation: Opening the Door to the International Court of Justice', in Burns and Osofsky (2009) 334).

97 CESCR (2003) [31]; (1999) [56]; McInerney-Lankford, Darrow and Rajamani (2011) 43; Knox (2009) 205–6 and 209.

98 Emphasis added.

99 As provided for in Article 56 of the UN Charter. See OHCHR (2009) [84]–[85]; Hall and Weiss, (2012) 343; Marc Limon, 'Human Rights and Climate Change: Constructing a Case for Political Action' (2009) 33 *Harv Envtl L Rev* 439, 454; Knox (2009) 201, 212–13.

100 McInerney-Lankford, Darrow and Rajamani (2011) 35.

The CESCR has recognised that the obligations to respect, protect and fulfil human rights have extraterritorial application, within a cooperative framework. With respect to the right to health, the Committee has indeed acknowledged that:

> To comply with their international obligations in relation to article 12 [ICESCR], States parties have to *respect the enjoyment of the right to health in other countries*, and to *prevent third parties from violating the right in other countries*, if they are able to influence these third parties by way of legal or political means, in accordance with the Charter of the United Nations and applicable international law. Depending on the availability of resources, *States should facilitate access to essential health facilities, goods and services in other countries, wherever possible, and provide the necessary aid when required.*[101]

The Committee has taken the same stand as to the rights to food and water.[102] With specific regard to climate change, this approach has been upheld by the UN OHCHR, paying particular attention to cooperation and the right to sustainable development.[103] More specifically, the Commissioner noticed the importance of 'the recognition of the extraterritorial [human rights] obligations of States [which] allows victims of . . . dangerous climate change to have access to remedies'.[104]

However, this view has not been clearly expressed with respect to civil and political rights and a literal interpretation of the language of Article 2(1) ICCPR would exclude State responsibility for not respecting such rights abroad.[105] To resolve this problem, a disjunctive interpretation of Article 2(1) ICCPR has been put forward, whereby the wording 'within [State] territory *and* subject to its jurisdiction' should be read as 'within [State] territory *or* subject to its jurisdiction'.[106] Nevertheless, within this context it is difficult to argue that transboundary GHG pollution, that is, mere factual conduct, not the exercise of specific power, brings individuals abroad within the jurisdiction of a State.[107]

Finally, the *erga omnes*, and, in some cases, peremptory[108] nature of human rights duties could support the stand that the obligations to respect, protect and

101 CESCR (2000) [39], emphasis added.

102 CESCR (1999) [36]; CESCR (2003) [34].

103 OHCHR (2009) 28, [86]–[88].

104 OHCHR, *Analytical Study of the Relationship between Human Rights and the Environment – Report*, UN Doc A/HRC/19/34 (24 December 2011) [72].

105 McInerney-Lankford, Darrow and Rajamani (2011) 40–1.

106 *Legal Consequences of the Construction of a Wall in the Occupied Palestinian Territory (Advisory Opinion)* [2004] ICJ Rep 136, 180, [111] (emphasis added).

107 For a different opinion, at least with respect to extreme environmental effects, see Knox (2009) 203.

108 HRCte, General Comment No 31, *Nature of the General Legal Obligation on States Parties to the Covenant*, UN Doc CCPR/C/21/Rev.1/Add.12 (26 May 2004) [2]; McInerney-Lankford, Darrow and Rajamani (2011) 23. According to the UN Sub-Commission on Prevention of Discrimination and Protection of Minorities, certain rights cannot be derogated from, even in time of emergency

fulfil human rights have extraterritorial application.[109] In this context, extraterritorial human rights obligations between a State and non-State entities (A–Bx, A–Cy, A–Dz) could possibly be asserted. This would also permit a more straightforward interpretation of Article 55 UN Charter: within such a legal framework, China would have a direct extraterritorial obligation to reduce GHG emissions in order to respect, protect and fulfil fundamental rights of citizens of the Maldives, such as self-determination and work. This approach underpins the Inuit and Athabaskan petitions to the IAComHR, respectively claiming that the US and Canada should abide by human rights obligations towards the Inuit and Athabaskan people living in the Arctic Circle. Indeed, the petitions allege that unsustainable GHG emissions in the US and Canada violate 'individual human rights', such as property, physical well-being and cultural life, of particular Inuit and Athabaskan natural persons, whose address and telephone number is explicitly listed.[110]

6.3. Causation

The causal link between anthropogenic GHG emissions and human rights breaches is not a direct one. The intermediate steps of rising atmospheric temperatures (climate change) and environmental changes make the breach of first and second generation human rights a secondary consequence (specific causation) of anthropogenic GHG emissions.[111]

This makes attribution of responsibility for the breach of the obligations to respect, protect and fulfil human rights particularly complex when the State is responsible because public agents, or private agents under its control, emit GHGs and thus eventually breach human rights. Indeed, it is not difficult to determine a connection and quantify the link between a State's GHG emissions and rising atmospheric temperature. By contrast, it is not easy to determine a further connection between rising temperatures and specific environmental events, such as tornados, which then affect human rights in particular situations. Think, for

threatening the life of the nation, including the right to life (UN Sub-Commission on Prevention of Discrimination and Protection of Minorities, *Siracusa Principles on the Limitation and Derogation of Provisions in the International Covenant on Civil and Political Rights*, Annex, E/CN.4/1984/4 (1984) [58]).

109 Human rights are, indeed, *universal*, inalienable, indivisible, interdependent and interrelated (World Conference on Human Rights, *Vienna Declaration and Programme of Action* (1993) [5]). In particular, as to the indivisible nature of fundamental claims, the right to self-determination of peoples is embedded in Article 1 of both the ICCPR and the ICESCR (see Susannah Willcox, 'A Rising Tide: the Implications of Climate Change Inundation for Human Rights and State Sovereignty' (2012) 9 *Essex Human Rights Law Review* 1).

110 *Inuit Petition* (2005) 10–12; *Arctic Athabaskan Council Petition* (2013) 8–10. See further Boumghar in this volume.

111 See Rachel Payne, 'Climate Change, Basic Rights and International Obligations' (2012) 2 *Yale Rev Int'l Stud* 64, 72; Limon (2009) 457; Richard SJ Tol and Roda Verheyen, 'State Responsibility for Climate Change Damages – A Legal and Economic Assessment' (2004) 32 *Energy Pol'y* 1109, 1112.

instance, about the loss of lives caused by Typhoon Haiyan.[112] Difficulties are due to the fact that several factors may have contributed to generate the Typhoon.

In other words, an individual or group is required to prove a nexus between specific conduct, namely, GHG emissions, and alleged human rights violations, in order to be able to engage a judicial process. A person must demonstrate to be 'injured' by GHG emissions, which provides standing to bring a case in courts. For instance, in *Aalbersberg and 2,084 Other Dutch Citizens v The Netherlands*, dealing with a complaint against the Dutch participation in the deployment of nuclear arms and their potential use as a violation of the right to life, the HRCte held:

> For a person to claim to be a victim of a violation of a right protected by the Covenant [ICCPR], he or she must show either that *an act or an omission* of a State party *has already adversely affected his or her enjoyment of such right*, or that *such an effect is imminent*, for example on the basis of existing law and/or judicial or administrative decision or practice.[113]

With regard to environmental pollution, for example, the owner of a property subject to certain emissions from a neighbouring estate could be entitled to bring an action against the State for not preventing these emissions.[114] However, from the perspective of climate change it seems particularly difficult to hold a State responsible for breaching individual human rights, alleging, for instance, that its GHG emissions spawned a typhoon subsequently causing the death of an individual and thus breaching a specific right to life. To what extent, nevertheless, is the applicant required to substantiate the indirect causal link between the harm suffered and State action or inaction causing it?

In *Asselbourg and 78 Others and Greenpeace Luxemburg v Luxemburg*, the ECtHR adjudicated upon Luxembourg regulation governing pollution generated by the production of steel from scrap iron, possibly in breach of the right to private and family life under Article 8 ECHR. The Court rejected the plaintiffs' allegations, based on the assumption that only in 'wholly exceptional circumstances' may the risk of a future violation 'confer the status of "victim" on an individual applicant', and only if there is 'reasonable and convincing evidence of the probability of the occurrence of a violation concerning him or her personally', whereas 'mere suspicions or conjectures' are insufficient.[115] However, in *Tătar v Romania*, the Court softened this stance, holding that the use of specific gold extraction methods in a

112 OHCHR (2009) [23]. See also Shelton, *Litigating* (2009) 17.

113 *Aalbersberg and 2,084 Other Dutch Citizens v The Netherlands*, Comm No 1440/2005 (HRCte, 14 August 2005) [6.3], emphasis added. Accordingly, see *Open Door and Dublin Well Woman v Ireland*, Appl No 14234/88 (ECtHR, 29 October 1992) [43]–[44]. See also Sarah Joseph and Melissa Castan, *The International Covenant on Civil and Political Rights* (OUP, 3rd edn, 2013) 71 ff; Stefan Kadelbach, 'Internationale Durchsetzung', in Oliver Dörr, Rainer Grote and Thilo Marauhn, *EMRK/GG Konkordanzkommentar zum europäischen und deutschen Grundrechtsschutz* (Mohr Siebeck, 2nd edn, 2013) Vol. II, 1893, 1911 ff, [30]–[37].

114 See, for instance, *Lopez Ostra v Spain* (9 December 1994) [42].

115 *Asselbourg and 78 Others and Greenpeace Luxemburg v Luxemburg*, Appl No 29121/95 (ECtHR, 29 June 1999) [1].

mine constituted a material risk to the right to private and family life of a nearby living applicant, which entailed a duty for State authorities to assess risks and take appropriate pre-emptive measures. The Court therefore held Romania in breach of Article 8 ECHR, despite the fact that the applicants failed to prove the existence of a causal link between exposure to sodium cyanide and a health condition characterised by asthma.[116] This approach is consistent with the precautionary principle and might attenuate evidence requirements as to the causal link between GHG emissions and human rights violations, based on the notion of 'threat' rather than 'actual damage'.

Evidence of the causal link may also be facilitated when claims for redressing human rights violations are submitted collectively by a group of citizens, which is often the case in the field of environmental and human rights damage. Significantly, in *Lago Agrio* a group of Ecuadorian citizens filed a lawsuit against Texaco (acquired by Chevron in 2001) in Ecuador, alleging severe environmental contamination of the Lago Agrio area in Amazonia, where Texaco conducted its oil operation activities. The plaintiffs alleged this led to increased cancer rates and other serious health problems for the residents of the region. The Provincial Court of Justice of Sucumbíos condemned Chevron to pay compensation for $18.2 billion.[117] Even more significantly, in *Gbemre v Shell Petroleum* the plaintiff, acting for himself and on behalf of the Iwherekan Community, sued the multinational oil corporation Shell for its practice of 'gas flaring' in the course of oil exploitation operations in the Niger Delta, causing, inter alia, excessive GHG emissions.[118] The Nigerian Supreme Court upheld a gross violation of the fundamental right to life, including a healthy environment, as well as of the dignity of the human person protected by the Constitution, and condemned Shell to shut down its 'gas flaring' activities.[119]

However, whereas in the cases of *Lago Agrio* and *Gbemre* environmental damage was localised, it is more difficult to prove a causal link when the impact of GHG emissions is not a direct one. In *Kivalina*, the District Court for the Northern District of California excluded that 'plaintiffs must show to a scientific certainty that defendant's effluent, and defendant's effluent alone, caused the precise harm suffered by the plaintiffs', and required the proof of a 'substantial likelihood that the defendant's conduct caused plaintiff's injury' or was the 'seed

116 *Tătar v Romania*, Appl No 6702/01 (EctHR, 27 January 2009) [25] ff.

117 *Aguinda v Chevron Corp*, No 002-2003 (Super Ct of Nueva Loja, 14 February 2011). Enforcement abroad is nevertheless proving problematic (see Manuel A Gómez, 'The Global Chase: Seeking the Recognition and Enforcement of the Lago Agrio Judgment outside Ecuador' (2013) 1 *Stanford Journal of Complex Litigation* 429).

118 As a consequence of this heavy carbon emitting activity, Nigeria produces more GHGs than the States of the sub-Saharan region together, amounting to almost 25 per cent of Africa's total GHG emissions (Eferiekose Ukala, 'Gas Flaring in Nigeria's Niger Delta: Failed Promises and Reviving Community Voices' (2011) 2 *Washington and Lee Journal of Energy, Climate, and the Environment* 97, 102–3).

119 *Gbemre (for Himself and Representing Iwherekan Community in Delta State, Nigeria) v Shell Petroleum Development Company Nigeria Ltd, Nigerian National Petroleum Corporation and Attorney-General of the Federation* (Federal High Court of Nigeria in the Benin Judicial Division, suit FHC/B/CS/53/05, 14 November 2005).

of the injury'.[120] In light of this, the distance of Kivalina City from the zone of discharge of GHGs, basically involving the entire world, and the combination of other environmental factors causing sea-level rise was considered to disrupt the causal link.[121] In the *Inuit* case, the IAComHR dismissed the Inuit petition by simply stating that 'the information provided does not allow [the Commission] to determine whether the alleged facts would tend to characterise a violation of rights protected by the American Declaration'.[122] This quite general statement might be read as a reference to the impossibility of establishing a causal link between GHGs emitted by the US and human rights breaches.[123]

Subject to the principle of exhaustion of domestic remedies, difficulties in proving the causal link also arise when taking into account the inter-State implications of human rights obligations. In fact, it is equally difficult for a State to demonstrate that the devastation caused by Typhoon Haiyan is ultimately due to climate change as it is for individuals.

6.4. Imputation

Besides the objective issue of causation, the subjective element of apportionment of responsibility arises. Indeed, several sources, in particular, more than one State and other persons, contribute in time to GHG emissions. In light of this premise, attributing responsibility based on an at least marginal contribution to the greenhouse effect by every GHG molecule emitted regardless of its origin is not persuasive.[124] It therefore seems *prima facie* difficult to divide liability between different States for breaching negative and positive human rights obligations, from both a spatial and temporal perspective.

In *Kivalina*, the District Court for the Northern District of California excluded causality not only based on objective elements, but also, and more specifically, with regard to imputation. Indeed, the Court held:

> Plaintiffs have not alleged that the 'seed' of their injury can be traced to any of the Defendants . . . Plaintiffs allege that the genesis of the global warming phenomenon dates back centuries and is a result of the emission of greenhouse gases by a multitude of sources *other than* the Defendants . . . Significantly, the source of the greenhouse gases are undifferentiated and cannot be traced to any particular source, let alone defendant, given that [GHGs] 'rapidly mix in the atmosphere' and 'inevitably merge' with the accumulation of emissions in California and the rest of the world.[125]

120 *Native Village of Kivalina* (30 September 2009) 16–18.

121 Ibid, at 21.

122 IAComHR, *Letter to Sheila Watt Cloutier et al*, Petition P-1413–05 (16 November 2006), <http://graphics8.nytimes.com/packages/pdf/science/16commissionletter.pdf>.

123 See Hari M Osofskiy, 'The Inuit Petition as a Bridge? Beyond Dialectics of Climate Change and Indigenous Peoples' Rights', in Burns and Osofsky (2009) 272, 283–4.

124 But see Hans-Joachim Koch, Michael Lührs and Roda Verheyen, 'Germany', in Richard Lord, Silke Goldberg, Lavanya Rajamani and Jutta Brunnée, *Climate Change Liability – Transnational Law and Practice* (CUP, 2012) 376, 401, [15.68].

125 *Native Village of Kivalina* (30 September 2009) 20.

The Court held that the impossibility of tracing any specific alleged effect of global warming to any particular emissions by any specific person, entity, or group at any particular point in time disrupted imputation. Attribution of responsibility was considered impossible because environmental phenomena affecting human rights have been caused individually and cumulatively over the span of centuries, thus preventing the determination of what emissions actually caused the erosion of the city of Kivalina.

In *Massachusetts v EPA*, the Court reached an opposite conclusion and held that, if unregulated by EPA, GHG emissions endanger public health.[126] In this case, nevertheless, issues of imputation focused on the exercise of regulatory power over GHG emissions by a public agency, not on the quantification of compensation for past emissions.

In order to address these problems, based on the principles established in the *Trail Smelter* arbitration it must be assumed that multiple sources of pollution do not exclude responsibility and the award of environmental damages.[127] Along these lines, a possible response to the issue of imputation in relation to causation and reparation of damage consists of prorata or joint and several liability. If liability is prorated, each emitter should be held proportionally responsible for the attributable percentage of the total damages. According to joint and several liability, emitters are individually and mutually responsible, and thus damage must be redressed by any and all of the parties in various amounts or in full by any one of the parties, who are then reciprocally liable.[128]

If it is difficult to apply these principles and quantify damage as to private legal persons, as in *Kivalina*, this does not seem impossible at the aggregate State level, given that there are currently only around 200 States in the international community and the respective emission trends are known.[129] Within this context, liability should preferably be attributed based on per capita emissions and the criterion of common but differentiated responsibility.[130] Proportionality seems

126 *Massachusetts v EPA* (2 April 2007).

127 *Trail Smelter* (16 April 1938 and 11 March 1941) 1905.

128 See Peter Butt *et al*, *Legal Dictionary* (LexisNexis, 2014) 112 ('Concurrent Liability'), 329 ('Joint and Several'), and 469 ('Proportionate Liability').

129 See section 1 above.

130 Correctly, the Inuit petition points out the disproportionate per capita emission trend of the US, which was the major GHG emitter at the time of the petition and is currently the second major GHG emitter in the world after China:

> With only 4.7% of the world's population, the United States produced 24% of global emissions in 2000. On a per-person basis, US emissions in 2000 were more than five times the global average. They were nearly two-and-a-half times the per capita emissions in Europe, and nine times those in Asia and South America. Only five countries exceeded the United States in per capita emissions in 2000—Qatar, the United Arab Emirates, Kuwait, and Bahrain—all of which have much smaller populations and huge reserves of highly carbon-intensive commodities. Yet among the countries with significant emissions, the United States had the highest level of per capita emissions.
>
> (*Inuit Petition* (2005) 69)

preferable to joint and several liability, because, in light of the principle of consensual jurisdiction,[131] it would be difficult for a State providing full compensation to recover damages from other jointly and severally responsible States. On the other hand, it could still be argued that universally dividing responsibility between all States of the international community is too widespread, and is thus tantamount to granting immunity. In this respect, it is interesting, but also deceiving, that the Inuit petition was dismissed by a general statement on the impossibility of determining a human rights violation as to unsustainable GHGs emitted by the US.[132] The IAComHR missed an exceptional opportunity to seriously discuss the question of the apportionment of responsibility among States for GHG emissions, and thus practice is currently lacking as to the possibility of applying proportionate or joint and several liability to States for climate change-related environmental effects. Hopefully, the Commission will shed light on the issue in dealing with the Athabaskan petition.

Shared liability was regarded by the ECtHR as a possible option to apportion responsibility in *Matthews*. In this case, the UK was held liable, together with all the other parties to the Maastricht Treaty, for failing to remedy the exclusion of Gibraltar citizens from the election of the European Parliament, thus breaching the principle of non-discrimination under Article 14 ECHR, although compensation was not in issue.[133] Along these lines, it must be noted that compensation is not the only means available to provide reparation, which also includes *restitutio in integrum* and satisfaction. Therefore, the impossibility of quantifying the exact liability of each and every GHG emitter should not exclude responsibility. Restitution for breaching human rights might entail the adoption of adequate GHG mitigation and adaptation policies and apologies.[134] For instance, GHG mitigation and adaptation measures were the main purpose of the petitions that 37 non-governmental organisations and individuals filed with the WHC between 2004 and 2006, requesting the addition of several world heritage sites to the List of World Heritage in Danger.[135] The WHC acknowledged the 'genuine concerns' of the petitions and the impact of climate change on cultural heritage, but did not inscribe any site on the List because of climate change, and only took action for adaptation measures as to the Waterton Glacier International Peace Park.[136] Unfortunately, these possibilities were also not adequately explored by the IAComHR in the *Inuit* case.

131 See Article 36 of the ICJ Statute.

132 See section 6.3 above.

133 *Matthews v the United Kingdom*, Appl No 24833/94 (ECtHR, 18 February 1999) [33].

134 Voigt (2008) 13, 15–17, 19.

135 *Waterton Glacier International Peace Park Petition* (16 February 2006), *Huascaran National Park Petition* (17 November 2004), *Sagarmatha National Park Petition* (15 November 2004), *Belize's Barrier Reef Petition* (15 November 2004), *Great Barrier Reef Petition* (21 September 2004) <www.climatelaw.org/cases/topic/unesco>. See the list of endangered sites at <http://whc.unesco.org/en/danger>.

136 WHC, *Report of Decisions Adopted at the Twenty-Ninth Session*, WHC-05/29.COM/22 (2005) Decision 29 COM 7B.a, 36–7, [4] and [5]; WHC, *Report of Decisions Adopted at Its Thirty-Third Session*, WHC-09/33.COM/20 (2009) Decision 33 COM 7B.22, [7] and [12].

The perspective is of course different when a State takes mitigation and adaptation policies that have a negative impact on fundamental rights. For instance, in the case of indigenous peoples forcibly evicted by reforestation policies, it should not be excessively complicated to identify the responsible States and take action before competent human rights bodies.[137]

7. The potential of the human right to a sustainable environment

So far, we have considered obstacles to the implementation of the State obligations to respect, protect and fulfil human rights mainly in light of first and second generation claims. However, would third generation human rights, in particular, the recognition of a human right to a sustainable environment, simplify these issues?[138] The question is sensible, especially in light of the fact that 'human rights are vulnerable to environmental degradation' and 'depend on a supportive environment'.[139]

Given that climate change is currently the most relevant threat to environmental sustainability, the right to environment should include a claim to the stabilisation of rising atmospheric temperatures below the unsustainable threshold of 2° Celsius. In this respect, scholars speak of a human right to 'climatic stability'.[140] The major advantage of the fundamental right to a (climatically) sustainable environment would be a simplification of causation (and imputation) problems. Indeed, the correlative obligations to respect, protect and fulfil a (climatically) sustainable environment would make it possible to hold a State responsible for the sole fact of releasing excessive GHGs into the atmosphere, without proving a breach of any further fundamental rights.[141] This would directly trigger human rights

137 Margaretha Wewerinke, *Climate Change, Human Rights and the International Legal Order: the Role of the UN Human Rights Council* (University of Cambridge WP, 2013) 9–12; McInerney-Lankford, Darrow and Rajamani (2011) 21; Owen Cordes-Holland, 'The Sinking of the Strait: the Implications of Climate Change for Torres Strait Islanders' Human Rights Protected by the ICCPR' (2008) 9 *Melbourne Journal Int'l L* 405, 414; Michael G Faure and André Nollkaemper, 'International Liability as an Instrument to Prevent and Compensate for Climate Change' (2007) 26A *Stan Envtl LJ* 123, 131.

138 McInerney-Lankford, Darrow and Rajamani (2011) 36–9; Limon (2009) 469.

139 In *Ogoniland*, the AComHPR concluded that, besides other human rights, including, in particular, the right to health under Article 16 of the Banjul Charter, the right to a general satisfactory environment under Article 24 of the Charter 'obligates governments to desist from directly threatening the health and environment of their citizens' (*Social and Economic Rights Action Center* (2001) [52]). The ICESCR (Article 12(b)) and the Convention on the Rights of the Child (Articles 24(2) and 29(e)) recognise that the enjoyment of human rights depends on a sustainable environment. See also John H Knox, *Report of the Independent Expert on the Issue of Human Rights Obligations relating to the Enjoyment of a Safe, Clean, Healthy and Sustainable Environment*, UN Doc A/HRC/22/43 (22 December 2012) 7, [19].

140 Bell (2013) 163–4.

141 Ole W Pedersen, 'The Janus-Head of Human Rights and Climate Change: Adaptation and Mitigation' (2011) 80 *Nordic Journal Int'l L* 403, 412.

protection mechanisms, in particular, UN universal procedures.[142] Furthermore, such obligations, which should naturally have extraterritorial application, would constrain the margin of discretion States enjoy in deciding climate change mitigation and adaptation policies.[143]

The Draft Statute of the International Environmental Agency and the International Court of the Environment presented by the ICE Coalition at the 1992 UNCED Conference in Rio de Janeiro provides a definition of the right to environment, also based on the necessity of containing GHG emissions.[144] Inter alia, the Draft Statute provides:

> Article 1
> Everyone has a fundamental right to the environment and an absolute duty to preserve life on earth for the benefit of present and future generations.
>
> Article 8
> States, in particular, shall . . . prohibit all activities that may cause irreversible damage to the basic natural processes of the biosphere and, as a precautionary measure, suspend those activities whose effects cannot be determined until all such uncertainty has been removed [. . .]

De lege lata, similar provisions are embedded in the Protocol of San Salvador Additional to the ACHR in the Area of Economic, Social and Cultural Rights,[145] and in the ACHPR.[146]

A human rights interpretation of the obligation to ensure a sustainable environment and level of GHG emissions has been put forward in the Inuit petition, both as an indigenous peoples' right and as a customary right under general international law.[147] The petition basically claims that the US obligations to respect, protect and fulfil the different first and second generation human rights of the Inuit are grounded in a specific claim of indigenous people to environmental protection.[148] Along the same lines, the Athabaskan petition invokes a State duty to protect the environment, inter alia, within the context of the Declaration

142 Limon (2009) 452 and 463–6; CIEL, *Climate Change and Human Rights: Practical Steps for Implementation* (Friedrich-Ebert-Stiftung, 2009).

143 See, for instance, *Saramaka v Suriname*, Pet No 12.338 (IAComHR, 28 November 2007) [128], [151]. See also Bell (2013) 167; Shelton, *Human Rights* (2009) 11 and 21.

144 ICE Coalition, *Draft Statute of the International Environmental Agency and the International Court of the Environment* (June 1992) Articles 1–8 and Preamble.

145 Protocol of San Salvador (1988) Article 11: '1. Everyone shall have the right to live in a healthy environment and to have access to basic public services. 2. The States Parties shall promote the protection, preservation, and improvement of the environment.'

146 ACHPR, Article 24 (Right to a General Satisfactory Environment): 'All peoples shall have the right to a general satisfactory environment favourable to their development.'

147 *Inuit Petition* (2005) 73–4. See also Shelton, *Human Rights* (2009) 14.

148 Ibid. See also Shelton, *Litigating* (2009) 221 ff.

on Human Rights and Climate Change in the Americas adopted by the OAS in 2008.[149]

More generally, it is possible to maintain that a human right to a sustainable environment is developing globally, and is already recognised as soft law.[150] However, despite the recognition of this right in around 120 constitutions, which could be considered to give rise to a general principle of law,[151] it is currently impossible to assume the existence of such an obligation within the framework of hard international law.[152] Also in this respect, the IAComHR missed the opportunity to clarify a practically relevant and theoretically fascinating question in quickly dismissing without proper motivation the Inuit petition on climate change and human rights.[153]

Alternatively, it is possible to think of a duty to protect the environment within the context of the right to sustainable development. Indeed, the concept of sustainable development entails the notions of 'poverty reduction' and 'environmental sustainability'.[154] Nonetheless, currently the right to development is not clearly established in international law either.[155]

For the time being, the well-established customary no-harm rule can be invoked as a substitute for the human right to a sustainable environment,[156] as complemented by the polluter pays principle and interpreted in light of the precautionary principle.[157] Indeed, the fundamental right to a sustainable environment would be nothing other than a human rights interpretation of the existing prohibition of

149 *Arctic Athabaskan Council Petition* (2013) 54–6; OAS, GA Res 2429 (XXXVIII-O/08).

150 See, in particular, Principle 1 of the 1972 Stockholm Declaration on the Human Environment and Principle 1 of the 1992 Rio Declaration on Environment and Development.

151 See John Knox (2012) [12] and [14]; Carel Bruch, Wole Cocker and Chris VanArsdale, 'Constitutional Environmental Law: Giving Force to Fundamental Principles in Africa' (2001) 26 *Colum J Envtl L* 131.

152 See the contribution by Francesco Francioni and Ottavio Quirico in this volume. See also Knox (2012) [14]; Limon (2009) 470.

153 For a critical analysis of the possibility of a right to environment, its structure, content and scope, see Chiara Feliziani, 'The Duty of Member States to Guarantee the Right to a Healthy Environment: a Consideration of *European Commission v Italy* (C-297/08)' (2012) 24 *JEL* 535.

154 See WCED, *Our Common Future*, Brundtland Report (1987) [1]:

> Sustainable development is development that meets the needs of the present without compromising the ability of future generations to meet their own needs. It contains within it two key concepts: the concept of "needs", in particular the essential needs of the world's poor, to which overriding priority should be given; and the idea of limitations imposed by the state of technology and social organisation on the environment's ability to meet present and future needs.

155 See further Varayudej in this volume.

156 *Trail Smelter* (16 April 1938 and 11 March 1941); *UNCLOS*, Article 194(2). See also Principle 21 of the Stockholm Declaration, Principle 2 of the Rio Declaration and the ILC's 2011 Draft Articles on Prevention of Trans-boundary Harm from Hazardous Activities. For a scholarly debate, see McInerney-Lankford, Darrow and Rajamani (2011) 21–2; Shelton, *Litigating* (2009) 258; Cordes-Holland (2009) 414–15; Roda Verheyen, *Climate Change Damage and International Law* (Martinus Nijhoff, 2005) 150–1.

157 See Principles 15 and 16 of the Rio Declaration. See also Shelton, *Human Rights* (2009) 42; Voigt (2008) 9.

(transboundary) pollution. More specifically, the right to a safe level of GHG emissions is already in place under the UNFCCC regime, which has quasi-universal application.[158] A strict application of this (*erga omnes*)[159] obligation, based on faulty excess emissions,[160] should facilitate the subjective and causal apportioning of responsibility for climate change damage, regardless of further environmental damage and human rights injury jointly caused.[161] A similar reasoning underpins the action threatened by the Pacific Island State of Tuvalu in the ICJ against the US and Australia for refusing to ratify the Kyoto Protocol.[162] Nevertheless, the universal effectiveness of the no-harm rule is hampered in the climate change field by the fact that only 40 States, excluding major emitters such as the US, China and India, have accepted binding GHG emissions constraints under the Kyoto Protocol, as amended in Doha in 2012.[163] Furthermore, customary and UNFCCC rules have the potential to attract an obligation to make full reparation for injury,[164] but, of course, do not trigger human rights protection mechanisms. Therefore, if imputation of responsibility is currently possible for States not respecting binding commitments jointly or proportionately based on the no-harm rule and the Kyoto Protocol, a further causal link must be demonstrated between excessive GHG emissions and specific breaches of first and second generation human rights in order to trigger human rights enforcement procedures.[165]

A human rights-based interpretation of the no-harm rule is thus suitable. Unlike first and second generation fundamental rights, a human right to a sustainable environment would generate an *erga omnes* procedural claim, that is, *actio popularis*,

158 See, in particular, Articles 2 and 4(2) UNFCCC, and Articles 2 and 3 of the Kyoto Protocol. Currently, the UNFCCC and the Kyoto Protocol respectively include 196 and 192 parties.

159 *Gabčíkovo-Nagymaros Project (Hungary v Slovakia)* [1997] ICJ Rep 7, Separate Opinion of Judge Weeramantry. See also Tol and Verheyen (2004) 1113; ILC, *Report to the General Assembly*, UN Doc A/31/10 (1976, vol II, part II) *Yearbook of the ILC* 75 (Draft Articles on State Responsibility, First Reading, Article 19).

160 Farber (2007) 40.

161 Pedersen (2011) 416; Voigt (2008) 15, 19–20; Karen E MacDonald, 'A Right to a Healthful Environment – Humans and Habitats: Re-thinking Rights in an Age of Climate Change' (2008) 17 *EEELR* 213, 220–2; Faure and Nollkaemper (2007) 142–4; Tol and Verheyen (2004) 1115. Of course, this would not exclude more compelling GHG emission limits under domestic legislation (Faure and Nollkaemper (2007) 153).

162 Rebecca Elizabeth Jacobs, 'Treading Deep Waters: Substantive Law Issues in Tuvalu's Threat to Sue the United States in the International Court of Justice' (2005) 14 *Pac Rim L & Pol'y J* 103.

163 The US has also expressed a negative view on the possibility of a human rights approach to climate change (US, *Observations on the Relationship between Climate Change and Human Rights* (2008) 1, [4], <www.ohchr.org/Documents/Issues/ClimateChange/Submissions/USA.pdf>). Theoretically, the peremptory nature of some human rights might also permit to extend binding GHG commitments to States that are not parties to the Kyoto Protocol, such as the US.

164 Article 31 of the ILC's 2001 Draft Articles on State Responsibility. See also Voigt (2008) 18; Faure and Nollkaemper (2007) 174–6.

165 See Caroline Dommen, 'Claiming Environmental Rights: Some Possibilities Offered by the United Nations' Human Rights Mechanisms' (1998) 11 *Geo Int'l Envtl L Rev* 1.

displacing the focus from bilateral claims to absolute rights.[166] Anybody would thus have a cause of action and could start proceedings to enforce the general right in issue.[167] With particular regard to States, besides regional human rights protection frameworks, in particular the European, Inter-American and African systems, it is currently possible to think of general petitions before the UN HRC. Otherwise, alternative mechanisms should be put in place, which has been discussed by the Coalition for an International Court for the Environment, particularly in the wake of the UNFCCC negotiations.[168] If the principle of joint and several liability is chosen to apportion responsibility, the latter solution is preferable, including compulsory jurisdiction to facilitate recovering damages.

Conclusion

Theoretically, the obligations to protect, respect and fulfil human rights may and should be interpreted as tools compelling States to take GHG mitigation and adaptation policies. Some obstacles nevertheless arise as to the implementation of State responsibility with respect to first and second generation fundamental rights, because of subjective and objective issues, including, in particular, policy discretion, the territorial scope of human rights obligations, and difficulties in proving causation and imputation.

Despite the seriousness of these problems,[169] some solutions can be envisaged. Specifically, the relatively small number of States composing the international community, where a few countries are responsible for most GHG emissions, might make imputation possible, based on per capita emissions under proportionate

166 On the distinction between bilateral, multilateral and absolute rights see Kelsen (1989) 127–8 and 191.

167 ICE Coalition (1992) Article 4: 'Everyone, whether an individual or an association, has the right to take legal action to prevent activities that are harmful to the environment and to seek compensation for any environmental damage.'

168 Article 10 (The International Court of the Environment) of the ICE Coalition Draft Statute of the International Environmental Agency and the International Court of the Environment provides:

> The functions of the Court shall be: a) to protect the environment as a fundamental human right in the name of the International Community; b) to decide any international environmental disputes involving the responsibility of States to the International Community which has not been settled through conciliation or arbitration within a period of 18 months; c) to decide any disputes concerning environmental damage, caused by private or public parties, including the State, where it is presumed that, due to its size, characteristics and kind, this damage affects interests that are fundamental for safeguarding and protecting the human environment on earth.'

See also ICE Coalition, *Creating the International Court for the Environment*, <http://icecoalition.com/the-solution>; Ole W Pedersen, 'An International Environmental Court and International Legalism' (2012) 24 *Journal of Environmental Law* 547; Amedeo Postiglione, 'A More Efficient International Law on the Environment and Setting up an International Court for the Environment within the United Nations' (1990) 20 *Envtl L* 321.

169 In the view of co-author Jürgen Bröhmer, obstacles are so serious that they are insurmountable.

or joint and several liability. Furthermore, the recognition of a human right to (a sustainable) environment would simplify causation, establishing a direct link between GHG emissions and fundamental rights, and at least reduce the margin of discretion States enjoy as to GHG mitigation and adaptation policies.

In practice, the IAComHR might have addressed these questions in the so far most relevant case in which State responsibility under tripartite human rights obligations has been raised, that is, the petition filed by the Inuit against the US for human rights violations caused by the environmental effects of GHG emissions in the Arctic. This was an exceptional opportunity to investigate an issue that underpins Article 19 of the 1976–1996 ILC's Draft Articles on State Responsibility and the 1992 proposal of the ICE Coalition for an International Environmental Agency and the International Court of the Environment. Since the petition was dismissed without clear motivations, the IAComHR missed a crucial chance to shed light on a vital and developing area of the law. For practical and doctrinal purposes, a review of the case would therefore be suitable. Hopefully, the Commission will shed light on fundamental issues concerning human rights and climate change dealing with the more recent 'twin' Athabaskan petition.

2 Balancing human rights in climate policies

Bridget Lewis

Introduction

The effects of anthropogenic GHG emissions are likely to impinge upon the ability of individuals and communities to enjoy the human rights that are guaranteed to them, such as the rights to life, health and self-determination. Consequently, States may be obliged to take particular action with respect to climate change in order to comply with their duties to respect, protect and fulfil human rights. These actions may comprise a range of mitigation and adaptation measures, and may involve an obligation on States to control the conduct of private actors.[1]

While the potential for climate change to impact upon human rights is significant, the steps taken by States in response to climate change also have the potential to affect the enjoyment of human rights. This chapter therefore considers the ways in which action on climate change might lead to other violations, and thus balances the necessity of mitigation and adaptation policies prompted by human rights obligations against potentially conflicting fundamental rights. The examination considers risks inherent in common mitigation and adaptation measures and the potential for human rights breaches in the implementation of those measures.

One of the key challenges of adopting a human rights-based approach to climate change is the requirement, in locating effective and practicable solutions to climate change, to balance the rights of all individuals and communities who are affected by the various aspects of a complex, widespread and cumulative problem. Given the potential for human rights violations to result from the failure to act on climate change and the taking of mitigation and adaptation measures, States are confronted with the need to balance possibly competing human rights imperatives. Furthermore, the transnational, intergenerational and cumulative impacts of GHG emissions create the potential for conflict not just between different substantive human rights obligations but also on the planes of geography and time, as different acts or omissions may result in variable impacts for different

1 See the contributions by Ottavio Quirico, Jürgen Bröhmer, Marcel Szabó and by Anna Riddell in this volume.

individuals and communities as well as diverse impacts on current and future generations. The complexities inherent in the nature of climate change therefore create significant challenges for States in terms of balancing competing human rights obligations. Human rights law provides some guidance in this respect, and this chapter considers how these principles might assist States in addressing the problem.

1. Tripartite human rights obligations and positive mitigation and adaptation measures

Before examining the human rights implications of specific mitigation and adaptation measures, some general points can be made on the relationship between human rights and climate change mitigation and adaptation. The first relates to the issue of climate justice, and the understanding that climate change is likely to affect most severely those who are least responsible for its causes and least capable of responding to its effects. This reality must be recognised in the way that climate change responses are developed, both as a moral consideration and also from a practical and economic perspective. Not only should the inequity of the effects of climate change be acknowledged, but so too the potential for disproportionate or discriminatory impacts of mitigation and adaptation measures. It is important that steps taken to mitigate the magnitude of climate change or respond to its effects do not result in the further impoverishment or marginalisation of already poor and disenfranchised groups.

A second aspect which should be acknowledged is the relationship between human rights, development and access to energy. Development is a key driver of the enjoyment of human rights, and reliable and affordable access to energy is an essential component of that development. GHG mitigation policies alter the distribution of energy and progress towards development.[2] Responses to climate change which diminish the availability of reliable energy sources, or which affect the economy of a developing State more broadly, have the potential to set back progress on the fulfilment of a wide range of human rights.[3] An undeniable part of this relationship is the fact that, at least in the context of current technological realities, it is not possible for developing States to pursue socio-economic growth without relying on fossil fuels as a source of energy.[4] There is therefore an unavoidable tension between the competing interests of development (and the positive human rights benefits it promises to bring) and reducing GHG emissions (in order to avoid the negative human rights impacts of climate change). Added to this is the fact that in many developing States the effects of climate change are

2 Javier de Cendra de Larragán, 'Tying the Knot of Energy Security and Climate Change Mitigation: a Tale of Solidarity?' (2011) 22 *Yb Int'l Env L* 76.

3 Dinah Shelton, *Human Rights and Climate Change* (Northwestern University, 2009) 23.

4 Margaux J Hall and David C Weiss, 'Avoiding Adaptation Apartheid: Climate Change Adaptation and Human Rights Law' (2012) 37 *Yale Journal of International Law* 309, 322.

already being experienced and adaptation measures are already required. Where resources are scarce this creates the additional challenge of balancing adaptation measures against the broader development agenda. This challenge can be understood in terms of potentially competing human rights, and human rights principles can be useful in identifying conflicts and possible resolutions.

These factors illustrate the complexity of issues relating to human rights and climate change mitigation and adaptation. While it is possible to consider the human rights impacts of any given mitigation or adaptation measure, the broader relationship between human rights and development dictates that a rights-based approach to climate change will inevitably involve a range of priorities and compromises which must be struck in a way which is equitable both within and between States.

As previously explained, human rights law imposes three levels of obligations on States. These are the obligations to respect, protect and fulfil human rights. While the duties to protect and fulfil human rights seem to suggest an obligation on States to take positive action in the face of climate change, the duty to respect human rights arguably has less direct application. This obligation requires that States refrain from acting in a manner which impacts negatively on human rights. The obligation to respect human rights is crucial if a State decides to take action, because the implementation of mitigation and adaptation measures may negatively affect human rights.[5] This must be taken into account in the decision-making process leading to such measures.[6] For instance, in the past, climate change-induced catastrophic floods in China, Thailand and the Philippines prompted logging bans that led to millions of people losing their occupations, thus negatively affecting their right to work. Similarly, as a consequence of increasing forest fires in the Amazonian and Southeast Asian rainforests caused by rising temperatures, governments have imposed blanket bans on burning forests, which might nevertheless negatively affect agricultural burning practices and thus the right to food.[7] Relocation of people caused by the fact that certain lands have become uninhabitable because of climate change is likely to impinge upon fundamental claims such as the right to work and adequate housing.[8] Reforestation

5 See Siobhan McInerney-Lankford, 'Climate Change and Human Rights: an Introduction to Legal Issues' (2012) 33 *Harv Envtl L Rev* 431, 436; Hall and Weiss (2012) 348, 357–62; John Knox, 'Climate Change and Human Rights Law' (2009) 50 *Va J Int'l L* 163, 198–9.

6 McInerney-Lankford (2012) 436; Shelton (2009) 37; David B Hunter, 'Human Rights Implications for Climate Change Negotiations' (2009) 11 *Or Rev Int'l L* 331, 360–1.

7 Frances Seymour, *Forests, Climate Change, and Human Rights: Managing Risk and Trade-offs* (CIFOR, 2008) 12.

8 OHCHR, *Report on the Relationship between Climate Change and Human Rights*, UN Doc A/HRC/10/61 (15 January 2009) [57]–[60]; Mariya Gromilova, 'Revisiting Planned Relocation as a Climate Change Adaptation Strategy: the Added Value of a Human Rights-Based Approach' (2014) 10 *Utrecht L Rev* 76, 80–94; John Von Doussa, 'Human Rights and Climate Change: a Tragedy in the Making' (2008) 31 *UNSW LJ* 953, 962–3; Siobhan McInerney-Lankford, Mac Darrow and Lavanya Rajamani, *Human Rights and Climate Change: a Review of the International Legal Dimension*, World Bank Study, World Bank (2011) 63–4.

policies can breach fundamental rights of indigenous peoples by causing forced evictions, destroying biodiversity, agricultural practices and cultural diversity.[9] In fact, the obligation to respect human rights requires that States consider the potential human rights consequences of any proposed action and ensure that, whatever steps they take, they do not cause further interference with human rights. It therefore operates as a minimum standard to which States must adhere.[10] In this respect, the HRC urged States to 'take human rights into consideration when developing their environmental policies'.[11]

The obligations to protect and fulfil human rights also apply to climate change mitigation and adaptation action. The obligation to protect is particularly relevant, given that many of the adaptation and mitigation activities commonly employed by States can in fact be carried out by private actors. In this context States will be required to supervise and regulate the activities of corporate entities to ensure human rights are not violated. The obligation to fulfil is also important to shape the way in which mitigation and adaptation measures are deployed, so as to permit full enjoyment of fundamental rights.

As will be discussed below, human rights obligations may also operate to restrict States from taking particular forms of mitigation and adaptation action altogether, creating specific challenges in terms of balancing the different objectives of addressing climate change and maintaining human rights standards.[12]

9 Ole W Pedersen, 'The Janus-Head of Human Rights and Climate Change: Adaptation and Mitigation' (2011) 80 *Nordic J Int'l L* 403, 408–9; Hunter (2009) 358; Seymour (2008) 11.

10 OHCHR (2009) [72], [80]; COP 9, UNFCCC, *Milan Declaration of the Sixth International Indigenous Peoples Forum on Climate Change* (29–30 November 2003) 6; *Australian Conservation Fund v Minister for Planning* (2004) VCAT 2029. See also Michael P Vandenbergh, Brooke A Ackerly and Fred E Forster, 'Micro-Offsets and Macro-Transformation: an Inconvenient View of Climate Change Justice' (2012) 33 *Harv Envtl L Rev* 303; McInerney-Lankford, Darrow and Rajamani (2011) 8; Hunter (2009) 331, 351; Knox (2009) 198–9; Owen Cordes-Holland, 'The Sinking of the Strait: the Implications of Climate Change for Torres Strait Islanders' Human Rights Protected by the ICCPR' (2008) *9 Melb J Int'l L* 405, 414.

11 HRC Resolution 16/11, *Human Rights and the Environment*, UN Doc A/HRC/RES/16/11, 12 April 2011. The ECtHR has highlighted the necessity of affording 'due respect' to human rights in the decision-making process in environmental matters, given that environmental pollution negatively affects fundamental rights (*Fadeyeva v Russia*, Appl No 55723/00 (ECtHR, 9 June 2005) [105]). In their petition to the Inter-American Commission on Human Rights, the Inuit requested the US to take into account the impact of GHG emissions on the Arctic environment, and thus on basic rights, such as the rights to culture, health, food, property, and self-determination, before approving major governmental measures (Sheila Watt-Cloutier, *Petition to the Inter-American Commission on Human Rights Seeking Relief from Violations Resulting from Global Warming Caused by Acts and Omissions of the United States* (7 December 2005) 116; Shelton (2009) 17). See also Arctic Athabaskan Council on Behalf of All Arctic Athabaskan Peoples of the Arctic Regions of Canada and the United States, *Petition to the Inter-American Commission on Human Rights Seeking Relief from Violations of the Rights of Arctic Athabaskan Peoples Resulting from Rapid Arctic Warming and Melting Caused by Emissions of Black Carbon by Canada* (23 April 2013) 86.

12 Pedersen (2011) 80.

2. Potential human rights violations associated with mitigation and adaptation measures

There are a range of climate change mitigation and adaptation approaches available to States, both within international frameworks and domestically. Mitigation measures may involve switching to alternative sources of energy, utilising more efficient technologies or increasing carbon sequestration. Adaptation strategies can incorporate a wide range of activities and may be targeted at the particular needs of affected communities or individuals. These can include changes to planning and development requirements, the construction of sea walls or other protective infrastructure, changes to agricultural crops and practices, altered water storage mechanisms and relocation of vulnerable communities.[13]

The variety of climate change mitigation and adaptation measures available to States gives rise to a wide range of ways in which human rights can potentially be affected. For instance, incentives and subsidies for farmers to switch from food cultivation to bio-fuel production may negatively affect food production, especially in developing countries.[14]

The necessity for taking human rights into account in the implementation of mitigation and adaptation measures has been recognised by the Conference of Parties to the UNFCCC, which has stated that, in all climate change-related activities, States parties should fully respect human rights.[15] This is true with respect to substantive human rights obligations as well as procedural guarantees that apply to decision-making processes.

2.1. Substantive human rights obligations

One of the key areas of concern in the implementation of climate change mitigation and adaptation relates to the use of land and the potential for individuals or communities to be displaced or to suffer interference with their rights to use or access land. Interference with land use can occur in a number of ways. For example, mitigation measures may be based upon changing land use practices, such as increasing forestry or the production of biofuels. Adaptation steps could involve switching to alternative forms of agriculture or relocating people who may be vulnerable to rising sea levels. Individuals, families or communities may be forced to move in order to make room for these new practices or to comply with relocation plans. These activities raise concerns in relation to a number of fundamental rights.

13 Lenny Bernstein *et al*, *Climate Change 2007: Synthesis Report*, Contribution of Working Groups I, II and III to the Intergovernmental Panel on Climate Change, Fourth Assessment Report (2008), Summary for Policy Makers, section 4.

14 OHCHR (2009) [65]–[67]; Shelton (2009) 27; Hunter (2009) 355–6.

15 UNFCCC Conference of Parties, Decision 1/CP.16, *The Cancun Agreements: Outcome of the Work of the Ad Hoc Working Group on Long-term Cooperative Action under the Convention*, UN Doc FCCC/CP/2010/7/Add.1 (15 March 2011) 4.

Changes to land use may have significant implications for those people who rely on such lands for their subsistence. Mitigation activities such as reforestation and forest management or the cultivation of biofuels may require the use of land which has previously provided a source of income or food for local people. Similarly, adaptation measures may aim to address the impacts of climate change by encouraging changes to new agricultural practices or crop varieties. This has the potential to interfere with the farming practices of people who rely on subsistence agriculture. More broadly it also has the potential to interfere with the food security of the wider population, particularly where staple varieties are replaced with biofuels.[16]

Under the ICESCR all people are guaranteed the right to an adequate standard of living, which incorporates the rights to adequate food and water.[17] These rights are in jeopardy where people are unable to use the lands and waters they have previously relied on for subsistence. Where the interference is serious enough, other rights, including the right to the highest attainable standard of health[18] and even the right to life may be at risk.[19] When States contemplate employing mitigation and adaptation techniques that involve the reallocation of land, the obligation to respect human rights would require that they ensure that such techniques do not adversely affect the means of subsistence for people living on or near those lands. Further, where private actors are involved in mitigation and adaptation mechanisms, the obligation to protect human rights will require that States regulate those activities.

Certain forms of mitigation and adaptation may raise issues for the human right to the highest attainable standard of health.[20] For example, mitigation may be pursued through the construction of alternative energy sources such as nuclear power plants or wind turbines. These facilities may present a risk to the health of those people living nearby. As noted above, changes to agricultural practices may have impacts on food security and nutrition. Reallocation of land may also mean that certain plants with medicinal value may no longer be available. This is a particular issue for indigenous peoples, whose rights will be examined in more detail below.

16 Naomi Roht-Arriaza, 'Human Rights in the Climate Change Regime' (2010) 1(2) *Journal of Human Rights and the Environment* 211, 213.

17 *International Covenant on Economic, Social and Cultural Rights*, opened for signature 16 December 1966, 993 UNTS 3 (entered into force 3 January 1976) Art 11; Committee on Economic, Social and Cultural Rights, General Comment No 12, *The Right to Adequate Food (Article 11 of the Covenant)*, UN Doc E/C.12/1999/5 (12 May 1999); Committee on Economic, Social and Cultural Rights, General Comment No 15, *The Right to Water (Articles 11 and 12 of the Covenant)*, UN Doc E/C.12/2002/11 (20 January 2003).

18 ICESCR, Art 12; CESCR, General Comment 14, *The Right to the Highest Attainable Standard of Health (Article 12 of the Covenant)*, UN Doc E/C.12/2000/4 (11 August 2000).

19 *International Covenant on Civil and Political Rights*, opened for signature 16 December 1966, 999 UNTS 171 (entered into force 23 March 1976) ('*ICCPR*') Art 6.

20 ICESCR, Art 12.

With regard to freedom of movement, Article 12 ICCPR guarantees that all individuals are entitled to it, especially within their country of nationality.[21] This includes the right of all persons to choose their place of residence within their State of nationality, and protects against forced displacement.[22] This right is clearly implicated in situations where families or communities are forced to relocate in order to make way for mitigation or adaptation activities, for instance, reforestation policies. Article 12 ICCPR contemplates that the right to freedom of movement may be lawfully restricted where it is necessary to achieve a legitimate purpose, including national security, public order or public health, or to protect the rights and freedoms of others.[23] Arguably, measures which are implemented to address climate change are likely to be justifiable on the basis of at least one of these grounds.[24] However, in order to avoid a claim that human rights have been violated, a State would need to show that the relocation was necessary, that is, that there was no reasonable alternative to relocating the affected people. This may be easy enough to establish in the context of rising sea levels, where relocation of vulnerable people would seem most appropriate. It may be more onerous an obligation for States where choices are made to use some portions of land and not others, or to implement particular mitigation or adaptation strategies where others might have been equally as effective. In these situations, States would need to be able to demonstrate that the decision to use particular land, and the resulting relocation of persons from that land, were necessary.

Where the land used to implement climate change mitigation or adaptation projects is inhabited or utilised by indigenous groups, then a number of specific rights are implicated. Indigenous peoples are guaranteed the right to self-determination under common Article 1 of the ICCPR and the ICESCR. Under this right, indigenous peoples can freely pursue their economic, social and cultural development, and freely dispose of their natural wealth and resources.[25] The United Nations Declaration on the Rights of Indigenous Peoples (UNDRIP) gives expression to a number of specific rights which are owed to indigenous peoples, drawing on the rights guaranteed under the ICCPR and ICESCR.[26] In particular, the UNDRIP enumerates a number of rights in relation to indigenous lands which are relevant to adaptation and mitigation measures. Indigenous peoples have a right to the lands, territories and resources which they have traditionally owned, and States are obliged to give legal recognition and protection to such lands.[27] The right of indigenous peoples to maintain their spiritual relationship with their

21 HRCte, General Comment No 27, *Freedom of Movement (Article 12 of the Covenant)*, UN Doc CCPR/C/21/Rev.1/Add.9 (18 October 1999).

22 HRCte (1999) [7].

23 Art 12(3).

24 Pedersen (2011) 422.

25 ICCPR, Art 1; ICESCR, Art 1.

26 United Nations, *Declaration on the Rights of Indigenous Peoples*, GA Res 61/295 (2 October 2007).

27 UNDRIP, Art 26.

lands is also protected,[28] as is the right to practice their cultural traditions, including through having access to religious, cultural and historic sites.[29] Building on the right to freedom of movement guaranteed in the ICCPR, the UNDRIP reinforces the right of indigenous peoples not to be forcibly removed from their lands and territories.[30] States are obligated to prevent any action which has the aim or effect of dispossessing indigenous peoples from their lands or resources, and, failing that, to provide effective redress for such dispossession.[31]

The impact of climate change mitigation and adaptation actions on the rights of indigenous groups is of particular concern.[32] Mitigation measures aimed at reducing emissions from deforestation have the potential to significantly affect the rights of indigenous groups who reside in or use forest areas.[33] In this regard, within the framework of the REDD, the Cancun Agreements specifically require 'respect for the knowledge and rights of indigenous peoples and members of local communities'.[34] The concern is that forestry-based mechanisms give increased control to governments and private actors over forest areas, and where governments stand to receive significant offset funding from other States and overseas corporations there may be incentives to overlook the rights of indigenous peoples.[35] Potential is created for land grabs and corruption at the expense of local and indigenous communities, and governments may be reluctant to recognise indigenous land tenure over potentially lucrative lands. Without the protection of tenure, however, indigenous communities are not guaranteed to benefit financially from any transaction relating to their lands.[36]

Overall, with regard to both mitigation and adaptation measures, States have an obligation to respect the principle of non-discrimination, which underpins all human rights, particularly based on the right to equal protection before the law.[37] This is relevant, for instance, to clean development mechanisms, common but differentiated responsibility and indigenous peoples, who may be disproportionately affected by global warming.[38] Climate change action, and in particular

28 Ibid, Art 25.

29 Ibid, Arts 11 and 12.

30 Ibid, Art 10.

31 Ibid, Art 8.

32 Pedersen (2011) 405; Peter Kostishack, *In Focus – Respecting Rights in Climate Change Mitigation: the Debate about REDD*, International Human Rights Funders Group (30 October 2009).

33 Pedersen (2011) 408; Roht-Arriaza (2010) 224; Rowena Maguire, 'Designing REDD+ to Be Just: Considerations for a Legally Binding Instrument' (2014) 4(1) *Asian Journal of International Law* 169.

34 UNFCCC COP, Decision 1/CP.16, Appendix 1, 26, [1(2)(c)]; OHCHR (2009) 23, [68].

35 Kostishack (2009); Peter Newell, 'Climate Change, Human Rights and Corporate Accountability', in Stephen Humphreys (ed), *Human Rights and Climate Change* (Cambridge University Press, 2010) 126.

36 Kostishack (2009); Pedersen (2011) 408–9; Frances Seymour, 'Forests, Climate Change and Human Rights: Managing Risks and Trade-Offs', in Humphreys (2010) 207.

37 Von Doussa (2008) 961.

38 OHCHR (2009) 5, [11], 26, [81]; HREOC, Australia, *Human Rights and Climate Change* (Background Paper, 2008); Jay Williams, 'The Impact of Climate Change on Indigenous People – The Implications for the Cultural, Spiritual, Economic and Legal Rights of Indigenous People' (2012)

adaptation projects, also has the potential to contribute positively to development objectives. For example, programmes to improve agricultural practices and increase yield are good for development, as well as allowing for adaptation to the impacts of climate change.[39] However, issues can arise where these benefits are not distributed equitably. Human rights law provides that States must respect, protect and fulfil fundamental rights without distinction as to race, religion, age, sex, property or any other status.[40] Where adaptation projects offer potential advantages, human rights law requires that these must be delivered in a way which is free from discrimination. Further, the obligations to protect and fulfil human rights have particular implications where certain individuals or groups are more vulnerable to the negative effects of climate change. It is well recognised that climate change has disproportionately severe impacts on some individuals, communities and States. Those with fewest resources or who rely most directly on natural resources for subsistence, or those who reside in low-lying or otherwise susceptible areas are likely to suffer the most serious consequences.[41] The obligations to protect and fulfil human rights require that States address these vulnerabilities, and ensure that adaptation measures are tailored to meet the needs of those people who are mostly likely to be affected.

2.2. Procedural rights

As well as the substantive rights identified above, including the rights to self-determination, food, water, freedom of movement, health and indigenous rights, a number of procedural 'environmental' human rights also apply to the taking of climate change mitigation and adaptation measures. These claims are created based on a systemically integrated interpretation of the rights of access to information, justice and participation embedded in both environmental and human rights instruments. They thus naturally complement the environmental implications of substantive human rights.[42]

14 *Int'l J L Hum Rts* 648; Timo Koivurova, Sébastien Duyck and Leena Heinämäki, 'Climate Change and Human Rights', in Erkki J Hollo, Kati Kulovesi and Michael Mehling (eds), *Climate Change and the Law* (Springer, 2013), 287, 322.

39 Hall and Weiss (2012) 323.

40 ICCPR, Art 2; ICESCR, Art 2. *Convention on the Elimination of All Forms of Discrimination against Women*, opened for signature 18 December 1979, 1249 UNTS 13 (entered into force 3 September 1981); *Convention on the Elimination of All Forms of Racial Discrimination*, opened for signature 7 March 1966, 660 UNTS 195 (entered into force 4 January 1969).

41 Naomi Roht-Arriaza (2010) 211; McInerney-Lankford (2012) 436; Edward Cameron, 'Human Rights and Climate Change: Moving from an Intrinsic to an Instrumental Approach' (2010) 38(3) *Georgia Journal of International and Comparative Law* 673, 708; Amy Sinden, 'Climate Change and Human Rights' (2007) 27 *Journal of Land Resources and Environmental Law* 255. Barnett explores the issue of vulnerability to climate change in detail, and points out that other human rights violations can be drivers of vulnerability to climate change, for example where people are forcibly relocated to unsafe areas or where their rights to health and social security are already poorly protected: Jon Barnett, 'Human Rights and Vulnerability to Climate Change', in Humphreys (2010) 257.

42 See the contribution by Francesco Francioni and Ottavio Quirico in this volume.

It has long been recognised that people have the right to access information about projects which may affect them and to participate in decision-making in relation to those projects, as well as to access legal remedies where they suffer harm as a result.[43] The Aarhus Convention on Access to Information, Public Participation in Decision-Making and Access to Justice in Environmental Matters sets out these three rights to information, participation and access to remedies.[44] The ICCPR also guarantees the right to seek and receive information.[45] These rights require that where States propose to implement a climate change mitigation or adaptation mechanism they must provide information about the project to those people who are likely to be affected and allow them to participate in the decision-making process. Overall, these claims can be regarded as correlative to the positive State obligation to take action in order to fulfil human rights.[46]

The jurisprudence of the ECtHR provides an example of how these procedural guarantees are relevant to breaches of substantive human rights. The Court has decided a number of cases in which States were found to be in violation of their obligations under the ECHR[47] because they had failed to adequately assess the human rights implications of particular projects, or had failed to provide opportunities for those affected to participate in the associated decision-making processes. In a series of cases where human rights violations have been alleged based on environmental harm, the Court has clarified that States are under an obligation to assess the potential risks to human rights and to provide information about those risks to those people who may be affected.[48]

These procedural obligations would apply to climate change mitigation and adaptation measures. For example, proposals to construct alternative energy facilities such as hydro-electric plants would need to incorporate an assessment of human rights impacts and that information would need to be made publically available. Furthermore, opportunities must be provided for affected persons to participate in the decision-making processes surrounding the proposal.[49]

43 Roht-Arriaza (2010) 216–17; Dinah Shelton, 'Human Rights, Environmental Rights and the Right to Environment' (1992) 28 *Stanford Journal of International Law* 103.

44 *Convention on Access to Information, Public Participation in Decision-Making and Access to Justice in Environmental Matters*, opened for signature 25 June 1998, 2161 UNTS 447 (entered into force 30 October 2001) ('*Aarhus Convention*') Art 1.

45 ICCPR, Art 19(2).

46 See Koivurova, Duyck and Heinämäki (2013) 306–19; Shelton (2009) 6.

47 *European Convention for the Protection of Human Rights and Fundamental Freedoms*, opened for signature 4 November 1950, 213 UNTS 221 (entered into force 3 September 1953) ('European Convention on Human Rights').

48 See, for example, *Tatar v Romania* (ECtHR, Appl No 67021/01, 27 January 2009); *Oneryildiz v Turkey*, 48939/99 (30 November 2004); *Guerra and Others v Italy*, 116/1996/735/932 (19 February 1998); Pedersen (2011) 413–15.

49 See, for instance, *Gray v Minister for Planning, Director-General of the Department of Planning and Centennial Hunter Pty Ltd* (Land and Environment Court of New South Wales, Australia, 27 November 2006) 152 LGERA 258.

These procedural safeguards are obviously similar to the right of indigenous peoples to free, prior and informed consent, although that requirement is arguably of a higher standard, as prior consent must be obtained before a project can proceed. In an effort to minimise the possible exploitation and abuse of indigenous lands, the UNDRIP guarantees a number of procedural rights to indigenous peoples, to ensure their full and informed participation in decision-making processes which may affect them. States are obliged to consult and cooperate in good faith with indigenous peoples and obtain their free, prior and informed consent before implementing any action which would affect their occupation or use of their lands.[50] Further, States are obliged to make restitution or pay compensation for any lands which have been taken, occupied or damaged without the free, prior and informed consent of the indigenous owners of those lands.[51] These provisions mean that, where States propose to implement mitigation or adaptation measures, they will first have to seek the consent of any indigenous populations who occupy or rely on the land which will be affected by those projects.

The procedural rights found in the Aarhus Convention would in theory still allow States to proceed with the proposed activity even without consent, if information is provided and opportunities are offered to participate.[52] This has been confirmed by the ECtHR, which has highlighted the importance of procedural guarantees in many cases.[53] Such an approach is prominent in the European context, where States are afforded a wide margin of appreciation.[54] It is one way of dealing with the complex challenge of balancing competing objectives, where interference with human rights might be an unavoidable consequence of pursuing broader legitimate objectives. This balancing act is particularly complex in relation to climate change mitigation and adaptation, where a number of potentially conflicting imperatives intersect.

3. Challenges to positive climate policies raised by human rights: a case-by-case assessment

As the previous section has demonstrated, there are a number of human rights which are potentially affected by the taking of climate change mitigation and

50 UNDRIP, Arts 10, 18, 19, 32; Tara Ward, 'The Right to Free, Prior and Informed Consent: Indigenous Peoples' Participation Rights within International Law' (2011) 10(2) *Northwestern Journal of International Human Rights* 52. The Inter-American Court and Commission on Human Rights have applied the right to free, prior and informed consent in a number of matters relating to indigenous land rights (see for example *Maya Indigenous Communities of the Toledo District v Belize*, Case 12.053, Report No 40/04 (IAComHR, 12 Ocotber 2004); *Saramaka People v Suriname*, Preliminary Objections, Merits, Reparations, Costs, Series C No 172 (IACtHR, 28 November 2007)).

51 UNDRIP, Art 28.

52 Pedersen (2011) 414.

53 See, for instance, *Branduse v Romania*, Appl No 6586/03 (7 April 2009).

54 John Knox, 'Climate Change and Human Rights Law' (2009–10) 50 *Virginia Journal of International Law* 163, 210.

adaptation action. States which decide to implement particular mitigation or adaptation strategies are bound by their obligations to respect, protect and fulfil human rights. However, the practical implications of this finding must be considered in the broader context of climate change, including the cumulative, transnational and long-term effects of global GHG emissions. Consideration must also be given to the interaction between action on climate change and other global issues, most notably development. A number of questions exist in relation to the enforceability of human rights obligations and the need to strike a balance where a particular activity may present both positive and negative human rights outcomes.

One set of issues relates to the enforceability of human rights guarantees and the possibility of proving a violation. States' human rights obligations are typically owed to people within their territory or subject to their jurisdiction or control.[55] However, the effects of GHG emissions are notoriously transnational, creating difficulties in establishing obligations and proving violations. Related issues arise with regard to mitigation and adaptation activities, as many of the formal processes established through the UNFCCC framework involve States funding or implementing activities within the territories of other States. Where a State has a high degree of involvement in the project it may be a relatively simple matter to show the requisite degree of control to establish responsibility for any consequent human rights violation, but as a State's role becomes more remote it may be the case that only the territorial State could be found liable.[56]

Another set of issues relates to how negative human rights impacts of mitigation and adaptation activities should be balanced against the positive outcomes they contribute to in terms of addressing climate change overall, or how they interact with other human rights issues, including those associated with development. There are a number of factors to consider in this regard. First, a given mitigation or adaptation mechanism may produce both positive and negative human rights consequences. Adaptation projects might work to improve agricultural processes and increase yield, generating positive outcomes for development, but they might at the same time create issues for other human rights, as noted above.[57] Second, while mitigation and adaptation measures have the potential to cause negative human rights impacts, failure to take action might lead to other human rights impacts, potentially with broader and more serious consequences. An approach which imposes strict human rights protections for mitigation and adaptation activities may result in less effective strategies for dealing with climate change overall, leading ultimately to more human rights issues in years to come.[58]

55 See ICCPR, Art 2; HRCte, General Comment No 31, *Nature of the General Legal Obligation on States Parties to the Covenant*, UN Doc CCPR/C/21/Rev.1/Add.13 (29 March 2004); Robert McCorquodale and Penelope Simons, 'Responsibility beyond Borders: State Responsibility for Extraterritorial Violations by Corporations of International Human Rights Law' (2007) 70(4) *Modern Law Review* 598, 602–3; Knox (2009–10) 210; Pedersen (2011) 420.

56 Pedersen (2011) 420.

57 Hall and Weiss (2012) 322–3; Roht-Arriaza (2010) 225.

58 Pedersen (2011).

The interaction between climate change mitigation and adaptation and development presents significant issues as well. Climate change mitigation is often at cross-purposes to development, as many developing States rely on sources of energy which emit GHGs.[59] Requiring developing States to reduce emissions therefore threatens to hinder their development, and to set back the various positive human rights outcomes which ought to flow from that development. States which do implement mitigation strategies must be mindful not only of the direct human rights implications that such projects might have, but also of the impact they might have on access to energy and development more broadly, and the flow-on effect this may have on the fulfilment of the full range of human rights.

Human rights law provides some mechanisms for balancing potentially competing fundamental rights. As noted above, legal, proportionate and temporary restrictions on the freedom of movement are permitted where they are necessary to achieve a legitimate purpose, such as national security, public health or the rights of others.[60] Similar principles also apply to the rights to liberty,[61] freedom of expression[62] and religious practice.[63] The ICCPR also permits States to derogate from certain rights in times of declared national emergency.[64] It is, however, difficult to establish whether climate change presents circumstances which justify derogation from human rights obligations, for instance, an 'exceptional and actual or imminent danger which threatens the life of the nation'.[65] Does climate change pose a threat to the 'power structure' or to the 'well-being of the community'?[66] Furthermore, certain human rights, including the right to life, cannot be derogated from even in time of emergency.[67]

Balancing rights is therefore a delicate issue.[68] It may be the case that the negative consequences of mitigation and adaptation activities will be justified on the grounds that they are necessary to achieve the broader benefits that come from taking action on climate change.

With regard to the fields of environment and human rights, for instance, in *Hatton and Others v UK* the ECtHR adjudicated upon an alleged breach of the rights to health and private and family life, because of an increase in noise near the house of the plaintiffs resulting from governmental policies regulating night flights at the airport of Heathrow, London. The Court held that the economic

59 Hall and Weiss (2012) 322.
60 ICCPR, Art 12.
61 Ibid, Art 9.
62 Ibid, Art 19.
63 Ibid, Art 18.
64 Ibid, Art 4.
65 UN Sub-Commission on Prevention of Discrimination and Protection of Minorities, *Siracusa Principles on the Limitation and Derogation of Provisions in the International Covenant on Civil and Political Rights*, E/CN.4/1984/4, Annex (1984) [39].
66 Christoph Schreuer, 'Derogation of Human Rights in Situations of Public Emergency' (1982) 9 *Yale Journal of World Public Order* 113, 122.
67 ICCPR, Art 4; UN Sub-Commission (1984) [58].
68 Schreuer (1982) 125.

interest in maintaining a full service of night flights overrode the rights of the claimants, since only a small percentage of people suffered from noise, housing prices had not dropped, and the applicants could move elsewhere without significant financial losses.[69]

In any case, States are bound to minimise any negative impacts of their mitigation and adaptation policies, in line with their obligation to respect human rights. This obligation can work in concert with the procedural rights to participate in decision-making and to have access to remedies in order to ensure that the impact on affected individuals and groups is no greater than necessary. Ultimately however these assessments must be carried out on a case-by-case basis and there is no single approach which will apply to all situations.

Conclusion

States have obligations to respect, protect and fulfil human rights which apply whenever they seek to implement climate change mitigation and adaptation measures. As this chapter has demonstrated, there are a number of ways in which mitigation and adaptation measures might negatively impact upon human rights. This is most obvious where measures involve the appropriation or reallocation of land, and especially where that land is owned or utilised by indigenous peoples, but the range of rights which could be affected is as broad as the variety of mitigation and adaptation techniques which States might employ. Human rights law requires that States at the very least take steps to minimise impacts on fundamental rights.

One of the factors which complicates the application of human rights law to climate change mitigation and adaptation is the recognition that, while such strategies may affect fundamental rights, the failure to take steps to address climate change will present other, potentially greater, human rights concerns. Further, any given mitigation or adaptation mechanism may present both positive and negative human rights outcomes, particularly when viewed in the broader context of development. The challenge of balancing competing human rights objectives in addressing climate change is therefore a significant one.

A rights-based approach to climate change offers many advantages in terms of identifying vulnerabilities and addressing the needs of those most likely to be affected, and it provides a useful set of minimum standards which can help to regulate the implementation of mitigation and adaptation strategies. However, it is also an approach which inevitably involves a number of compromises, and it is not always possible to avoid negative human rights impacts entirely. The principles outlined above relating to procedural safeguards and the balancing of competing rights can be useful in determining the appropriate compromise in any given context, and it is argued that, if States are committed to upholding their obligation to respect human rights, then negative consequences can be kept to a minimum.

69 *Hatton and Others v the United Kingdom*, Appl No 36022/97 (ECtHR, 2 October 2001).

3 Human rights responsibility of private corporations for climate change?

The State as a catalyst for compliance

Anna Riddell

Introduction

In recent years it has been acknowledged that the climate change debate must involve a consideration of the role of the private sector. This is true not only at the domestic level, but also in the international domain, which no longer belongs solely to States. In fact, the rise of the private corporation[1] has created another entity capable of acting globally, which in some respects is coming to be the dominant player.[2] The dramatic changes in global actors came about in parallel with alterations in the world trade order resulting from the Second World War and later the oil crisis. The reduction of trade barriers and public control on foreign investment, increased privatisation of public enterprises, and companies expanding their geographical reach thanks to technology and communications innovation also contributed. Historically, foreign investment was discouraged or even prohibited, but the realisation that it could bring beneficial technology transfers, job creation and strong currency changed this attitude and nowadays it is encouraged, particularly by developing countries that see the opportunity to gain the development benefits of being associated with a private corporation.[3]

As to environment and climate change, it is possible to perceive both benefits and detriments from the activities of private corporations. They have the potential to be a driving force behind sustainable development, through financial and technology transfers and increasingly 'green' processes. This is the case, for instance, of CDMs under Article 12 of the Kyoto Protocol and voluntary carbon offsetting schemes.[4] At the same time, private corporations pose a likely threat to

1 There are several terms used to describe such entities, which, although they have slight variance in meaning, are often used interchangeably. These are: private corporation; transnational corporation; multinational corporation; and multinational enterprise. In recent UN documents and literature there is also a trend for referring to the 'private sector' more generally, or to 'business' or 'global business'. The term 'private corporation' will be used throughout for simplicity, but it is intended to encompass all these various meanings.

2 Karl P Sauvant (ed), *The Rise of Transnational Corporations from Emerging Markets: Threat or Opportunity?* (Edward Elgar, 2008).

3 World Bank, *World Trade Report* (2008) 15 ff.

4 See World Bank, *State and Trends of Carbon Pricing* (2014) 43.

the environment through production processes or methods that are inherently harmful,[5] and often both aspects manifest themselves in the same project. However, any benevolence is far outweighed by a lack of regard for environmental protection; private companies have free rein to act internationally, with little legally binding regulation because the rules relating to their conduct have not kept pace with their increasing global influence.[6] Cutting corners on employment, environmental or operational standards is possible, and thus private corporations do so with the aim of making the largest possible profit. It is therefore clear that addressing climate change in a meaningful way involves engaging these behemoths of industry and development and persuading them to significantly reduce their harmful practices and GHG emissions. The difficult question remains how this can be achieved via human rights, given that relevant legally binding measures are aimed at States as opposed to private actors.

1. Private corporations as subjects of human rights obligations

At Johannesburg in 2002 the private sector was finally acknowledged as being critical in the protection of the environment. This paradigm shift was noted by then Secretary-General of the UN, Kofi Annan, who stated that 'more and more we are realising that it is only by mobilising the corporate sector that we can make significant progress.'[7] The goal became to promote the activities of businesses and to encourage good practice, rather than focusing on regulating their activities or subjecting them to increased control. This stand was confirmed at Rio+20.[8] However, despite this approach of encouragement rather than coercion, perhaps because of the difficulty of making such firms legally accountable, thoughts on the legal personality of private corporations have also been evolving, so that these entities are now recognised as subjects of not only domestic law, but also international law.[9]

While international law is still mainly addressed to States, a series of developments since the Second World War have expanded the categories of actors on the international scene. A forerunner of this was the decision of the ICJ in which it decided that the UN is a subject of law, thus opening the potential for other

5 For an assessment of GHGs emitted by key private corporations see John Morhead and Tim Nixon, *Global 500 GHGs Performance 2012–13: 2014 Report on Trends*, Financial and Risk White Paper (Thomson Reuters, 2014).

6 See, for instance, Juliette Jowie, 'World's Top Firms Cause $2.2tn of Environmental Damage, Report Estimates', *The Guardian* (19 February 2010).

7 Kofi Annan, Speech at Lekgota Business Day, Johannesburg (1 September 2002), as reported in *Earth Time Johannesburg Summit: Big Business Gets Green Light from Kofi Annan* (2 September 2002).

8 Rio+20, UN Conference on Sustainable Development (2012) <www.uncsd2012.org>.

9 See, for instance, Peter Muchlinski, 'Corporations in International Law', in Rüdiger Wolfrum (ed), *Max Planck Encyclopaedia of Public International Law* (OUP, 2009).

new types of legal persons.[10] This was closely followed by the UDHR, which put individuals at the centre of the (human) rights–duties relationship, and meant that individuals, international organisations, non-governmental organisations and (somewhat controversially) companies have (human) rights and obligations. The preamble to the UDHR calls on 'every individual and every organ of society' to protect and respect human rights.[11] As Henkin observes, 'every individual includes juridical persons. Every individual and every organ of society excludes no one, no *company*, no market, and no cyberspace. The Universal Declaration applies to them all'.[12] This has been reiterated many times, for example in 2002 a Report by the International Council on Human Rights stated that 'there is a clear basis in international law for extending international legal obligations to companies in relation to human rights'.[13]

However, despite these advances it remains the case that private actors cannot easily be held accountable for their actions through binding human rights provisions both domestically and internationally. How then is it possible to influence their actions and encourage human rights responsibility for GHG emissions?

2. Fundamental climate change-related obligations of private corporations: 'soft' responsibility

The vast majority of instruments which make reference to environmental protection and human rights obligations of private corporations are 'soft law', with all the resultant limitations this brings in terms of applicability. In addition to the usual debate about the efficacy of soft law, there is an extra dimension to the question when one talks of applying it to private corporations. The driving forces behind respect of soft law instruments are accountability and acceptance of the provisions. Large and influential private corporations are unlikely to be worried

10 *Reparation for Injuries Suffered in the Service of the United Nations, Advisory Opinion* [1949] ICJ Rep 174, 179.

11 *Universal Declaration of Human Rights*, Adopted and proclaimed by UN General Assembly Res 217 A (III) of 10 December 1948, Articles 29 and 30.

12 Louis Henkin, 'The Universal Declaration at 50 and the Challenge of Global Markets' (1999) 25 *Brooklyn Journal of International Law* 17, 25 (emphasis added).

13 International Council on Human Rights, *Beyond Voluntarism: Human Rights and the Developing International Legal Obligations of Companies* (February 2002). Currently, full legal personality is still only possessed by States, but it is clear that non-State actors such as private corporations do possess at least limited international legal personality, without the capacity to modify their own legal obligations (Donald K Anton and Dinah Shelton, *Environmental Protection and Human Rights* (CUP, 2011) 863). Higgins argues that the notion of legal personality is a construct which does not always reflect reality, and suggests adopting the term 'participants' to describe actors in the international legal process, that is, State participants and individual participants, the latter including private corporations (Rosalyn Higgins, *Problems and Process: International Law and How We Use It* (OUP, 1999) 50).

about the impact of a slight deviation from soft obligations on their accountability, and if a private corporation does not agree to be bound by these duties, it simply continues to operate outside them. Additionally, soft law cannot provide redress for those affected by violations. In the reality of corporate actions, even with soft agreements, environmental and human rights abuses will occur, and it is of critical importance to have a means of providing redress for the victims. Since soft law creates no legal grounds for seeking compensation, hard or binding obligations are a necessary component of regulation.[14] However, even without binding legal force, in practice soft instruments have legal value, given the establishment of compliance monitoring mechanisms for many of them, which demonstrates a certain commitment to observe their provisions.

2.1. Sustainable companies

The first global instrument to deal with the topic of regulating private corporations was the 1976 OECD Guidelines for Multinational Enterprises. This was a multilaterally agreed instrument originally conceived of as a tool for facilitating foreign investment, and was addressed to developed countries. Initially, the Guidelines did not make any provision regarding social concerns or human rights. Fast-forward to their revision in 2000,[15] and the emphasis had shifted considerably to being principles and standards for responsible business conduct, including human rights and environmental protection. The Guidelines apply to private corporations operating in or from the 30 OECD participating countries and 12 non-member countries have also adopted them, which will contribute to the application of the Guidelines across developed and developing States not members of the OECD. There is also an enforcement mechanism whereby National Contact Points (NCPs) are endowed with the task of raising awareness of the Guidelines and solving disputes about their interpretation, as well as conducting enquiries into company behaviour.

The 2000 Guidelines required in Section II(2) that private corporations respect the human rights obligations of their host State, thus holding the private corporation to the same standards as that State.[16] It was not until the 2011 revision, however, that human rights became a key focus of the Guidelines.[17] The most notable addition was an entire section on human rights to bring the Guidelines into line with other developments, and making them the first intergovernmental agreement on corporate human rights standards.[18] As regards climate change, the Preface to the

14 See, for instance, Alexandre Flückiger, *Keeping Domestic Soft Law Accountable: towards a Gradual Formalization* (OUP, 2012); Francesco Francioni, 'Soft Law: a Contemporary Assessment', in Vaughan Lowe *et al* (eds), *Fifty Years of the International Court of Justice* (CUP, 1996) 167.

15 OECD, *Guidelines for Multinational Enterprises*, DAFFE/IME/WPG (2000) 15/FINAL (31 Oct 2001).

16 Ibid, Section 2 (General Policies).

17 OECD, *Guidelines for Multinational Enterprises* (29 September 2011).

18 Ibid, Chapter 4 (Human Rights).

Guidelines lists among its objectives the need to 'enhance the contribution to sustainable development made by multinational enterprises',[19] and stresses that the common aim of adhering governments is to encourage positive contributions from private corporations to economic, environmental, and social progress, while minimising the difficulties to which their various operations may give rise.[20] The General Principles also feature a particular section focusing on the environment, which calls on private corporations to 'take due account of the need to protect the environment, public health and safety, and generally to conduct their activities in a manner contributing to the wider goal of sustainable development'.[21] This has to take place 'within the framework of laws, regulations and administrative practise in the countries in which [multinational corporations] operate, and in consideration of relevant international agreements, principles, objectives and standards'.[22] Section VI (Environment), subsection 6(b) of the most recent 2011 update further specifies that enterprises should 'reduce greenhouse gas emissions' and '[be] efficient in their consumption of energy and natural resources'.[23]

An example of how the Guidelines and system of NCPs function in practice is provided by the referral to the Dutch and UK NCPs of the activities of Shell in Nigeria. A group of NGOs including Amnesty International and Friends of the Earth International submitted a complaint about breaches of the OECD Guidelines because of oil spills on 25 January 2011.[24] The Dutch NCP issued a report in 2013 which recommended that Shell further implement the OECD Guidelines, but did not establish a breach of them.[25] Upon criticism of this, and allegations that the NCP had allowed Shell to dictate the course of proceedings,[26] the NCP stated that 'based on the wording of the OECD Guidelines and the Procedural guidance for NCPs', their policy is to 'focus on improving possible negative behaviour of the company, so it will comply with the Guidelines in the near future',[27] and not to issue a decision on whether there has been a breach. The Dutch NCP also re-stated this position in addressing a subsequent request on the same issue.[28] The description of NCPs as an enforcement mechanism may

19 Ibid, Preface, [1].

20 Ibid, Preface, [9].

21 Ibid, Chapter VI (Environment).

22 Ibid.

23 Ibid, 6(b).

24 Amnesty International and Friends of the Earth International, *Royal Dutch Shell in the Niger Delta*, Complaint Submitted under the 'Specific Instance Procedure' of the OECD Guidelines (25 January 2011).

25 OECD NCP, Netherlands, *Shell – Friends of the Earth/Amnesty International*, Final Report (21 March 2013).

26 Amnesty International, *Nigeria: Oil Giant Shell Criticized Over Niger Delta Pipelines 'Sabotage' Claims*, Press Release (19 June 2013).

27 Dutch OECD NCP, *Reply to Concerns Raised in Press Release by Complainants* (26 June 2013).

28 OECD NCP, Netherlands, *Shell in the Niger Delta (II)*, Final Statement (21 September 2013) 6; Amnesty International and Friends of the Earth International, *Royal Dutch Shell in the Niger Delta*, Complaint Submitted under the 'Specific Instance Procedure' of the OECD Guidelines (30 December 2011).

thus be a little strong, as it appears these bodies are reluctant to hold corporations to account, preferring to appeal to them to uphold the Guidelines of their own volition. Shell has demonstrated the futility of this by continuing its controversial activities in Nigeria without attempting to address compliance with the Guidelines.

2.2. Caring for climate

In 1999 the UN Secretary-General launched the Global Compact,[29] asking businesses to support and adopt its ten core principles relating to human rights, core labour standards, protection of the environment and avoidance of corruption.[30] According to this initiative, duties on private corporations are partially shared with States, which are primarily responsible for the respect of human rights. Thousands of companies participate in the Global Compact on a voluntary basis and report publicly on steps they take to comply with the ten principles. However, as the Compact itself states, it is 'not a performance or assessment tool . . . nor does it make judgments on performance'.[31]

The Global Compact contains significant provisions relating to environmental accountability and has even been interpreted as implying the direct applicability of environmental principles to private enterprises.[32] However, the parameters of the environmental framework which it seeks to persuade private corporations to uphold are unclear, since, despite a mention of the Rio Declaration, there is no reference to any specific treaty. The OECD Guidelines and the UN Norms on the Responsibilities of Transnational Corporations and Other Business Enterprises with Regard to Human Rights (discussed below) are far clearer about the rules they seek to enforce.

In 2007, an additional intitive was launched by the Global Compact in conjunction with UNEP and the UNFCCC called 'Caring for Climate', aimed at defining a framework for engaging businesses in the climate change discussion.[33] It provides a platform to help business leaders advance practical solutions, share experiences, inform public policy and educate stakeholders. To date, 335 private corporations in 56 countries have signed up to the five commitments the initiative requires: reducing emissions, setting targets and reporting annually; devising a business strategy to address climate risks; engaging with policymakers to encourage scaled-up climate action; working with other enterprises to combat climate change; and becoming a climate-friendly business champion with stakeholders.[34] Annual reports analyse whether or not the signatories have met these criteria.

29 UN Global Compact, <www.unglobalcompact.org>.

30 Ibid, *The Ten Principles*, <www.unglobalcompact.org/AboutTheGC/TheTenPrinciples/index.html>.

31 UN Global Compact, Frequently Asked Questions, <www.unglobalcompact.org/AboutTheGC/faq.html>.

32 UN Global Compact, *The Ten Principles*, Principle 7; Charles Pearson, *An Environmental Code of Conduct for Multinational Companies?* (Durham: Duke University Press, 1987) 3, 155.

33 See <http://caringforclimate.org>.

34 UN Global Compact Office, *Caring for Climate: Progress Report 2013* (November 2013) 5.

Though the Global Compact itself may have had little impact in terms of addressing climate change, Caring for Climate has made real inroads into engaging corporations with the necessity of prioritising GHG emissions reductions and preventing climate change more broadly. The strict analysis from independent data has seen 50 countries delisted since 2007 for not complying with the commitments, and companies ranked on performance and disclosure aspects.[35] A Leadership Index produced annually can provide companies with a very real way of demonstrating their credentials in this field, if being climate-change aware is an important consideration for their reputation. Caring for Climate is thus resulting in considerable GHG emissions reductions, and successful engagement of private corporations in the responsibility for climate change.

2.3. A general framework for environmental protection

The ambitious 2003 Norms adopted by the UN Sub-Commission on the Promotion and Protection of Human Rights attempted to codify many key principles in the fields of fundamental rights, labour law, environmental protection, consumer protection and prevention of corruption. The main author of the Norms, Sub-Commission member David Weissbrodt, has hailed them the 'first non-voluntary initiative' in this field and described them as a soft law instrument deriving its authority from binding sources.[36] These norms state:

> Transnational corporations and other business enterprises shall carry out their activities in accordance with national laws, regulations, administrative practices and policies relating to the preservation of the environment of the countries in which they operate, as well as in accordance with relevant international agreements, principles, objectives, responsibilities and standards with regard to the environment as well as human rights, public health and biosafety, bioethics and the precautionary principle, and shall generally conduct their activities in a manner contributing to the wider goal of sustainable development.[37]

This would permit the application of tripartite human rights duties to private corporations, including their consequences as to GHGs.[38] Unfortunately, the Norms were received with mixed reviews. Strong opinions both in support and

35 Ibid.

36 David Weissbrodt and Muria Kruger, 'Norms on the Responsibilities of Transnational Corporations and Other Business Enterprises with Regard to Human Rights' (2003) 97 *AJIL* 901, 903 and 913.

37 UN Sub-Commission on the Promotion and Protection of Human Rights, *Norms on the Responsibilities of Transnational Corporations and Other Business Enterprises with Regard to Human Rights*, UN Doc E/CN.4/Sub.2/2003/38/Rev.2 (13 August 2003) Section 14.

38 On the application of tripartite human rights obligations to States see the contribution by Ottavio Quirico, Jürgen Bröhmer and Marcel Szabó in this volume.

against the extension of human rights responsibilities to businesses were expressed. The descriptive part of the Norms was largely uncontroversial, but on issues such as direct responsibility and accountability it was unclear what the consequences would be.[39] The main problem was that States had little appetite for a document which entangled the duties of States and private corporations so closely. The Norms appeared to assign State obligations to non-State actors, and it was not easily understood how this would work in practice.[40] In particular, whereas it is theoretically straightforward how the obligation to respect human rights could compel private corporations, particularly energy companies, to reduce GHGs, it is less clear how the obligations to protect and fulfil human rights would work. The UN Commission on Human Rights considered the Norms, and thanked the Sub-Commission for its work, saying they contained 'useful elements and ideas for consideration', but it did not approve them and concluded that they had 'no legal standing'.[41]

2.4. Environment and project financing

The Equator Principles represent a rather different attempt at providing rules for environmental conduct of corporations. They were created in 2003 by private sector banks, led by Citigroup, ABN AMRO and Barclays, as a voluntary set of standards for determining, assessing and managing social and environmental risk in project financing.[42] As of June 2013, 79 financial institutions had adopted the Principles, and the revised 2013 version (EP3) has become the de facto standard for banks and investors to assess major development projects around the world.[43]

A first revision of the Principles in 2006 added human rights to the agenda,[44] and in 2013 a second revision introduced climate change, meaning both are now key considerations in identifying, assessing and managing environmental and social risk in large and expensive financing projects.[45] The 2013 revision also appended an Annex on GHG emissions, requiring consideration of less carbon-

39 See, for instance, Canada, *Submission to the High Commissioner for Human Rights on the Responsibilities of Business Enterprises with Regard to Human Rights* (2004) [2.6].

40 Ibid, [2.2].

41 OHCHR, *Responsibilities of Transnational Corporations and Related Business Enterprises with regard to Human Rights*, Decision 2004/116 (20 April 2004).

42 *Equator Principles* (4 June 2003) <www.griequity.com/resources/AdminandGovernance/Equator_Principles.pdf>.

43 *Equator Principles* (4 June 2013) <www.equator-principles.com/resources/equator_principles_III.pdf>.

44 *Equator Principles* (4 June 2006) <www.equator-principles.com/resources/equator_principles_II.pdf>.

45 *Equator Principles* (2013) 2.

intensive fuel sources and technologies for new projects, as well as reporting for all projects with GHG emissions over a certain level.[46]

The Principles represent a significant global initiative, but, although they have generally been well received, some NGOs say they do not go far enough in their provisions.[47] However, consultation and grievance mechanisms under Principles 5 and 6 alone constitute a significant step forward in the attitude of private corporations towards the physical environment in which they invest, and those living in it.[48]

2.5. Applying tripartite human rights obligations to private corporations

In 2005 the UN appointed Professor John Ruggie to the newly created position of Special Representative on Business and Human Rights. Following the 2003 Norms, Ruggie wanted to reinvigorate the debate on the responsibility of private corporations, from a pragmatic, evidence-based perspective. He spent some time conducting extensive research and consulting experts and representatives of governments, business and civil society in various regions of the world. His first action was to propose the 'Protect, Respect and Remedy' policy framework in 2008.[49] This consisted of the three pillars of the State duty to protect against human rights abuses, the corporate responsibility to respect human rights, and the shared duty to facilitate greater access by victims to effective remedies, shedding light on the applicability of tripartite human rights obligations to private corporations. This was welcomed by the UN HRC, which further mandated Ruggie to make recommendations on how to operationalise and strengthen the three pillars of the framework.[50] A final report was presented in March 2011, which in June 2011 was fully endorsed by the Council: the 'Guiding Principles on Business and Human Rights' (GPs) is the most comprehensive document in existence purporting to give responsibility to private corporations for human rights.[51] It notes 'the role of business enterprises as specialized organs of society performing specialized

46 Ibid, 12, Annex A.

47 See Bank Track, *Equator Principles*, <www.banktrack.org/show/pages/equator_principles>.

48 *Equator Principles* (2013) 5 (Stakeholder Engagement) and 6 (Grievance Mechanisms).

49 John Ruggie, *Protect, Respect and Remedy: a Framework for Business and Human Rights*, Report by the Special Representative of the Secretary-General on the Issue of Human Rights and Transnational Corporations and Other Business Enterprises, UN Doc A/HRC/8/5 (7 April 2008).

50 HRC, Resolution 8/7 (18 June 2008).

51 John Ruggie, *Guiding Principles on Business and Human Rights: Implementing the United Nations 'Protect, Respect, Remedy' Framework*, Report of the Special Representative of the Secretary-General on the Issue of Human Rights and Transnational Corporations and Other Business Enterprises, UN Doc A/HRC/17/31 (21 March 2011). By means of Resolution A/HRC/RES/17/4, the HRC also established a Working Group to promote the effective and comprehensive dissemination and implementation of the Principles.

functions, required to comply with all applicable laws and to respect human rights'.[52]

GP 11 provides that 'business enterprises should respect human rights. This means they should avoid infringing on the human rights of others and should address adverse human rights impacts with which they are involved'.[53] The official commentaries included in the Principles state that 'the responsibility to respect human rights is a global standard of expected conduct for all business enterprises wherever they operate . . . [It] exists over and above compliance with national laws and regulations protecting human rights'.[54] It is also encouraging that GP 12 comprehensively requires respect for all fundamental rights,[55] given the possible broad range of human rights impacts which climate change may have.

These 'foundational principles' are accompanied by some 'operational principles', namely policy commitment (GP 16), human rights due diligence (GPs 17–21) and remediation (GP 22). The second of these is the core of the obligation on private corporations, involving 'assessing *actual and potential human rights impacts*, integrating and acting upon the findings, tracking responses and communicating how impacts are addressed'.[56] Crucially when one considers climate change human rights impacts, GP 17 requires that human rights '*due diligence*' cover 'adverse human rights impacts that the business enterprise may cause or contribute to through its own activities, or which may be directly linked to its operations, products or services by its business relationships'.[57] The combination of referring to 'actual or potential' impacts, and impacts the business 'may cause or contribute to' together go some way to address the tricky issue of causation which characterises the climate change field.[58] The clever use of the term 'due diligence', which is already familiar in a CSR context, perhaps facilitated acceptance and understanding of the Principles by private corporations. However, the concept has a rather different meaning in human rights than in CSR, which has led to some confusion about what the Principles actually require.[59] To be consistent with the human rights meaning of the term, GPs 18–21 elaborate objective legal criteria laying down the standard of conduct required,[60] with respect to identification and assessment of actual and potential human rights impacts,[61] effective integration

52 Ruggie (2011) 6.
53 Ibid, Principle 11.
54 Ibid, para II.A.11.
55 Ibid, Principle 12 and commentary.
56 Ibid, Principle 17 (emphasis added).
57 Emphasis added.
58 See the contribution by Quirico, Bröhmer and Szabó in this volume.
59 For a full discussion on the issue of due diligence in the GPs, see Jonathan Bonnitcha and Robert McCorquodale, *Is the Concept of 'Due Diligence' in the Guiding Principles Coherent?* (SSRN Paper, 29 January 2013).
60 Ruggie (2011) 16 ff.
61 Ibid, Principle 18.

and appropriate action,[62] tracking responses,[63] and external communication and reporting.[64]

Overall, despite detailed requirements, the Principles can go no further in obliging private corporations to comply than any of the above instruments, because they have no binding legal force, even if they do have the support of the UN. The lack of a legally enforceable grievance mechanism means compliance with the Principles is, as with the other instruments, reliant upon private corporations voluntarily fulfilling the requirements.

3. Enforcing responsibility via States: the pivotal role of the obligation to protect human rights

3.1. Lack of responsibility of private corporations for GHG emissions and State protection

As can be seen from the above analysis, there are mechanisms and principles which seek to control the damaging behaviour of private corporations in relation to human rights and climate change. However, given their soft law nature and lack of ability to compel observance, it is clear that whatever improvements have resulted from private companies adhering to these rules have been thanks to the voluntary compliance of these actors. For instance, Coca-Cola has a progressive CSR policy, particularly with regard to energy:

> we recognise that climate change may have long-term direct and indirect implications for our business and supply chain. As a responsible multinational company, we have a role to play in ensuring we use the best possible mix of energy sources, improve the energy efficiency of our manufacturing processes, and reduce the potential climate impact of the products we sell.[65]

However, in the past Coca-Cola has been the subject of much criticism for its operations in India, where several different plants have contributed to water shortages in drought stricken areas, and polluted groundwater and soil, harming both the environment by destroying farms, and also the livelihoods of the human population.[66] Therefore, despite the apparent positive developments and the enthusiasm for a more socially and environmentally responsible private corporation, it is clear that if the adherence to human rights standards is left on

62 Ibid, Principle 19.

63 Ibid, Principle 20.

64 Ibid, Principle 21.

65 The Coca-Cola Company, *Position Statement on Climate Protection* (2014) <www.coca-colacompany.com/position-statement-on-climate-protection>.

66 India Resource Centre, <www.indiaresource.org/index.html>.

a voluntary basis, it is unlikely that the majority of corporations will place human rights and climate change considerations high on their agenda, unless they can see a 'business case' for doing so. Some have even argued that the proliferation of codes is improving corporate images while preventing discussion on the development of necessary hard law measures.[67]

Today, if one thinks of private corporations the many incidents of questionable conduct in breach of the obligation to respect human rights by even some of the largest companies come to mind. The Union Carbide-owned plant in Bhopal, India, allowed poisonous gases to escape, killing and permanently injuring hundreds of thousands, in an incident which is now a world-famous example of a subsidiary of a private corporation disregarding the impact of poor maintenance and cost-cutting measures on the local population.[68] The oil extraction process undertaken by Shell in Nigeria also highlights the disregard which private corporations can have for the environment and human rights. Large-scale oil spills into the Niger delta are common, with huge spills in 2008 and 2011 far exceeding the Deepwater Horizon incident.[69] For decades the Ogoni people, native to the delta, have suffered as a result of the pollution, but despite repeated legal attempts to hold Shell accountable they have yet to be successfully challenged.[70] Procter & Gamble are another company with a poor environmental record, having spent years polluting the Fenholloway River in Florida from a cellulose plant which pumped out up to 50 million gallons of waste water per day, contaminating the river and nearby wells, and making the fish population inedible.[71] Unilever too on numerous occasions has been held liable for discharging polluting substances into water sources.[72] The list goes on.

67 Justine Nolan, 'All Care, No Responsibility? Why Corporations Have Limited Responsibility and No Direct Accountability for Human Rights Violations under International Law', in Gretchen Bellamy, Lara Blecher and Nancy Kaymar-Stafford (eds), *Corporate Responsibility for Human Rights Impacts* (American Bar Association, 2014) 3, 4; Corporate Watch, *What's Wrong with Corporate Social Responsibility? The Arguments against CSR* (2006) <www.corporatewatch.org>. On 25 June 2014, by means of Resolution A/HRC/26/L.22/Rev.1, the HRC mandated an open-ended intergovernmental working group to elaborate a legally binding instrument regulating the activities of transnational corporations and other business enterprises under international human rights law. Interestingly, the Resolution was approved with the support of developing countries, despite the opposition of developed States (see International Justice Resource Centre, *In Controversial Landmark Resolution, Human Rights Council Takes First Step toward a Treaty on Transnational Corporations' Human Rights Obligations* (15 July 2014) <www.ijrcenter.org>).

68 *Union Carbide Company v Union of India* (1991) SCC (4) 584 (Indian Supreme Court, 3 October 1991).

69 See, for instance, Friends of the Earth Netherlands, *Oil Spills in the Niger Delta* (2008) <https://milieudefensie.nl/publicaties/factsheets/oilspills-in-the-niger-delta>.

70 See the discussion on the cases before the Dutch OECD NCP and the AComHPR above and below.

71 *Hodges v Buckeye Cellulose Corporation*, Case No F-370, 174 So 2D 565 (District Court of Appeal of Florida, First District, 29 April and 21 May 1965).

72 *USA v Conopco Inc d/b/a/ Unilever Home and Personal Care USA*, Case No 3:13-CR-00223, in the US District Court for the District of Connecticut.

Examples of cases against private corporations in relation to climate change and human rights are difficult to find. Paradoxically, it is not easy to hold the main actors responsible for breaching human rights through GHG emissions independently of States. Internationally, this is due to the absence of a compulsory substantive and procedural human rights framework for corporations. However, since it has long been accepted that they have a duty to take action to prevent breaches of human rights law, States are responsible for GHGs emitted by private corporations in violation of fundamental rights.[73] In fact, a State has a legal obligation to adopt effective laws and practices that control actions and omissions not only by State agents, but also by non-State actors, such as private corporations.[74] Due diligence requires States to do all they can to protect human rights, including by regulating such entities where necessary.[75] Thus, for example, in *Ogoniland* irresponsible oil development practices in breach of human rights by private corporations and the government of Nigeria were in issue before the AComHPR. According to its jurisdictional competence, the Commission only held the Nigerian government responsible for breaching human rights, including, most significantly, the rights to health and to a healthy environment under Articles 11 and 24 of the Banjul Charter.[76] However, the Commission emphasised that 'Governments have a duty to protect their citizens, not only through appropriate legislation and effective enforcement but also by protecting them from damaging acts that may be perpetrated by private companies'.[77] With specific regard to climate change, the duty to prevent human rights breaches underpins the decision of the European Committee of Social Rights in *Marangopoulos*, whereby Greek energy policy not preventing the operation of GHG-emitting lignite mines was considered in breach of the right to health under the European Social Charter.[78] The Inuit and Athabaskan petitions to the IAComHR against the US and Canada are also based on the State failure to prevent GHG emissions by private

73 In the *Velázquez Rodríguez* case, the IACtHR noted: 'If the State apparatus acts in such a way that the violation goes unpunished and the victim's full enjoyment of such rights is not restored as soon as possible, the State has failed to comply with its duty to ensure the free and full exercise of those rights to persons within its jurisdiction. *The same is true when the State allows private persons or groups to act freely and with impunity to the detriment of the rights recognised in the Convention* [ACHR]' (*Velásquez Rodríguez v Honduras* (IACtHR, 29 July 1988) [176] (emphasis added)).

74 For a full discussion on the State duty to protect human rights see the contribution by Quirico, Bröhmer and Szabó in this volume.

75 However, there has been a political reluctance on the part of both home and host States to take this obligation seriously (see Surya Deva and David Bilchitz, *Human Rights Obligations of Business: beyond the Corporate Responsibility to Respect?* (CUP, 2013)).

76 *Social and Economic Rights Action Center and Center for Economic and Social Rights v Nigeria*, Communication 155/96 (IAComHR, 27 October 2001) [69]. See also the contribution by Faustin Ntoubandi and Roland Adjovi in this volume. As we have seen, subsequent action under the OECD Guidelines for Multinational Enterprises against Royal Dutch Shell proved unfruitful (see section 2.1 above).

77 *Social and Economic Rights Action Center* (2001) [57].

78 *Marangopoulos Foundation for Human Rights (MFHR) v Greece*, Collective Complaint No 30/2005 (ECteSR, 26 December 2005).

corporations in violation of several human rights, although the outcome is so far controversial.[79]

Domestically, private corporations could be held responsible for excessive GHG emissions in breach of human rights obligations before national courts. Along these lines, in *Gbemre* a claim was submitted with the Federal High Court of Nigeria against the Shell Petroleum Development Company of Nigeria Limited, the Nigerian National Petroleum Corporation and the Attorney-General of the State of Nigeria. The Court held the practice of gas flaring by these companies in the course of their oil production activities, including GHGs, in breach of the fundamental rights to life and dignity under Articles 33 and 34 of the Nigerian Constitution, which were considered intertwined with the rights to health and environment under Articles 16 and 24 ACHPR. The State was also held responsible for not adopting adequate legislation preventing oil pollution.[80] In this case, besides State responsibility for failing to protect human rights, corporate responsibility for violating the obligation to respect human rights was made possible by the fact that emissions were localised and thus easily imputable.[81] However, since private corporations are innumerable, when GHG emissions are delocalised it is very difficult to establish a causal nexus between their polluting activity and climate change-related human rights violations. This is proved by the *Kivalina* case, where the US District Court for the Northern District of California had to decide on alleged damages caused by ExxonMobil Corporation and other energy companies for the destruction of the city of Kivalina, Alaska, by flooding caused by climate change. The Court dismissed the claims in light of the impossibility of establishing a causal link between GHGs emitted by the defendant corporations and the destruction of Kivalina. In fact, according to the Court GHGs have been emitted over centuries by corporations dispersed throughout the world and multiple environmental factors contribute to causing sea level rise.[82]

Thus, it is only through States fully accepting their responsibility to protect human rights that the most effective route for controlling the activities of private corporations is to be found. Because of its centralised nature, the State might permit to aggregate the responsibility of dispersed private corporations. The duty to

79 Inuit, *Petition to the Inter-American Commission on Human Rights, Seeking Relief from Violations Resulting from Global Warming Caused by Acts and Omissions of the United States* (IAComHR, 7 December 2005) 107–8; Arctic Athabaskan Council on Behalf of All Arctic Athabaskan Peoples of the Arctic Regions of Canada and the United States, *Petition to the Inter-American Commission on Human Rights, Seeking Relief from Violations of the Rights of Arctic Athabaskan Peoples Resulting from Rapid Arctic Warming and Melting Caused by Emissions of Black Carbon by Canada* (IAComHR, 23 April 2013) 18 ff. See also the contribution by Mouloud Boumghar in this volume.

80 *Gbemre v Shell Petroleum Development Company Nigeria Limited and Others*, Case No FHC/B/CS/53/05 (NgHC, 14 November 2005).

81 See further the contribution by Tineke Lambooy and Hanneke Palm in this volume.

82 *Native Village of Kivalina, City of Kivalina v ExxonMobil Corporation and Others*, Order Granting Defendants' Motions to Dismiss for Lack of Subject Matter Jurisdiction, Case No C 08–1138 SBA (US District Court for the Northern District of California, Oakland Division, 30 September 2009) 20 and 21.

protect human rights involves preventing and remedying the actions of private corporations, for example, by requiring non-financial disclosure obligations which could include due diligence and accountability as to human rights, or by imposing emissions caps or fuel-efficiency regulations. In this sense, the OECD 'Common Approaches' compel Member States to conduct an environmental review of supported credits for export of capital goods and services.[83] More generally, UN GPs 8–10 provide that States should pursue policy coherence, by ensuring their climate change regulations are in line with human rights obligations.

3.2. Barriers to the implementation of the obligation to protect human rights

Although in principle it sounds straightforward – States can control the activities of private corporations domiciled within their territory by introducing domestic legislation with which enterprises must comply – in practice the difficulty of balancing the interests of society with economic and policy considerations has meant that few States have really taken the opportunity to introduce appropriate legislation.[84] This is facilitated by jurisdictional immunity enjoyed by States under the political question doctrine.[85]

Furthermore, from a global perspective State control of transnational activities has a fundamental flaw: the biggest GHG emitters do not act in one State, but worldwide and in many guises. Often subsidiaries will be carrying out the harmful activity, and the parent company based in a State attempting to enforce domestic emissions caps is thus able to absolve itself of legal responsibility. If both 'home' and 'host' States applied similar controls, this issue would be somewhat alleviated, but as noted above, some States, such as Burma, keep the regulation of private corporations activities to a minimum in order to encourage them to operate there.[86] Tighter regulations in some States can also cause 'emissions dumping' in those States which have not introduced emissions caps.[87]

One possible way to avoid this problem would be to allow the extraterritorial application of human rights law. This is a particularly attractive concept in relation to climate change, where the polluting activity may have been carried out in one State, but its harmful impacts may be felt in another country or countries. The UN GPs take a conservative approach to this issue, leaving unsettled the question of whether there is an extraterritorial dimension to the duty to protect human rights. The commentary to GP2 states that 'at present States are not generally required under international human rights law to regulate the extraterritorial

83 OECD, *Recommendations on Common Approaches for Officially Supported Export Credits and Environmental and Social Due Diligence (The 'Common Approaches')* (28 June 2012) [13].

84 See the contribution by Bridget Lewis in this volume.

85 *Native Village of Kivalina* (2009) III.A and Conclusion.

86 World Bank (2014) 91.

87 See the contribution by Olivier De Schutter in this volume.

activities of businesses domiciled in their territory and/or jurisdiction. Nor are they generally prohibited from doing so, provided there is a recognized jurisdictional basis'.[88] Amnesty International has criticised this as being one of the weak points of the Principles and has asked that it be the focus of future work, though this suggestion has not yet been taken up.[89] However, even if the Principles had further engaged with this, the case law on the extraterritoriality of human rights mainly requires the 'occupation or control' of a territory,[90] which, when viewed in the climate change context, is problematic, given the transboundary nature of global warming.

Though direct extraterritorial application of human rights in relation to GHG emissions remains controversial, there is a possibility for States to impose indirect extraterritoriality on private corporations, by requiring that parent companies fulfil due diligence obligations in relation to their subsidiaries. This could take the form of reporting duties, which could help to foster a culture of respect for human rights both at home and abroad.[91] However, the likelihood of States actually imposing such requirements is even slimmer than the chance that they will translate their duty to protect into binding domestic legal obligations on private corporations.

Conclusion

Climate change is the field of law which is most crucial for our continued existence on the planet, yet relevant obligations are difficult to apply against the most polluting entities, that is, private actors. Both internationally and domestically, human rights responsibility of private corporations is centred around the State, which should be ultimately answerable for GHGs emitted by private corporations. Internationally, this is due to the fact that there are no substantive and procedural human rights mechanisms effectively binding upon private corporations. Whether or not these should be in place with regard to GHG emissions is nonetheless uncertain, given that even domestically, where human rights duties are enforceable, it is in practice very difficult, or even impossible, to demonstrate a causal link and attribute delocalised GHG emissions to private companies. As a consequence, the obligation to protect human rights emerges as the main tool in keeping private corporations responsible, whereby the State is answerable for failure to prevent their GHG emissions. The effectiveness of this duty is nevertheless problematic, because of issues of policy discretion and extraterritoriality.

88 Ruggie (2011) Commentary to Principle 2.

89 Amnesty International, *United Nations: a Call for Action to Better Protect the Rights of Those Affected by Business-related Human Rights Abuses*, Public Statement (14 June 2011).

90 *Cyprus v Turkey*, Appl Nos 6780/74 and 6950/75 (ECtHR, 26 May 1975).

91 Stefanie Ricarda Roos 'Climate Change and Human Rights: What Follows for Corporate Human Rights Responsibility?', in Oliver C Ruppel, Christian Roschmann and Katharina Ruppel-Schlicting (eds), *Climate Change: International Law and Global Governance* (Nomos, 2013) 299, 311.

Part II

Specific rights

4 Climate change and right to life

Limits and potentialities of the human rights protection system

Christine Bakker

Introduction

It is increasingly being recognized that climate change has a significant negative impact on the enjoyment of human rights.[1] Within this ambit, the right to life is the most fundamental; it is a prerequisite for the enjoyment of all other human rights. Indeed, '[l]ife is tantamount to human existence'; it is 'the foundation of man as a citizen and as a member of society'.[2]

The right to life has been recognized as a legal principle since the 1776 Virginia Bill of Rights, and, besides its systematic inclusion in national constitutions, it has become the cornerstone of all international and regional human rights instruments.[3] The States that have ratified these instruments have committed themselves to respect, protect and fulfil the right to life for all persons within their jurisdiction. Besides the 'negative obligation' to refrain from violating this right themselves, this commitment also entails a number of 'positive obligations' for States parties, such as the adoption of adequate legislative, administrative and other measures, and the provision of an effective remedy in case of violation of these rights.[4]

At the same time, climate change puts the enjoyment of the right to life at risk, both directly and indirectly. The human right to life is *directly* affected when people die in hurricanes or floods caused by global warming. While this most often

1 See John H Knox, *Report of the Independent Expert on the Issue of Human Rights Obligations relating to the Enjoyment of a Safe, Clean, Healthy and Sustainable Environment, Mapping Report*, A/HRC/25/53 (30 December 2013); Human Rights Council, Res 18/22, *Human Rights and Climate Change* (30 September 2011); Decision 1/CP.16, *The Cancun Agreements: Outcome of the Work of the Ad Hoc Working Group on Long-term Cooperative Action under the Convention*, COP-16, Cancun (29 November–10 December 2010) [8].

2 Christian Tomuschat, 'The Right to Life: Legal and Political Foundations', in Christian Tomuschat, Evelyne Lagrange and Stefan Oeter (eds), *The Right to Life* (Leiden, Koninklijke Brill, 2010) 3.

3 Article 3 UDHR (1948), Article 1 ADRDM (1948), Article 6 ICCPR (1966), Article 6 CRC (1989), Article 2 ECHR (1950), Article 4 ACHR (1969), Article 4 ACHPR (1982), and Article 5 ArCHR (2004). All these instruments protect the right to life from arbitrary deprivation.

4 See also Knox (2013).

happens in developing countries, recent extreme weather events in the US and Europe have confirmed that such risks are increasing in other parts of the world as well.[5] The right to life is *indirectly* at risk when the rights to health, to food and to clean water are threatened as a result of rising sea levels, floods or extreme climatological phenomena. In other words, the negative impact of climate change on people's health and living conditions also entails risks for the enjoyment of their right to life.

In this chapter, the question will be considered to what extent the current human rights system, both in normative terms and its enforcement mechanisms, is sufficiently equipped to protect the right to life in the face of global threats such as climate change. In other words, the chapter will consider the potentialities and limits of bringing claims before human rights bodies, alleging a violation of the right to life as a result of, for example, the failure of a State to take adequate measures to prevent or mitigate climate change. To this end, the chapter will examine the main judgments and decisions of the regional and international human rights courts and monitoring bodies, considering how they have interpreted States' obligations to respect, protect and fulfil the right to life, especially in cases that concern acts or omissions in the environmental field. Where a link is established in these cases with other human rights and in particular the right to health, this will also be considered.

The analysis will first consider the judgments and recommendations of the regional human rights bodies: the ECtHR, the IACtHR and 1AComHR, and the AComHPR (Section 1). Subsequently, it will discuss the views, decisions and general comments from the international bodies: the HRCte and the CteRC, and those related to the right to health of the CESCR (Section 2). Based on the conclusions drawn from this analysis, some reflections will be put forward on the question whether there is a need to re-conceptualize the human rights obligations of States in the light of climate change, in a more collective, future perspective (Section 3).

The chapter will demonstrate that despite the progressive interpretation of States' obligations not to commit and to prevent environmental damage and natural disasters, in particular by the regional human rights courts, the possibilities for successful climate change-related human rights claims based on the right to life are extremely limited. Limiting factors are the individual nature of the right to life and the related 'victim requirement', restrictions to extraterritoriality, and the condition of causality between harm and the acts or omissions of a particular State. On the other hand, the case law and decisions of human rights monitoring bodies increasingly recognize a linkage between climate change and, in particular, the rights to health and to an adequate standard of living.

5 See Bernice Lee and Diarmuid Torney, 'New Drivers of US Climate Action? The Politics of Extreme Weather and Adaptation', in Christine Bakker and Francesco Francioni (eds), *The EU, the US and Global Climate Governance* (Ashgate, 2014) 163.

Therefore, the chapter concludes by suggesting that climate change-related claims should, where possible, be founded on *both* the right to life *and* on other fundamental rights (the rights to health, to an adequate standard of living, or, where available, the right to a healthy environment), particularly when a human rights instrument covers both individual and collective rights.[6] Moreover, it is suggested that a reflection is required on how human rights obligations should, *de lege ferenda*, be re-conceptualized in order to also protect these rights from the far-reaching threats of climate change. Considering that the protection of life on earth is a public interest of all people, it is argued that States should guarantee the right to life of their people in the broadest sense, and that human rights courts and monitoring bodies should interpret the obligations of States to protect people from environmental harm, in a more collective perspective. [7]

1. Regional standpoints on the obligations to respect, protect and fulfil the right to life

At the regional level, human rights courts and other monitoring bodies have developed an 'environmental dimension' of human rights, including the right to life, through an expansive interpretation of the obligations of States that are parties to the relevant human rights conventions. After an examination of the main conclusions that can be drawn from this regional case law, their possible applicability to the threats posed by climate change to the enjoyment of the right to life will be considered.

1.1. European Court of Human Rights: due diligence obligations and environmental threats

In the ECHR the right to life is enshrined in Article 2:

> 1. Everyone's right to life shall be protected by law. No one shall be deprived of his life intentionally save in the execution of a sentence of a court following his conviction of a crime for which this penalty is provided by law.

The ECtHR has developed a substantial jurisprudence on the obligations of States to respect, protect and fulfil the right to life.[8] Moreover, it has expressed

6 See Section 3.

7 See Francesco Francioni, 'International Common Goods: an Epilogue', in Federico Lenzerini and Ana Vrdoljak (eds), *International Law for Global Common Goods: Normative Perspectives on Nature, Culture and Human Rights* (Hart Publishing, 2014) 449.

8 *Streletz, Kessler and Krenz v Germany*, Appl Nos 34044/96, 35532/97 and 44801/98 (22 March 2001) [87]; *McCann and Others v UK*, Appl No 18984/91 (27 September 1995) [149]; *Andronicou and Constantinou v Cyprus*, Appl No 86/1996/705/897 (9 October 1997) [171]; *Bubbins v UK*, Appl No 50196/99 (17 March 2005) [135]; *Salatkhanovy v Russia*, Appl No 17945/03 (16 October 2008) [70]. See also the contribution by Quirico in this volume.

itself on the scope of such obligations in several cases where the risks to the right to life were directly related to environmental concerns. The main cases are *LCB v The United Kingdom*, *Oneryıldız v Turkey*, *Budayeva and Others v Russia* and *Murillo Saldias v Spain*,[9] where the Court considered the obligation of States to protect the right to life, and in particular the obligation to prevent violations of this right. In *LCB v The UK*, the Court did not uphold the alleged violation of Article 2, since the applicant had not established a causal link between her father's exposure to radiation from nuclear tests while serving in the Army, and the leukaemia from which she died. However, in *Oneryıldız v Turkey*, a violation of Article 2 was confirmed, because the State failed to take any measures to prevent an explosion that occurred at a municipal rubbish tip, killing thirty-nine people who had illegally built their dwellings around it. In *Budayeva and Others v Russia*, the Court ruled that Russia had not complied with its positive obligation to warn the local population of the risks posed by a severe mudslide after heavy rainfalls, despite the foreseeable threat to life. In *Murillo Saldias v Spain* the Court also confirmed the existence of such positive obligations in the event of natural disasters.[10] Another environmental case currently pending before the ECtHR in which a violation of the right to life is invoked is *Smaltini v Italy*,[11] involving a claim that the air pollution emitted by the Ilva steelworks factory in the city of Taranto was the cause of the leukaemia from which the claimant died.

In these cases, the Court has laid down a number of criteria for assessing if a State has complied with its positive obligations in a specific situation.[12] These obligations are both of a substantive and of a procedural nature, and can be summarized as follows. First, Article 2 of the ECHR does not only concern deaths resulting directly from the actions of the agents of a State; public authorities also have a duty to guarantee the rights of the Convention when they are threatened by other (private) persons or activities.[13] Second, States' obligations apply to dangerous activities (for example nuclear tests or the operation of chemical factories with toxic emissions), either carried out by public authorities or by private companies. The extent of these obligations depends on factors such as the *harmfulness of the dangerous activities* and the *foreseeability of the risks to life*.[14] Third, the obligation to prevent the loss of life also applies in cases of natural disasters, even

9 *LCB v UK*, 14/1997/798/1001 (9 June 1998) (hereafter: *LCB*); *Paul and Audrey Edwards v UK*, Appl No 46477/99 (14 March 2002); *Oneryıldız v Turkey [GC]*, Appl No 48939/99 (18 July 2002) and *Budayeva and Others v Russia*, Appl Nos 15339/02, 21166/02, 20058/02, 11673/02 and 15343/02 (30 November 2004).

10 *Murillo Saldias v* Spain, Appl No 76973/01 (28 November 2006); *Manual on Human Rights and the Environment* (Council of Europe Publishing, 2nd edn, 2012) 37.

11 Appl No 43961/09, lodged on 7 August 2009.

12 For a comprehensive analysis of this case law, see Riccardo Pavoni, *Interesse pubblico e diritti individuali nella giurisprudenza ambientale della Corte europea dei diritti umani* (Editoriale scientifica, 2013).

13 *LCB* (1998) [36]; *Oneryıldız* (2002) [71]; *Budayeva* (2004) [128].

14 *Oneryıldız* (2002) [73]; *LCB* (1998) [37]–[41].

though they are as such beyond human control.[15] Therefore, public authorities have the primary duty to put in place a legislative and administrative framework including regulations taking into account the special features of a situation or an activity and the level of potential risk to life, to fulfil the public's right to information concerning such activities, and to maintain an adequate defence and warning infrastructure in the case of natural disasters.[16] Moreover, where an infringement of the right to life has occurred, the State must provide an effective remedy, ensuring inter alia, that an independent and impartial investigation is undertaken, and that breaches of the right to life are prosecuted and punished as appropriate.[17]

When considering the relevance of this jurisprudence for the protection of the right to life from threats posed by climate change, the following points emerge. On the one hand, the Court's extensive interpretation of the above-mentioned positive obligations is a clear confirmation that States have a far-reaching due diligence obligation with respect to environmental threats. However, while dangerous activities, especially high levels of CO_2 and other GHG emissions, contribute to global warming, it is impossible to establish any measurable connection between the emissions of a specific factory – or even of all industries and other emitters within one State combined – and the overall phenomenon of climate change.[18] In other words, there are too many uncertainties regarding the *causality* between the dangerous activities carried out within the borders of an individual State on the one hand and the possible loss of life on the other to hold this State responsible for any deaths resulting from the effects of climate change, or for failing to take the necessary measures to prevent such deaths. Therefore, in addition to the causality requirement, the criterion of *foreseeability of the risks to life* cannot be fulfilled.

Furthermore, the limited scope of the jurisdiction of the ECtHR is another barrier to successful climate change disputes, since cases can only be heard with regard to dangerous activities or natural disasters that occur within the territory of the States parties to the ECHR, that is, within Europe, and which entail risks for the life of people within the jurisdiction of these same States. The case law of the ECtHR on the extraterritorial application of the ECHR has evolved from the *territorial* notion of the '*espace juridique*' in *Bankovic*,[19] according to which a State is responsible for extraterritorial human rights violations when it exercises overall effective control over the territory or physical space where those violations occur, to a broader, *personal* basis, which makes extraterritorial human rights responsibility dependent on whether a State exercises physical power and control over the individuals who allege to be victims of violations, even if such harm occurred

15 *Murillo Saldias* (2006).
16 *Oneryıldız* (2002) [89–90]; *Budayeva* (2004) [129], [132].
17 *Oneryıldız*, (2002) [91]–[94]; *Budayeva* (2004) [138]–[142].
18 See Alan Boyle, 'Human Rights and the Environment: Where Next?' (2012) 23 *EJIL* 613, 640–2.
19 ECtHR, *Banković and Others v Belgium and 16 Other Contracting States*, Appl No 52207/99 (19 December 2001).

abroad.[20] However, the ECtHR has not gone so far as to extend this extraterritorial application to environmental damage caused by industries or companies in States parties to the ECHR, affecting the rights of people in other States. Even though in *Tatar v Romania* the Court referred to principle 21 of the Stockholm Declaration and principle 14 of the Rio Declaration,[21] which both stipulate the duty of States to ensure that local industrial activities do not cause any transboundary harm, it did not explicitly extend the scope of the State's positive obligations under the ECHR not to cause and to prevent environmental harm to people in neighbouring States. Consequently, the chances for successful climate change-related claims before the ECtHR are extremely limited.

Finally, it is interesting to note that the ECteSR, which monitors the implementation of the European Social Charter (ESC), has recognized a linkage between the right to life, as laid down in Article 2 ECHR, and the right to health, included in Article 11 ESC. In *Marangopoulos Foundation for Human Rights v Greece*,[22] the Committee found a violation of the right to health arising from the failure of the respondent State to take adequate measures to prevent and counter atmospheric pollution caused by lignite mining and power stations fuelled by lignite. Although lignite is not a fossil fuel emitting the main greenhouse gases covered by the Kyoto Protocol targets, the Committee did take the international and European legal frameworks relating to climate change into account, and gave substantial weight to Greece's non-compliance therewith. This is a significant precedent, in the first place because it transcends, as it were, the boundaries between different human rights by explicitly recognizing the connection between the right to life and the right to health, which may be one of the possible options for more effective human rights litigation in climate-related cases, as will be further discussed below.[23] Second, the Committee recognized the relevance of the general attitude of a State in respecting environmental norms and commitments when considering compliance by that State with its positive human rights obligations.

1.2. Inter-American human rights protection system: connecting the right to life with people's physical environment

According to Article 1 ADRDM:

> Every human being has the right to life, liberty and the security of his person.

20 ECtHR, *Al-Skeini and Others v United Kingdom*, Appl No 55721/07 (7 July 2011) [130]–[42]; Riccardo Pavoni, *Climate Change and Human Rights: a Viable and Unavoidable Relationship*, paper presented at the International Conference on Mitigation of Climate Change: Law, Policy and Governance, University of Delhi (25–27 April 2014) 8.

21 ECtHR, *Tatar v Romania*, Appl No 67021/01 (27 January 2009).

22 ECteSR, *Marangopoulos Foundation for Human Rights (MFHR) v Greece*, Complaint No 30/2005 (6 December 2006).

23 The ECteSR has reiterated this connection in the case *FIDH v Greece*, Complaint No 72/2011 (23 January 2013) [50].

Article 4 ACHR stipulates:

> 1. Every person has the right to have his life respected. This right shall be protected by law and, in general, from the moment of conception. No one shall be arbitrarily deprived of his life.

However, as in the European system, the IAComHR and the IACtHR have also not limited their interpretation of the right to life to the 'protection against arbitrary killing'. The IAComHR affirmed that the 'realization of the right to life, and to physical security and integrity is necessarily related to and in some ways dependent on one's physical environment.'[24] Moreover, in *Sawhoyamaxa Indigenous Community v Paraguay*, the IACtHR held that Paraguay failed to respect the right to life of the members of the Sawhoyamaxa Community, affirming that 'States have the duty to guarantee the creation of the conditions that may be necessary in order to prevent violations of such inalienable right',[25] and that 'States must adopt any measures that may be necessary to create an adequate statutory framework to discourage any threat to the right to life.'[26] The question then arises what kind of measures are required to protect the right to life from environmental threats, including those resulting from climate change.

The IAComHR and IACtHR have long been at the forefront of the environmental rights jurisprudence.[27] The most important cases in this regard are *Yanomami*,[28] *Mayagna (Sumo) Awas Tigny Community v Nicaragua*,[29] *Yakye Axa Indigenous Community v Paraguay*,[30] *Kawas-Fernández v Honduras*,[31] *Maya Indigenous Community of the Toledo District v Belize*[32] and *The Kichwa Peoples of the Sarayaku Community and Its Members v Ecuador*.[33] Although these cases primarily concern violations of the land and property rights of indigenous peoples resulting from the granting of logging concessions by the State, these violations were also considered to threaten their fundamental rights. In the Inter-American case law, the procedural obligations of States related to environmental threats are also particularly well defined.[34]

24 IAComHR, *Report on the Human Rights Situation in Ecuador*, OEA/Ser.L/V/II.96, Doc.10Rev1 (1997) Ch VIII.

25 *Sawhoyamaxa Indigenous Community v Paraguay*, Merits, Reparations, Costs, Series C No 146 (IACtHR, 29 March 2006) [145](a).

26 Ibid, [153].

27 For a comprehensive overview of these cases, see Cathrin Zengerling, *Greening International Jurisprudence: Environmental NGOs before International Courts, Tribunals, and Compliance Committees* (Martinus Nijhoff Publishers, 2013) 109–12.

28 *Yanomami Indians v Brazil*, Case No 7615 (IAComHR, 3 March 1985). See also Boumghar in this volume.

29 *The Mayagna (Sumo) Awas Tigny Community v Nicaragua* (IAComHR, 31 August 2001).

30 *Yakye Axa Indigenous Community v Paraguay* (IAComHR, 17 June 2005).

31 *Kawas-Fernández v Honduras* (IACtHR, 3 April 2009).

32 *Maya Indigenous Community of the Toledo District v Belize*, Case No 12.053 (IAComHR, 12 October 2004).

33 *Kichwa Peoples of Sarayaku v Ecuador* (IACtHR, 27 June 2012).

34 See Riccardo Pavoni, 'Environmental Jurisprudence of the European and Inter-American Courts of Human Rights: Comparative Insights', in Ben Boer (ed), *Environmental Law Dimensions of Human Rights* (OUP, 2015).

In two environmental cases, the right to life was at the heart of the dispute. In the *Yanomami* case, the IAComHR upheld the claim by the Yanomami Indians against Brazil that their right to life and, inter alia, their right to health, included in the American Declaration on the Rights and Duties of Man, were violated by the construction of the Trans-Amazonian highway that crosses their homelands and forced them to resettle without any compensation.[35] In *Sarayaku Community v Ecuador*, the IACtHR held that Ecuador had violated the right to life by allowing a private oil company to operate within the ancestral territory of the Kichwa people of Sarayaku.[36] The Court recognized that a State cannot be held responsible for all situations in which the right to life is at risk. It held that for such a responsibility of the State to arise, it must be determined that:

> at the time the events occurred, the authorities knew or should have known about the existence of a situation that posed an immediate and certain risk to the life of an individual or of a group of individuals, and that they did not take the necessary measures available to them that could be reasonably expected to prevent or avoid such risk.[37]

However, when considering to what extent this case law can also be applied to the risks posed by climate change to the right to life, two main difficulties are immediately clear. Indeed, the conditions formulated by the Court that 'the authorities knew or should have known about the existence of a situation that posed (i) *an immediate and certain risk* to the life of (ii) *an individual or of a group of individuals*' are problematic. Even though it could be argued that sufficient scientific data and analyses are available to establish that climate change is causing severe risks to life on earth, the *immediacy* and *certainty* of the risks for the life of identified people cannot be ascertained. According to the IACtHR, this is only the case when the authorities do not take the necessary measures *available to them* that could be *reasonably expected to prevent or avoid such risk.* It depends on the specific circumstances what measures can be considered to fulfil these conditions. In any case, according to the strict terms of this jurisprudence, the obligation to protect the right to life would be related to the immediate risk posed by, for example, a particular hurricane or flood, and not to the risks posed by climate change in general.

The question of whether States can be held responsible, based on human rights law, for causing or failing to prevent climate change, and the question of the causality between GHG emissions of a specific State on the one hand, and the effects of climate change for individual groups of people on the other, were addressed in the first case concerning climate change brought before a human rights body, the *Inuit* case.[38] The Inuit Circumpolar Conference filed this petition

35 *Yanomami* (1985) [3].

36 Zengerling (2013) 111.

37 *Kichwa Peoples of Sarayaku v Ecuador* (IACtHR, 2012) [155] and [166].

38 *Petition to the Inter-American Commission on Human Rights Seeking Relief from Violations Resulting from Global Warming Caused by Acts and Omissions of the United States* (7 December 2005).

in 2005, claiming that the human rights of the Inuit people, a population group living in the Arctic region, were violated as a result of climate change, caused by GHG emissions from the US. Since the US has not ratified the ACHR, the petitioners relied on the ADRDM. With regard to the rights to life (Article I ADRDM) and health (Article IX ADRDM), the claimants held that:

> [c]limate change has damaged the Arctic environment to such an extent that the damage threatens human life . . . Not only are harvested species becoming scarcer as the climate changes, the Inuit's access to these foods is diminishing due to difficulties in travel and changes in game location.[39]

Therefore, reduced access to the Inuit's traditional means of subsistence is among the main factors constituting an alleged violation of the right to life. As is well known, the IAComHR rejected the admissibility of the petition without providing any specific reasons,[40] but a special hearing was held to investigate the relationship between global warming and human rights.[41] Despite this rejection, the Inuit petition still constitutes a significant precedent in itself, and provides valuable arguments substantiating the linkages between climate change and human rights, including the right to life. Similar arguments have been advanced in the more recent petition on black carbon submitted by the Arctic Athabaskan Council to the IAComHR.[42]

Finally, it may seem surprising that, save the Inuit and Arctic Athabaskan peoples,[43] claimants in environmental cases have not relied on the right to a healthy environment included in the San Salvador Protocol.[44] However, the reason why this right has never been invoked is that the Protocol does not recognize it as a justiciable right.[45] Had this been the case, it could arguably broaden the scope for climate-related claims in the Inter-American system.

39 Ibid, 91.

40 In a letter dated November 16, 2006, the IAComHR informed the petitioners that the Commission would not consider the petition because the information it provided was insufficient for making a determination. See Jessica Gordon, 'Inter-American Commission on Human Rights to Hold Hearing after Rejecting Inuit Climate Change Petition' (2006–7) 7(2) *Sustainable Development Law and Policy* 55. The decision of inadmissibility is not available on the website of the IAComHR.

41 The testimonies given at this hearing are available at <www.ciel.org/Climate_Change/IACHR_Inuit_5Mar07.html>.

42 Arctic Athabaskan Council on Behalf of All Arctic Athabaskan Peoples of the Arctic Regions of Canada and the United States, *Petition to the Inter-American Commission on Human Rights Seeking Relief from Violations of the Rights of Arctic Athabaskan Peoples Resulting from Rapid Arctic Warming and Melting Caused by Emissions of Black Carbon by Canada* (2013) 5, 54–7, 74.

43 Ibid, 55; *Inuit Petition* (2005) 74.

44 *Additional Protocol to the American Convention on Human Rights in the Area of Economic, Social and Cultural Rights* ('San Salvador Protocol'), adopted on 17 November 1988, Art 11.

45 See Riccardo Pavoni, 'Environmental Jurisprudence of the European and Inter-American Courts of Human Rights: Comparative Insights', in Ben Boer (ed), *Environmental Law Dimensions of Human Rights* (OUP, 2015), referring to Article 19(6) of the San Salvador Protocol. According to this provision, only Art 8(1)(a) (Right to Unionization) and Art 13 (Right to Education) can be the object of individual petitions.

1.3. African Commission on Human and Peoples' Rights: linking the rights to life, to health and to a satisfactory environment

Article 4 ACHPR provides:

> Human beings are inviolable. Every human being shall be entitled to respect for his life and the integrity of his person. No one may be arbitrarily deprived of this right.

To date, the ACtHPR, which came into being in 2004, has decided only two cases on the merits.[46] However, the AComHPR (African Commission), which became operational in 1987, has adopted several decisions concerning the right to life.[47] In particular, in 2001 the African Commission delivered a landmark decision in human and environmental rights law in the *Ogoniland* case.[48] This case concerned alleged violations by Nigeria of the right to life (Article 2 ACHPR), the right to health (Article 16) and the right to a general satisfactory environment (Article 24), claiming that the Nigerian government, through a State company, has been directly involved in oil production operations, which caused the contamination of the environment and serious health problems. Aiming to suppress popular protests, the government ordered military operations including the destruction of Ogoni villages and food sources.[49] The African Commission upheld the claims and appealed to the Nigerian government to ensure protection of the environment, health and livelihood of the people of Ogoniland by, inter alia:

> [e]nsuring that . . . the safe operation of any further oil development is guaranteed through effective and independent oversight bodies for the petroleum industry, and by [p]roviding information on health and environmental risks . . . to communities likely to be effected by oil operations.[50]

The fact that the African Commission upheld a claim for violation of the right to a general satisfactory environment, together with violations of the rights to life and health is an encouraging precedent. It could be argued that human rights instruments that protect the individual right to life, the right to health and the collective right to a healthy environment, such as the African Charter, provide a more solid basis for climate change-related claims.[51]

46 Specifically, in *Norbet Zongo v Burkina Faso*, Appl No 13/2011 (March 2014), the Court ruled that Burkina Faso had failed in its obligations to protect journalists, including their right to life.

47 See, for example, the cases 276/03 *Centre for Minority Rights Development (Kenya) and Minority Rights Group (on behalf of Endorois Welfare Council) v Kenya*; 204/97 *Mouvement burkinabé des droits de l'homme et des peuples v Burkina Faso*; and 266/03 *Kevin Mgwanga Gunme et al v Cameroon.*

48 *The Social and Economic Rights Action Center and the Center for Economic and Social Rights v Nigeria (Ogoniland case)*, Case No 155/96 (AComHPR, 27 October 2001). See also the contribution by Faustin Ntoubandi and Roland Adjovi in this volume.

49 Idem, [1]–[9]. See also Zengerling (2013) 113.

50 *The Social and Economic Rights Action Center* (2001) Findings.

51 On this point, see also section 2.3 below.

2. The positions of universal human rights monitoring bodies

2.1. UN Human Rights Committee: adhering to the 'victim-requirement'

Article 6(1) ICCPR states:

> Every human being has the inherent right to life. This right shall be protected by law. No one shall be arbitrarily deprived of his life.

The UN HRCte, tasked with the monitoring of the implementation of the ICCPR, has adopted a General Comment, describing the right to life as a 'supreme' and non-derogable human right that should not be interpreted in a narrow way.[52] States are urged to take positive measures in order to protect human life, including the reduction of infant mortality, increase in life expectancy, and eradication of malnutrition and epidemics.[53] The Committee has also mentioned the need to adopt such measures in its recommendations to States based on their periodic reporting on the implementation of the ICCPR.[54]

In the context of its Communications procedure, the HRCte has, to date, not developed a systemic practice of identifying States' obligations with respect to the right to life. Nevertheless, in a few cases concerning the right to life of prisoners,[55] a political activist[56] and persons who were the victims of forced disappearances,[57] the Committee has confirmed a State's due diligence obligations to take appropriate measures to protect the life of these persons.[58] However, in *André Brun v France*, concerning an alleged violation of the right to life resulting from the cultivation of transgenic plants, the Committee observed that the facts of the case do not show that:

> the position of the State party on the cultivation of transgenic plants in the open field represents, in respect of the author, an actual violation or an imminent threat of violation of his right to life and his right to privacy, family and home.[59]

52 HRCte, General Comment No 6, *Article 6*, UN Doc HRI/GEN/1/REV.6 (1982) [1].
53 Ibid, [5].
54 For example, *Concluding Observations on Paraguay*, UN Doc CCPR/C79/Add.48, and on *Zambia*, UN Doc CCPR/C/79/Add.62.
55 *Lantsov v Russia*, Communication No 763/1997 (26 March 2002) 96–102.
56 *Roger Chongwe v Zambia*, Communication No 821/1998 (25 October 2000) UN Doc A/56/40 (Vol II) 137–43.
57 *Celis Laureanu v Peru*, Communication No 540/1993 (22 March 1996) 108–15.
58 For more details, see *Leading Cases of the Human Rights Committee*, Compiled by Raija Hanski and Martin Scheinin, Institute for Human Rights, Turku/Abo (Åbo Akademi University, 2003).
59 Communication No 1453/2006, UN Doc CCPR/C/88/D/1453/2006 (23 November 2006).

This statement, which confirms the 'victim requirement', clearly illustrates the obstacles for bringing claims related to environmental concerns based on an alleged violation of the right to life under the ICCPR.

2.2. Committee on the Rights of the Child: explicitly linking climate change with the rights to life and health

Article 6(1) CRC specifies:

> States Parties recognize that every child has the inherent right to life.

In the context of the periodic reporting of States on the implementation of the CRC, the CteRC has asked the Pacific island State Tuvalu to provide information on the measures taken to minimize the impact of climate change on the enjoyment of children's rights, in particular measures to protect children in emergency situations and to secure safe drinking water and sanitation.[60]

With regard to the relationship between various rights, the CteRC affirmed in its General Comment No 5,[61] that the '(e)njoyment of economic, social and cultural rights is inextricably intertwined with enjoyment of civil and political rights.' In particular, the CteRC has recognized a clear link between children's right to health and their right to life. Moreover, in its General Comment No 15, the CteRC refers to climate change, drawing attention to the relevance of the environment, beyond environmental pollution, to children's health. It states:

> Environmental interventions should include addressing climate change as this is one of the biggest threats to children's health and to exacerbating health disparities. States should, therefore, put children's health concerns at the center of their climate change adaptation and mitigation strategies.[62]

The CteRC may help clarify the State's obligations related to the protection of children's right to life from climate change-related risks, for the following reasons. First, the CRC includes both the right to life and the right to health, and the CteRC has explicitly recognized the linkages between these two rights, as well as the interconnectedness between the right to health and climate change. Second, the CteRC can use the reporting exercise of States more systematically to ask what measures they are taking to protect childrens' rights to life (and other, directly related rights, in particular the right to health and the right to adequate living conditions), from climate change-related threats. Finally, through their representatives, including NGOs, children can present a complaint to the CteRC,

60 *List of Issues to Be Taken up in Connection with the Consideration of the Initial Report of Tuvalu*, CRC/C/TUV/Q/1 (8 February 2013) [10].

61 CteRC, *General Comment No 5*, CRC/C/GC/2003/5 (November 2013) [6].

62 CteRC, *General Comment No 15*, CRC/C/GC/15 (April 2013) [50].

if they consider that a State has failed to take the necessary measures to protect their rights.[63]

2.3. Committee on Economic, Social and Cultural Rights: right to health, collective rights and related State obligations

Some of the rights included in the ICESCR that are related to the right to life may also be adversely affected by climate change, in particular the right to health. In this respect, the CESCR can issue recommendations based on the periodic reports of the States parties; it can issue general comments, and since the entry into force of the Optional Protocol on a complaint procedure in May 2013, it can also hear complaints from States, individuals or NGOs. The Committee has provided a detailed analysis of the relevant positive obligations of States in its General Comment No 14.[64] Taking account of the different availability of resources among States, a distinction is made between 'progressive' and 'immediate' obligations. General Comment No 14 provides that 'progressive realization means that States parties have a specific and continuing obligation to move as expeditiously and effectively as possible towards the full realization of Article 12.'[65] Moreover, the Committee has clarified the obligations to respect, to protect and to fulfil the right to health, also distinguishing between obligations at the national level, and obligations of international assistance and cooperation.[66] The UN Special Rapporteur on the Right of Everyone to the Enjoyment of Physical and Mental Health, Paul Hunt, has outlined his vision on the link between climate change and the right to health, arguing that the right to health also entails obligations of States to take measures, through international cooperation, to stop and reverse climate change.[67] Furthermore, in the context of the periodic reporting by States parties to the Covenant, the ICESCR has expressed its concern on the effects of climate change on the enjoyment of the right to health and recommended a State party to 'intensify its efforts to address issues of climate change, including through carbon reduction schemes'.[68]

It could thus be argued that the right to health – as well as certain other social, economic and cultural rights, such as the rights to an adequate standard of living and to a healthy environment – provides a more suitable basis for climate change-

63 *Optional Protocol to the Convention on the Rights of the Child on a Complaint Procedure*, entered into force on 14 April 2014, <http://treaties.un.org/doc/source/signature/2012/CTC_4-11d.pdf>.

64 General Comment No 14, UN Doc E/C.12/2000/4 (2000).

65 Ibid, [31]. Article 12(1) ICESCR provides: 'The States Parties to the present Covenant recognize the right of everyone to the enjoyment of the highest attainable standard of physical and mental health.'

66 CESCR (2000) [34]–[37] and [38]–[42] respectively.

67 Paul Hunt and Roda Khosla, 'Right to the Highest Attanable Standard of Health', in Stephen Humphreys (ed), *Human Rights and Climate Change* (CUP, 2010) 252–3.

68 See, for example, *Concluding Observations of the Committee on Social, Economic and Cultural Rights: Australia* (2009) [27].

related claims than the individual right to life. Considering that at least some of these rights can be conceived in a more collective manner, the obligations of States to respect, protect and fulfil these rights also have a wider scope than those related to the right to life.[69] This idea is confirmed by the express recognition by the ECteSR of the linkage between the right to health, included in the ESC, and the right to life, enshrined in the ECHR, in two cases concerning States' obligations related to the prevention of environmental harm.[70]

3. Re-conceptualizing human rights obligations related to climate change: some initial reflections

The analysis in the preceding sections has revealed that the possibilities of invoking the right to life before human rights bodies to hold States accountable for harm caused by climate change are extremely limited. Therefore, the question can be asked whether, in the face of the challenges posed by climate change, this most fundamental of human rights can still be guaranteed based on its current interpretation and enforcement.

There are several ways to look at this *impasse*. It can either be argued that human rights are not 'suitable' to protect people from environmental harm and from the risks of climate change, because these risks do not fit the existing normative framework. However, it can also be suggested that if the existing normative framework is insufficient to respond to 'new' challenges, the norms themselves need to be progressive, and through the appropriate means of interpretation by the competent judicial bodies, be adjusted to the changing global context and threats. In this regard, Francesco Francioni has convincingly argued:

> [I]t does not make much sense to engage human rights language to combat environmental degradation only when such degradation affects the rights to life, property, and the privacy of certain directly affected individuals. This reductionist use of human rights may even be counter-productive in that it tends to reduce environmental values to the very limited sphere of individual interest, thus adulterating their inherent nature of public goods indispensable for the life and welfare of society as a whole.[71]

This view is also particularly pertinent when considering the global threats caused by climate change and the risks it entails for the living conditions on our planet.[72]

69 Except the ICESCR, the right to health is also included, inter alia, in the ACHPR, the Protocol of San Salvador in the Inter-American system, the CRC, and the ECteSR.

70 See section 1.1 above.

71 Francesco Francioni, 'International Human Rights in an Environmental Horizon' (2010) 21 *EJIL* 41, 55, also cited by Riccardo Pavoni, 'Public Interest Environmental Litigation and the European Court of Human Rights: No Love at First Sight', in Lenzerini and Vrdoljak (2014) 331, 333.

72 See also the chapter by Francesco Francioni and Ottavio Quirico in this volume.

Francioni therefore pleads for:

> a more imaginative and courageous jurisprudence which takes into consideration the collective dimension of human rights affected by environmental degradation and adapts the language and technique of the human rights discourse to the enhanced risk posed by global environmental crises to society and, indeed, to humanity as a whole.[73]

In this same sense, Riccardo Pavoni has put forward that the ECtHR should also take account of the collective dimension of environmental damage, beyond the harm suffered by individual persons:

> The ECtHR, as an institution overseeing a 'living instrument which . . . must be interpreted in the light of present day conditions'[74] should be more receptive to well-founded claims calling attention to serious shortcomings in the environmental law and practice of respondent States which potentially affect the well-being of entire communities of people, while not necessarily disclosing a special and immediate link with the enjoyment of ECHR rights by the complainants.[75]

This idea could arguably also be advanced with respect to the other existing human rights monitoring bodies, in particular the Inter-American system, in which the Court and Commission have already confirmed the obligations of States to respect, protect and fulfil fundamental individual rights, including in a more collective dimension, particularly in connection with the right to health.[76] Such an approach would entail a progressively expansive interpretation of States' obligations, recognizing, for example, that in order to prevent violations of the individual right to life from environmental harm, the adoption of collective measures may also be required. Such measures could include, in principle, the adoption of laws prohibiting activities with particularly heavy GHG emissions, or at least the adoption of measures to comply with States' regional and/or international commitments to reduce such emissions.

Moreover, there may be a need for regional human rights courts and other monitoring bodies to progressively reconsider their approach to extraterritoriality in the context of environmental harm. In this regard, principles that have developed in the framework of international environmental law, such as the

73 Francioni (2010) 55.

74 *Tyrer v United Kingdom*, Appl No 5856/72 (25 April 1978) [31].

75 Pavoni, 'Public Interest' (2014) 333.

76 On the collective dimension of the right to health in light of globalization see Benjamin Mason Meier and Larisa M Mori, 'The Highest Attainable Standard: Advancing a Collective Human Right to Public Health' (2005) 35 *Columbia Human Rights Law Review* 101 ff. This collective dimension is also highlighted within the context of the 'right to a healthy environment' (see, for instance, Principle 1 of UN GA Resolution 45/94 of 14 December 1990 on the *Need to Ensure a Healthy Environment for the Well-being of Individuals*: 'All individuals are entitled to live in an environment adequate for their health and well-being').

precautionary principle, the 'polluter pays principle' and especially the principle of transboundary harm prevention, could be more systematically taken into account in the identification of States' obligations.

Finally, it has been scientifically established that the consequences of climate change not only affect the lives of the people living today, but that they will also have negative effects for future generations. The question whether the current generation owes a duty to future generations to preserve the global environment, in particular in the era of climate change, is the subject of a continuing scholarly debate.[77] In this context, it has been argued that such a duty is inherent in the concept of human rights itself, affirming that the main international human rights instruments:

> reveal a fundamental belief in the dignity of all members of human society and in an equality of rights that extends in time as well as space. Indeed, if we were to license the present generation to exploit our natural and cultural resources at the expense of the wellbeing of future generations, we would contradict the purposes of the United Nations Charter and international human rights documents.[78]

This would mean that States also have a duty to take the right to life of future generations into account, which is a rather abstract concept. However, it could be argued that the principle of *erga omnes* obligations, as first enunciated by the International Court of Justice in the *Barcelona Traction* case and in subsequent decisions[79] could be considered to support this idea. The question could be asked whether the concept of *erga omnes* obligations can be considered to apply also with regard to future generations. In other words, does the concept of *erga omnes*, which is generally understood to have a 'horizontal' applicability, namely, as an obligation of all States towards the international community as a whole, also have a 'vertical' dimension, stretching from the present to the future? In this perspective, the term '*omnes*' would refer to all of humankind, not only as it is represented today, but

77 See, Peter Lawrence, *Justice for Future Generations: Climate Change and International Law*, (Edward Elgar, 2014); Malgosia Fitzmaurice, 'Intergenerational Equity: a Reappraisal', in Malgosia Fitzmaurice (ed), *Contemporary Issues in International Environmental Law* (Edward Elgar, 2009); Edith Brown Weiss, 'Intergenerational Equity: a Legal Framework for Global Environmental Change', in Edith Brown Weiss (ed), *Environmental Change and International Law: New Challenges and Dimensions* (United Nations University Press, 1992); Anthony D'Amato *et al*, 'Agora: What Obligation Does Our Generation Owe to the Next? An Approach to Global Environmental Responsibility' (1990) 84 *Am J Int'l Law* 190.

78 Edith Brown Weiss, 'Our Rights and Obligations to Future Generations for the Environment', in 'Agora' (1990) 201.

79 *Barcelona Traction (Belgium v Spain)* [1970] ICJ Rep 3, 32, [33]–[34]; *Application of the Convention on the Prevention and Punishment of the Crime of Genocide (Bosnia and Herzegovina v Serbia and Montenegro)* [1996] ICJ Rep 595, 616, [31]; Id, [2007] ICJ Rep 43, 65 [147]; *Advisory Opinion on Legal Consequences of the Construction of a Wall in the Occupied Palestinian Territory* [2004] ICJ Rep 136, 170, [88], 199 [155]–[156].

also by future generations.[80] What this means in practical terms, and in terms of enforcement, would require substantial further thought. However, States, as the primary guarantors of human rights, may need to broaden the protection of these rights, and especially of the right to life, taking account of the challenges resulting from climate change.

Evidently, there is no explicit basis in international human rights law for an expansive interpretation of States' obligations as outlined above. This would entail a profound change in the way of thinking about human rights, which could only be brought about, if at all, by a gradual process, that is, *de lege ferenda*. Nevertheless, the enormity and the irreversibility of the threats posed by climate change would certainly justify such a fundamental reflection.

Conclusion

Even though it is clear that climate change adversely affects the enjoyment of the right to life, this chapter has shown that the possibilities of invoking this most fundamental human right in climate change-related claims are limited.

The ECtHR has increasingly broadened the scope of States' obligations to respect, protect and fulfil the right to life, also establishing a positive obligation to prevent violations of this right as a result of dangerous activities or natural disasters. At the same time, the strict adherence to the 'victim requirement' and the related prohibition of *actio popularis*, the limited acceptance of extraterritorial jurisdiction, and the criteria of causality and of foreseeable harm constitute significant obstacles for bringing climate change-related claims based on the right to life before the Strasbourg Court. In the Inter-American system, the prospects for such claims are somewhat better, considering that, contrary to the European system, NGOs also have standing before the Inter-American Commission and Court of Human Rights in their own right, so that they can lodge a petition either on behalf of victims or in the 'interest of the public'. Moreover, the recognition of economic, social and cultural rights, including the right to a healthy environment in the San Salvador Protocol, provides some scope for supporting claims based on the right to life, by also referring to other fundamental rights, often having a collective nature, even though these are not officially 'justiciable'. This possibility exists more explicitly in the African system, since the African Charter particularly emphasizes collective rights.

At the international level, the approach adopted by the UN HRCte, monitoring the implementation of the ICCPR, is less expansive than that of the regional courts, and virtually excludes a successful claim alleging the failure of a State to comply with its obligations to respect, protect and fulfil the right to life from threats resulting from climate change. However, some possibilities are offered through

80 See also Francioni, 'International Common Goods' (2014), for some further reflections on the material, spatial and temporal dimensions of human rights in the context of the protection of global common goods.

the CRC, which includes both the right to life and other fundamental rights, particularly the right to health. Thus, complaints against a State can be brought before the CteRC, based on a combination of individual and group rights. Moreover, it has been shown that collective rights, in particular the right to a healthy environment, may provide a more suitable basis for climate-related claims.

Therefore, it is submitted that claims based on the rights to life and to a healthy environment may be the most effective solution when applying existing human rights to climate change.[81] However, the limits of this approach are that: (1) only a few of the existing human rights instruments include the rights to life, to health and to a healthy environment; and (2) these are not the instruments which benefit from the most effective and legally binding enforcement mechanisms. Thus, more creative solutions should also be sought, suggesting that, in the face of the challenges posed by climate change, obligations of States should be progressively interpreted in a more collective, forward-looking perspective.

81 *Marangopoulos* (2006).

5 Climate change and interdependent human rights to food, water and health

The contest between harmony and invention

Alessandra Franca

Introduction

The relationship between weather, food and water has been known for a long time. Vivaldi's most famous work, the Four Seasons, which was made public around 1720, perfectly illustrates such a synergic interaction. This set of popular baroque concertos, both in the melody and sonnets, shows some characteristics of the seasons changing which highlight the nexus between weather, water and food. In the first concerto of the composition, No 1 in E major (Spring), the sonnet mentions the 'murmuring streams' which are 'softly caressed by the breezes'. Concerto No 2 in G minor brings the summer sounds, where 'beneath the blazing sun's relentless heat, men and flocks are sweltering, pines are scorched', but then 'the heavens roar and great hailstones beat down upon the proudly standing corn'. When autumn arrives in Concerto in F Major, it is with speed allegro that 'the peasant celebrates with song and dance the harvest safely gathered in', and 'the hunters emerge' to their chase. Finally, Concerto in F minor, Winter, describes with allegro non molto 'those outside drenched by pouring rain'.

Despite time-honoured knowledge about synergic interactions between the natural elements, it is not easy to legally deal with them given the disturbing interference of human action, particularly anthropogenic GHG emissions. Since global warming has a highly negative effect on food and water, fundamental rights covering these elements are crucial tools in this endeavour. This is all the more true when considering the interdependence of all human rights,[1] whereby the fundamental claims to food and water are basic to the enjoyment of all other rights, particularly the rights to health and life.[2] This chapter explores the interrelationship between climate change and the fundamental rights to food and water, as well as

1 See Daniel J Whelan, *Indivisible Human Rights: a History* (University of Pennsylvania Press, 2010).

2 CESCR, General Comment No 14, *The Right to the Highest Attainable Standard of Health (Art 12 of the Covenant)*, E/C.12/2000/4 (2000).

further implications for the right to health,[3] via a systemically integrated approach to the UNFCCC regime and international human rights protection systems.

1. The impact of climate change on food and water availability and access

Climate change is projected to negatively affect water and food availability. The 2014 Fifth Assessment Report (AR5) of the UN IPCC, which integrates the work of three Working Groups, points out threats posed by climate change to food and water availability and access.

Among other worrying conclusions, the IPCC holds that in light of an average global temperature increase by 0.85°C from 1880 to 2012, for each 1 degree of temperature increase, grain yields decline by about 5 per cent, and maize, wheat and other basic crops have experienced global yield reductions of 40 megatonnes per year between 1981 and 2002 because of warmer climate.[4] More generally, the Report of Working Group II considers the vulnerability of human and natural systems, the impacts and future risks of climate change, and potential limits to adaptation. In addition to a list of issues directly and indirectly related to water and food, the report devotes a chapter to freshwater resources and another to food security and food production systems.

The chapter on freshwater resources highlights the negative impacts of climate change on freshwater worldwide, with significant regional differences. Indeed, climate change reduces renewable surface water and groundwater resources, particularly in dry subtropical regions. Furthermore, global warming negatively impacts freshwater ecosystems by changing stream flow and water quality and is projected to reduce raw water quality, posing risks to drinking water. In addition, it is expected that water resources will be subject to change over the course of the 21st century, and continued warming may lead to further unforeseen change; for example, in glacier-fed rivers, total melt water yields from stored glacier ice may increase in many regions during the next decades, but then decrease thereafter.[5] Overall, it is possible to conclude that 'climate change can alter the availability of water and therefore threaten water security.'[6]

The chapter on food security and food production systems highlights, with a high level of agreement and robust evidence, that climate change affects both production and non-production elements. Indeed, the effects of climate change on crop and food production are evident in several regions of the world. More specifically, crop (wheat, rice and maize) yields are sensitive to extreme daytime temperatures around 30°C and to elevated tropospheric ozone. Furthermore,

3 The chapter does not develop additional inferences as to the right to life, which is dealt with by Christine Bakker elsewhere in this volume.

4 IPCC WGII AR5 (2014).

5 Ibid, 229 ff.

6 Ibid, 19.

climate trends are affecting the abundance and distribution of harvested aquatic species, both freshwater and marine, and aquaculture production systems in different parts of the world. Changes in climate and CO_2 concentration enhance the distribution and increase the competitiveness of agronomical invasive weeds. In the long term, this will result in increased global food prices between 3 and 84 per cent by 2050. All aspects of food security are thus affected by climate change, including food production, access, utilisation and price stability.[7]

2. The rights to food and water in the climate change regime and human rights protection systems

When the UNFCCC was adopted in 1992, even in the absence of scientific certainty, it was already known that dangerous anthropogenic interference with the climate system could impact on food production and water management and access. On the one hand, Article 2 sets the ultimate aim 'to stabilise greenhouse gas concentrations at a level that would prevent dangerous anthropogenic interference with climate system . . . within a time-frame sufficient to allow ecosystems to adapt naturally to climate change, to ensure that *food production is not threatened*.'[8] On the other, Article 4 takes into account the specific needs and concerns of developing country parties, based, inter alia, on 'areas liable to *drought* and desertification.'[9] No mention is nonetheless made of the human rights to food and water. Along the same lines, the Kyoto Protocol, which is supposed to complement and implement the UNFCCC, also provides no specific references to the rights to food and water. Furthermore, its effectiveness is doubtful.[10]

Instead, within the international human rights protection system, the claims to food and water have undergone an important development. Indeed, as stressed by Andrew Clapham, many subscribe to a so-called 'full belly thesis', according to which 'subsistence rights to food and water have to be secured before turning to civil and political rights relating to political participation, arbitrary detention, freedom of expression, or privacy'.[11] Nonetheless, only recently have these rights been explicitly recognised as independent claims. A clear definition of their content

7 Ibid, 485 ff.

8 Emphasis added.

9 Emphasis added.

10 On this issue, Sandrine Maljean-Dubois considers:

> [. . .] de fait, le Protocole de Kyoto a donné naissance à la procédure de non-respect la plus aboutie à ce jour. Elle s'inspire des procédures déjà éprouvées pour d'autres problèmes environnementaux, tout en les dépassant largement: elle est en réalité à la fois plus innovante et plus élaborée. Le Comité de contrôle du respect des dispositions figure parmi les plus puissants et les plus indépendants des comités de ce type institués par les conventions environnementales.

See Sandrine Maljean-Dubois, 'L'enjeu du contrôle dans le droit international de l'environnement et le Protocole de Kyoto en particulier', in Sandrine Maljean-Dubois (ed), *Changements climatiques. Les enjeux du contrôle international* (Centre d'Études et de Recherches Internationales et Communautaires, Université Paul Cézanne Aix-Marseille, 2007) 17, 27–8.

11 Andrew Clapham, *Human Rights: a Very Short Introduction* (OUP, 2007) 119.

has in fact been built slowly and fragmentarily through a range of instruments whose legal status is often debatable.

The right to food was first embedded in Article 25(1) UDHR, which was adopted in 1948 at the first session of the UN General Assembly,[12] within the context of the right to an adequate standard of living. However, the UDHR is not a treaty and thus its binding force is limited. Subsequently, the 1949 Geneva Conventions explicitly recognised the rights to both food and water.[13]

Latterly, the right to food has been recognised in the 1966 ICESCR, which spells out in Article 11:

> 1. The States Parties to the present Covenant recognise the right of everyone to an adequate standard of living for himself and his family, *including adequate food*, clothing and housing, and to the continuous improvement of living conditions. The States Parties will take appropriate steps to ensure the realisation of this right, recognising to this effect the essential importance of international co-operation based on free consent.
> 2. The States Parties to the present Covenant, recognising the fundamental right of everyone to be *free from hunger*, shall take, individually and through international co-operation, the measures, *including specific programmes*, which are needed:
>   (a) To *improve methods of production, conservation and distribution of food* by making full use of technical and scientific knowledge, by *disseminating knowledge of the principles of nutrition* and by developing or *reforming agrarian systems* in such a way as to achieve the most efficient development and utilisation of natural resources;
>   (b) Taking into account the problems of both food-importing and food-exporting countries, to *ensure an equitable distribution of world food supplies* in relation to need.[14]

12 Article 25(1):

> Everyone has the right to a standard of living adequate for the health and well-being of himself and of his family, including *food*, clothing, housing and medical care and necessary social services, and the right to security in the event of unemployment, sickness, disability, widowhood, old age or other lack of livelihood in circumstances beyond his control (emphasis added).

13 Articles 20, 26, 29, 46 of the Third Geneva Convention relative to the Treatment of Prisoners of War; Articles 85, 89, 127 of the Fourth Geneva Convention relative to the Protection of Civilian Persons in Time of War; Articles 54 of 1977 Protocol I relating to the Protection of Victims of International Armed Conflicts; Articles 5, 7 and 14 of 1977 Protocol II relating to the Protection of Victims of Non-International Armed Conflicts. Along these lines, it is also important to mention the 1979 Convention on the Elimination of All Forms of Discrimination against Women (Preamble and Article 14(2)(h)) and the 1989 Convention on the Rights of the Child (Article 24(2)(c)).

14 Emphasis added.

This provision is crucial, since it outlines a comprehensive stand on the right to food, which goes beyond the basic obligation to respect human rights. Indeed, it demands a proactive approach that encompasses an obligation to fulfil the right to food, with regard to its different components, including the obligations to facilitate, provide and promote the enjoyment of fundamental claims.

By taking into consideration how interconnected the natural elements of food and water are, as well as the corresponding rights, Article 11 ICESCR is also crucial to the recognition of the right to water. This was explicitly acknowledged by the CESCR in 2003; indeed, CESCR Comment No 15, on the Right to Water, states that Article 11 ICESCR encompasses the claim in issue.[15] The Committee pointed out that Article 11(1) 'specifies a number of rights' essential for the realisation of the right to an adequate standard of living, '*including* adequate food, clothing and housing'.[16] This expression signals that the catalogue is non-exhaustive and encompasses the right to water, which 'clearly falls within the category of guarantees essential for securing an adequate standard of living, particularly since it is one of the most fundamental conditions for survival.'[17]

The CESCR has also established a link between the right to water and the right to food within the context of sustainablity, noting the importance of 'ensuring sustainable access to water resources for agriculture to realise the right to adequate food.'[18]

This approach has spawned widespread recognition of the rights to food and water by the international society, NGOs, international bodies and organisations, as well as by international law scholars.[19]

15 CESCR, General Comment No 15, *The Right to Water*, UN Doc E/C.12/2002/11 (2003).

16 Emphasis added.

17 CESCR (2003) [3]. This stance is also based on paras 5 and 32 of CESCR General Comment No 6 (1995) on the Economic, Social and Cultural Rights of Older Persons.

18 CESCR (2003) [7].

19 Some instruments deserve to be mentioned, in particular, the 2004 *Voluntary Guidelines to Support the Progressive Realisation of the Right to Adequate Food in the Context of National Food Security* prepared by an Intergovernmental Working Group under the auspices of FAO and two *Resolutions on the Right to Water* adopted in 2010, one by the Human Rights Council (Doc A/HRC/RES/15/9) and another by the General Assembly (Doc A/RES/64/292). For a scholarly opinion, see Mara Tignino, *International Law and Freshwater* (Edward Elgar Publishing, 2013); Laurence Boisson De Chazournes, 'Le droit à l'eau et la satisfaction des besoins humains: notions de justice', in Denis Alland *et al* (eds), *Unité et diversité du droit international – Mélanges en l'honneur du Professeur Pierre Marie Dupuy* (Brill, 2013) 967 ff; Smita Narula, 'The Right to Food: Holding Global Actors Accountable under International Law' (2006) 44 *Columbia Journal of Transnational Law* 691; Claire Mahon, 'The Right to Food: a Right for Everyone', in Christopher Rosin, Paul Stock and Hugh Campbell (eds), *Food Systems Failure: the Global Food Crisis and the Future of Agriculture* (Routledge, Earthscan, 2012); Hans Morten Haugen, 'Food Sovereignty: an Appropriate Approach to Ensure the Right to Food?' (2009) 78 *Nordic Journal of International Law* 263; Danilo Zolo, 'The Water Right as a Social and Collective Right (2005) 1 *Jura Gentium* <www.juragentium.org/topics/palestin/en/water.htm>; Pierre-Marie Dupuy, *Le droit à l'eau, un droit international?* (EUI Working Paper, 2006).

3. Interdependent human rights: lack of food and water and right to health

The rights to food and water are a fundamental prerequisite to the enjoyment of the human right to health. Positively, Article 24(2) of the Convention on the Rights of the Child provides that 'States Parties recognise the right of the child to the enjoyment of the highest attainable standard of health and to facilities for the treatment of illness and rehabilitation of health.' On this basis, States commit to take appropriate measures 'to combat disease and malnutrition . . . through the provision of *adequate nutritious foods and clean drinking-water, taking into consideration the dangers and risks of environmental pollution.*'[20] In this respect, the CESCR has pointed out that, among core obligations implied by the right to health, there is a basic duty to ensure access to an adequate supply of the 'underlying determinants' of food and safe and potable water.[21]

The link between food, water and health is also normatively stressed by the fact that the ICESCR recognises the latter immediately after the former, under Article 12 ICESCR:

> 1. The States Parties to the present Covenant recognise the right of everyone to the enjoyment of the highest attainable standard of physical and mental health.
> 2. The steps to be taken by the States Parties to the present Covenant to achieve the full realisation of this right shall include those necessary for:
>   a. The provision for the reduction of the stillbirth-rate and of infant mortality and for the healthy development of the child;
>   b. The improvement of all aspects of *environmental* and industrial hygiene;[22]
>   c. The prevention, treatment and control of epidemic, endemic, occupational and other diseases;
>   d. The creation of conditions which would assure to all medical service and medical attention in the event of sickness.

The drafting history of this provision confirms that the reference to 'the highest attainable standard of physical and mental health' is not confined to the right to health care, but embraces a wide range of socio-economic factors indispensable to a healthy life, extending to underlying determinants, including adequate supply of safe food and nutrition, access to safe and potable water and adequate sanitation, and a healthy environment.[23]

20 Article 24(2)(c), Emphasis added.
21 CESCR (2000) [12] and [43].
22 Emphasis added. Along these lines, the Convention on the Rights of the Child (Articles 24(2) and 29(e)) recognises that that the enjoyment of human rights depends on a sustainable environment.
23 CESCR (2000) [3]–[4].

Among the different sub-headings of paragraph 2, a link between food, water and health can be established under letters (b) and (c). Indeed, not fulfilling the rights to food and water is tantamount to not preventing, treating and controlling diseases, thus breaching the right to health. Along these lines, Article 12(2)(b) ICESCR comprises, inter alia, preventive measures so as to ensure an 'adequate supply of safe and potable water and basic sanitation' and an 'adequate supply of food and proper nutrition', within the context of the necessity of hindering 'detrimental environmental conditions that directly or indirectly impact upon human health.'[24] This entails taking steps on a non-discriminatory basis to prevent threats to health from unsafe and toxic water conditions.[25]

Within this context, the right to water emerges as preliminary to both the rights to health and food. In fact, it is 'inextricably related' to the right to the highest attainable standard of health under Article 12 ICESCR, and to the right to adequate food under Article 11(1) of the Covenant.[26] Indeed, as to the right to food, water is necessary to food production, so as to prevent starvation.[27] With regard to the right to health, in its General Comment No 15, citing a report of the WHO, the CESCR pointed out that 'over one billion persons lack access to a basic water supply, while several billion do not have access to adequate sanitation, which is the primary cause of water contamination and diseases linked to water.'[28] The Committee has further signalled that States have, among other core duties, a basic obligation to 'take measures to prevent, treat and control diseases linked to water, in particular ensuring access to adequate sanitation.'[29] A sufficient amount of safe water is indeed necessary to 'prevent death from dehydration, to reduce the risk of water-related disease and to provide for consumption, cooking, personal and domestic hygienic requirements.'[30] More precisely, the quantity of water available for each person is to be defined according to the guidelines established by the WHO as well as relevant health, climate and work conditions.[31] As to quality, water for personal or domestic use must be safe and free from elements that constitute a threat to a person's health.[32]

24 Ibid, [15].

25 CESCR (2003) [8].

26 A further 'conjunction' with the right to life and human dignity is also highlighted (CESCR (2003) [3]).

27 CESCR (2003) [6].

28 Ibid, [1].

29 Ibid, [37(i)].

30 Ibid, [2].

31 See Jamie Bartram and Guy Howard, *Domestic Water Quantity, Service Level and Health* (WHO, 2003); Peter H Gleick, 'Basic Water Requirements for Human Activities: Meeting Basic Needs' (1996) 21 *Water International* 83–92.

32 CESCR (2003) [2].

4. Assessing the climate change regime in light of the fundamental rights to food, water and health: mitigation and adaptation

Since it is demonstrated that climate change has a negative impact on food, water and health conditions,[33] can the interrelated human rights to food, water and health contribute to prescribe and enforce compliance with the climate change regime?

Although the UNFCCC and the Kyoto Protocol do not explicitly refer to food and water from a human rights perspective, a general connection between the UNFCCC regime and human rights has been established by the Cancun Agreements, which provide that the 'Parties [to the UNFCCC] should, in all climate change-related actions, fully respect human rights.'[34]

According to the CESCR, the rights to food, water and health impose on States obligations to respect, protect and fulfil human rights, with the latter incorporating an obligation to facilitate, provide and promote.[35] This compels States to provide human rights when individuals or groups are unable, for reasons beyond their control, to realise them themselves by the means at their disposal,[36] which is likely to happen because of global warming.[37]

As to the right to food, States have a core obligation to take necessary action to mitigate and alleviate hunger.[38] However, the right to food must not be interpreted in a narrow or restrictive sense, simply equating it with a minimum package of calories, proteins and other specific nutrients. States thus have a positive obligation to grant access to available food, and sustainable accessibility is essential so as not to interfere with the enjoyment of other human rights.[39] More specifically, the core content of this right implies availability of food 'in a quantity and quality sufficient to satisfy the dietary needs of individuals, free from adverse substances, and acceptable within a given culture.'[40] Along these lines, the CESCR has held that the obligation to facilitate this claim means that the State must 'pro-actively engage in activities intended to progressively realise food security' and

33 See OHCHR, *Report on the Relationship between Climate Change and Human Rights*, UN Doc A/HRC/10/61 (15 January 2009) [25]–[27], [28]–[30], and [31]–[34]; Ingrid Ahlgren, Seiji Yamada and Allen Wong, 'Rising Oceans, Climate Change, Food Aid, and Human Rights in the Marshall Islands' (2014) 1 *Health and Human Rights Journal* 69 ff, stressing the impact of droughts on crop production, entailing poor health outcomes.

34 UNFCCC COP, Decision 1/CP.16, *The Cancun Agreements: Outcome of the Work of the Ad Hoc Working Group on Long-term Cooperative Action under the Convention*, UN Doc FCCC/CP/2010/7/Add.1 (15 March 2011) 4, [8] (emphasis added).

35 CESCR, General Comment No 12, *The Right to Adequate Food (Art 11 of the Covenant)* (1999) [6] and [15]; Comment No 15, [20].

36 CESCR (2003) [25].

37 See IPCC (2014); OHCHR (2009) [25]–[27], [28]–[30], and [31]–[34].

38 CESCR (1999) [6].

39 Ibid, [8].

40 Ibid.

should 'provide that right directly'.[41] Article 11(2)(a) ICESCR also sets out an obligation to promote the right to food, requesting dissemination of the 'principles of nutrition'.

Concerning the right to water, an overall definition has been provided by the 1977 Mar del Plata Action Plan, stating that all people 'have the right to have access to drinking water in quantities and of a quality equal to their basic needs.'[42] Along these lines, the CESCR has pointed out that 'the human right to water entitles everyone to sufficient, safe, acceptable, physically accessible and affordable water for personal and domestic uses.'[43] Based on Article 11 ICESCR, the CESCR has assumed that States must take positive measures to assist individuals and communities in enjoying this right.[44]

Similar considerations have been developed with regard to Article 12 ICESCR, which interlinks health and environment.[45] The obligation to fulfil the right to health thus requires States parties, inter alia, to give sufficient recognition to the right to health in national systems, preferably by way of legislative implementation, and to adopt a national policy including a detailed plan for realising such a right. States must not only ensure provision of health care, including immunisation programmes against major infectious diseases, they must also grant equal access for all to the underlying determinants of health, including nutritiously safe food and potable drinking water, basic sanitation and adequate housing and living conditions.[46] Ensuring that everyone has access to adequate sanitation is not only fundamental for human dignity and privacy, but is one of the principal mechanisms for protecting the quality of drinking water supplies and resources.[47]

In light of these premises, the negative impact of climate change on food, water and health conditions generates State obligations to adopt GHGs mitigation policies.[48] It also generates a duty to take positive adaptation measures to fulfil the human rights to food, water and health, as a response to environmental effects caused by climate change.[49] For instance, the obligation to fulfil the human rights to food, water and health should compel States to build water supply systems in

41 Ibid, [15].

42 UN, *Report of the Water Conference*, Mar del Plata (14–25 March 1977).

43 CESCR (2003) [2].

44 CESCR (2003) [25]. The Committee maintains that 'States parties are obliged to fulfil (provide) the right [to water] when individuals or a group are unable, for reasons beyond their control, to realise that right themselves by the means at their disposal'. Furthermore, according to the Committee the 'obligation to promote [human rights] obliges the State party to take steps to ensure that there is appropriate education concerning the hygienic use of water, protection of water sources and methods to minimise water wastage'.

45 CESCR (2000) [33].

46 Ibid, [36].

47 CESCR (2003) [29].

48 Catarina De Albuquerque (Independent Expert on the Issue of Human Rights Obligations related to Access to Safe Drinking Water and Sanitation), *Climate Change and the Human Rights to Water and Sanitation*, UN Position Paper (2009) 4.

49 Ibid, 5.

progressively desert areas and to promote recycling water practices, among other measures. A range of potential adaptation options across all food system activities, including innovations in food processing, packaging, transport, storage and trade can also be envisaged.[50] FAO studies on the relationship between climate change, water and food security foresee adaptive responses in cropping and hydrological systems.[51] This is clearly pointed out by the Inuit petition to the IAComHR, requesting the US to 'establish and implement, in coordination with petitioner and the affected Inuit, a plan to protect Inuit culture and resources, including, *inter alia*, the land, *water*, snow, ice, and *plant and animal species . . . and mitigate any harm to these resources caused by US greenhouse gas emissions*.'[52]

It is therefore suitable that effective GHG mitigation and adaptation strategies are integrated in the comprehensive policies that the CESCR advocates for the rights to food, water and health. In particular, violations occur not only via direct State action, but also by means of conduct of non-State entities insufficiently regulated by States.[53] Thus, with regard to the right to health, the CESCR has held that the State has an obligation to enact a strategy and laws to prevent the pollution of water, air and soil by extractive and manufacturing industries.'[54] Furthermore, States are to adopt appropriate legislative, administrative, budgetary, judicial, promotional and other measures towards full realisation of the right to health.[55] Also, in order to create a favourable climate for the realisation of this right, States parties are requested to take appropriate steps to ensure that the private business sector and civil society consider the importance of the right to health in pursuing their activities.[56]

Moreover, although the CESCR has recognised that they have a margin of discretion in choosing their policies, States have to take 'whatever steps are necessary to ensure that everyone is free from hunger and as soon as possible can enjoy the right to adequate food.'[57] This requires the adoption of a national strategy to ensure food and nutrition security, based on human rights principles.[58] In parallel with measures in the field of health, such a strategy should address the 'sustainable management and use of natural and other resources for food at the

50 IPCC WGII AR5 Chapter 7 (2013) 2–4; Olivier De Schutter (Special Rapporteur on the Right to Food), *Contribution on the Right to Food to the Meeting Convened by the Friedrich-Ebert-Stiftung with the Committee on Economic, Social and Cultural Rights* (2010) 2 ff.

51 Hugh Turral, Jacob Burke and Jean-Mark Faurès, *Climate Change, Water and Food Security* (FAO Water Reports, 2011) 77.

52 Sheila Watt-Cloutier, *Petition to the Inter American Commission on Human Rights* (7 December 2005) 118, emphasis added. See also Arctic Athabaskan Council on Behalf of All Arctic Athabaskan Peoples of the Arctic Regions of Canada and the United States, *Petition to the Inter-American Commission on Human Rights Seeking Relief from Violations of the Rights of Arctic Athabaskan Peoples Resulting from Rapid Arctic Warming and Melting Caused by Emissions of Black Carbon by Canada* (23 April 2013) 5 and 86–7.

53 CESCR (2003) [48].

54 CESCR (2000) [51] and [53].

55 Ibid, [33].

56 Ibid, [55].

57 CESCR (1999) [21].

58 Ibid, [21].

national, regional, local and household levels.'[59] Contrary to judicial assumptions,[60] the right to food thus sets a limit to State discretion in climate change policies.

With regard to the right to water, 'comprehensive and integrated strategies' must be adopted to ensure water 'sustainability for present and future generations.'[61] Such strategies and programmes include a set of measures which fit with GHG mitigation and adaptation policies, particularly monitoring water reserves; ensuring that proposed developments do not impinge upon access to adequate water; evaluating the 'impacts of actions that may impinge upon water availability and natural-ecosystems watersheds, such as climate changes, desertification and increased soil salinity, deforestation and loss of biodiversity'; increased efficient use of water by end-users; reduction of water distribution wastage; and response mechanisms in emergency situations.[62]

These policies must be adequately monitored and should be established by ensuring participation of the public in decision-making processes, according to procedural environmental rights, giving specific attention to disadvantaged or marginalised groups.[63]

The human rights to food, water and health are also particularly interesting with respect to climate change as a transboundary phenomenon, even if it is still in its infancy, to the extent that, according to the case law of the ICESCR, they tend to be considered extraterritorially applicable. Thus, international institutions and States are committed to take steps to respect, protect and fulfil the enjoyment of these rights abroad, which requires enforced political, normative and institutional cooperation.[64] Along the same lines, the Constitution of Uruguay as reformed in 2002 prompts the State to provide water to foreign countries that face shortages.[65]

A clear link between the right to health and climate change was established by the ECteSR in *Marangopoulos*.[66] In this case, the ECteSR adjudicated upon the Greek energy policy allowing the operation of lignite mines, emitting, inter alia, GHGs, without adequately reducing their environmental impact. Based on the UNFCCC regime as implemented in the EU, the Committee upheld the plaintiff's claim both substantively and procedurally, holding that Greece breached Article 11(1) ESC, providing for an obligation 'to remove as far as possible the causes of ill-health', Article 11(3) ESC, requiring 'to prevent as far as possible epidemic, endemic and other diseases, as well as accidents', and Article 11(2) ESC, requesting

59 Ibid, [25].

60 See, in particular, *Native Village of Kivalina, City of Kivalina v ExxonMobil Corporation and Others*, Order Granting Defendants' Motions to Dismiss for Lack of Subject Matter Jurisdiction, Case No C 08-1138 SBA (United States District Court for the Northern District of California, Oakland Division, 30 September 2009) III.A and Conclusion.

61 CESCR (2003) [11] and [28].

62 Ibid, [28].

63 CESCR (2003) [37].

64 Ibid, [31]; Comment No 12, [36].

65 Constitution of Uruguay (1830, lastly amended in 2002) Article 47.

66 *Marangopoulos Foundation for Human Rights (MFHR) v Greece*, Collective Complaint No 30/2005 (ECteSR, 26 December 2005).

'advisory and educational facilities for the promotion of health', for lack of access to information and decision-making.[67] Moreover, the Committee recalled the case law of the ECtHR, the IACtHR, the AComHPR and the UN CESCR and held that the right to health under Article 11 ICESCR includes the 'right to a healthy environment'.[68] This connection, which must be understood within the framework of the indivisibility and interdependence of human rights, establishes a further crucial link between the right to health and climate change, although the status of the right to environment is currently uncertain in international law.[69]

Therefore, considering the initial question as to whether the rights to food, water and health contribute to defining the effectiveness of the climate change regime, not only the ICESCR but also emerging judicial practice provide the basis for a positive answer. On the other hand, can the UNFCCC regime contribute to implementing the rights to food, water and health? How can these regimes interact?

5. The rights to food, water and health in light of the climate change regime: prospective responses

Almost twenty years after its establishment, we need to assess whether the climate change regime has evolved so as to recognise, respect, protect and fulfil the fundamental rights to food, water and health.

Theoretically, the UNFCCC regime was established as a binding system to deal with the dangerous impact of anthropogenic GHG emissions on climate and its consequences.[70] In practice, the answer is disappointing, since the regime itself does not show steady progress as to compliance and effectiveness. In particular, the 2009 Copenhagen Accord, which was designed to reach a new agreement for a post-2012 Kyoto Protocol commitment period, was far from reinforcing recommended legally binding measures. The Accord substitutes top-down mechanisms, such as normatively quantified emission caps, reduction commitments and compliance regimes, with bottom-up ones, where countries voluntarily determine and notify their own targets and actions.[71] However, in 2012 the Doha Amendment to the Kyoto Protocol established a more compelling second commitment period (from

67 Ibid, [194] ff.

68 Ibid, [160].

69 Particularly, in *Ogoniland* the AComHPR concluded that, besides other human rights, specifically including the right to health under Article 16 of the Banjul Charter, the right to a general satisfactory environment under Article 24 of the Charter 'obligates governments to desist from directly threatening the health and environment of their citizens' (*Social and Economic Rights Action Center and the Center for Economic and Social Rights v Nigeria* (2001) [57]). On this issue, see the contribution by Francesco Francioni and Ottavio Quirico in this volume.

70 As Tarlock observes, this system was built based on two main components. A first element 'seeks to find the most efficient and equitable means to reduce the root cause of anthropocentric climate change, increasing greenhouse gas emissions, in an effort to mitigate projected temperature increases'; by contrast, a second component 'accepts the projected increases and seeks to understand both the effects of global climate change and the impacts of those effects in order to adapt to them' (Dan Tarlock, 'How Well Can International Water Allocation Regimes Adapt to Global Climate Change?' (2000) 15 *J Land Use & Transnational Environmental Law* 423, 424).

71 See UNFCCC, *Copenhagen Accord* (18 December 2009).

2013 to 2020).[72] Ratifications are nonetheless still largely insufficient for binding targets to become compulsory. Furthermore, the 2011 UN Ministerial Conference in Durban decided to launch a new round of negotiations in order to reach an agreement in Paris on a more ambitious instrument in December 2015 'to take the form of a protocol, another legal instrument or an agreed outcome with legal force', and to be implemented from 2020 onwards. Nonetheless, considering the 'disjuncture between climate change science (urging immediate action) and the international political process',[73] what can be expected from Paris 2015 concerning imminent challenges to the rights to food, water and health?

Actually, based on the IPCC Fifth Assessment Report and its alarming conclusions, the fundamental rights to food, water and health should lead to an ambitious agreement, in the face of the failure of climate change mitigation efforts to reverse GHG emissions. In particular, in addition to GHG mitigation measures these rights should stimulate adaptation measures providing food, water and health threatened by climate change.[74] For instance, it is estimated that the average benefit of adapting crop management to GHG emissions is equivalent to about 15 to 18 per cent of current yields.[75]

In order to correctly address the rights to food, water and health within the climate change regime, a systemically integrated ('trans-juridical')[76] strategy should be adopted. In other words, a transnational and cross-sectoral approach seems imperative to regulate climate change in light of the transboundary impact of GHGs on the interrelated rights to water, food and health.[77] This should ultimately

72 UNFCCC, *Kyoto Protocol Including the Doha Amendment* (2013) <http://unfccc.int/kyoto_protocol/doha_amendment/items/7362.php>.

73 Jacqueline Peel, Lee Godden and Rodney Keenan, 'Climate Change Law in an Era of Multi-Level Governance' (2012) *Transnational Environmental Law* 245, 247.

74 Gupta and Lebel highlight the different nature of climate change and water allocation: 'while access to water is essentially a local issue although it can be internationalised, access in climate change is essentially a global issue' (Joyeeta Gupta and Louis Lebel, 'Access and Allocation in Earth System Governance: Water and Climate Change Compared' (2010) 10 *Int Environ Agreements* 377, 390).

75 See IPCC WG II (2014) Chapter 7.

76 By 'trans-juridicity' we mean a trans-disciplinary approach to the law, which requires 'trans-reading', 'trans-interpretation' and 'trans-application' of the UNFCCC regime and the rights to food, water and health.

77 Along these lines, considering the need to bring together the legal regimes on trade, climate change, food security, humanitarian crises and migration, Garcia and Gostin state:

> It is time for international law to catch up with global reality. Despite the synergies between trade, climate change, food, mass migrations, and armed conflict, the international community lacks the vision to see how these legal regimes could work together for the global common good – a classical formula of "One World, One Health". Steps are already being taken to move in this direction. For example, in January 2011 the Report of the United-Nations Secretary-General's High Level Panel on Global Sustainability, aptly titled, Resilient People, Resilient Planet: a Future Worth Choosing, called for a long-term vision "to eradicate poverty, reduce inequality, and make growth inclusive, and production and consumption more sustainable while combating climate change", and recognised multiple intersections between different areas of law and the effect of those areas on global development, health and the environment.

See Kelly K Garcia and Lawrence O Gostin, 'One Health, One World – The Intersecting Legal Regimes of Trade, Climate Change, Food Security, Humanitarian Crisis and Migration' (2012) 1 *Laws* 4, 27.

lead to expressly including the rights to food, water and health in a multilevel regulatory network for global warming, and to the establishment of new instruments aimed at implementating these rights in the framework of the climate change regime.

Such a necessity is clearly illustrated by the problem of vulnerable groups, particularly climate migrants, for whom the provision of international medical aid, distribution and management of safe and potable water and food and medical supplies is crucial. Furthermore, supranational and transnational organisations play a fundamental role in implementing these rights, in particular the WHO, the Office of the UN High Commissioner for Refugees, the International Committee of the Red Cross/Red Crescent, UNICEF and national medical associations.[78] Similar considerations apply, *mutatis mutandis*, to women, children, minority groups, and indigenous peoples,[79] which highlights further interconnections between climate change, the rights to food and water and other fundamental rights.[80]

Along these lines, the CESCR pointed out that organisations such as the WHO, FAO, ILO, UNICEF, UNEP, UNDP, UNWFP, IFAD, IMF, the World Bank, the UN Population Fund, the UN Centre for Human Settlements (Habitat), and regional development banks should cooperate more effectively, with respect to the rights to food, water and health.[81] With particular regard to the right to food, States members of international financial institutions, notably the IMF, the World Bank and regional development banks, are asked to pay attention to the protection of the right to health in influencing lending policies.[82]

Of course, climate change policies should only be constrained by the rights to food, water and health within the limits of the best efforts standard.[83] Therefore, it is important to distinguish the inability from the unwillingness of a State to comply with its obligations. Indeed, Article 2(1) ICESCR obliges a State to take the necessary steps 'to the maximum of its available resources',[84] so that, if

78 CESCR (2000) [65].

79 See Rhys Jones, Hayley Bennett, Gay Keating and Alison Blaiklock, 'Climate Change and the Right to Health for Māori in Aotearoa/New Zealand' (2014) 16 *Health and Human Rights Journal* 54 <www.hhrjournal.org>.

80 CESCR (2003) [16].

81 CESCR (2003) [53] and [60]; CESCR (1999) [40]. Among current initiatives, it is worth recalling major events reconciling governmental and non-governmental action, such as the 'world food summits' and the 'world water forums', some of which produced interesting documental outcomes, including the *Rome Declaration on World Food Security and the World Food Summit Plan of Action* (1996), the *Ministerial Declaration of The Hague on Water Security in the 21st Century* (2000), and the *Istanbul Water Consensus and Declaration of Heads of States on Water* (2009). As to NGOs, it is worth mentioning the Food First Information and Action Network (FIAN), the International Alliance against Hunger and the World Water Council, as well as the International Committee for the World Water Contract and its *Water Manifesto*.

82 CESCR (2000) [39].

83 Siobhán McInerney-Lankford, Mac Darrow and Lavanya Rajamani, *Human Rights and Climate Change: a Review of the International Legal Dimensions* (World Bank, 2011) 35.

84 CESCR (2003) [41].

resource constraints make it impossible to adopt GHG policies compliant with the rights to food, water and health, particularly in light of the developmental conditions of a State, that State then has the burden of proving that every effort has nonetheless been made to use all available resources and satisfy these obligations, as a matter of priority.[85] In light of the different interests at stake in mitigation policies and the difficulties of containing rising temperatures below 2°C, this ultimately means that States are particularly compelled by the rights in issue to take adaptation measures. Further limits to GHG mitigation policies aiming to protect the rights to food, water and health are intrinsic in the necessity of balancing different interests at stake, since, for instance, producing energy from clean sources such as ethanol or biodiesel may infringe upon food security.[86] Finally, this substantively and procedurally integrated regulatory process should take place by respecting fundamental procedural rights,[87] particularly access to judicial remedies.[88]

Conclusion

While the Four Seasons illustrates synergetic interaction between natural elements, what can this masterpiece tell as to the interaction between the fundamental rights to water, food, health and climate change?

Since food, water, and therefore health and other fundamental rights are inseparable and interdependent, they are affected by climate change as a whole. This requires the adoption of tight GHG mitigation policies, within the limits of the best effort standard, and even more compelling adaptation measures, via substantively and procedurally networked cross-sectoral regulation.

For the moment, however, what we can expect from Paris 2015 is quite uncertain. Actually, it is revealing that Vivaldi's famous masterpiece was composed as a part of a larger work entitled 'the contest between harmony and invention'. This contest is a good allegory for both the idea of sustainable development and the necessity of achieving a balance between climate change, water, food and health. According to Pablo Picasso 'we all know that Art is not truth. Art is a lie that makes us realise truth, at least the truth that is given us to understand'; thus, even if making the ideals of art a reality is a difficult undertaking, we should seriously consider trying to make progress in this way.

85 CESCR (2000) [47].

86 De Schutter (2010) 2 ff; International Bar Association, *Achieving Justice and Human Rights in an Era of Climate Disruption* (2014) 183.

87 Margaux J Hall, 'Advancing Climate Justice and the Right to Health through Procedural Rights' (2014) 1 *Health and Human Rights Journal* 8, 14 ff.

88 CESCR (2003) [59].

6 Waterworld

Climate change, Statehood and the right to self-determination

Cameron Moore

Introduction

The right to self-determination in international law grew out of the period of decolonisation from European imperialism. It has since developed a new life in respect of the emergence of States, particularly in Europe, as well as indigenous peoples and minorities more generally, specifically in Latin America and Africa. This chapter argues that the right to self-determination should further develop so as to address the new plight of small island States facing the prospect of inundation by rising sea levels resulting from climate change.

There are two difficulties that small island States face. The first is the physical, practical problem of progressively becoming uninhabitable because of submersion or sea encroachment causing scarcity of fresh water, arable land or protection from waves. The second difficulty is legal: to what extent can an uninhabitable group of islands, or possibly reefs, still claim to be a State? To what extent can their people claim ownership of natural resources in accordance with the law of the sea, especially in projected maritime zones?[1]

The right to self-determination did not develop to address these questions and does not provide a neat solution to the issue of States faced with inundation. This right may however provide a means to argue that the peoples of such States still have a right to exist and govern their natural resources, particularly in respect of their maritime zones, even if they become uninhabitable.

This chapter first considers the 2014 Fifth Assessment Report of the IPCC as the most comprehensive scientific assessment of global warming and particularly the plight of small island States as a result.[2] For those disinclined to accept this scientific assessment, evidence of inundation of some islands and the concern of their governments in itself warrant consideration of this issue. The investigation then considers the place of self-determination in the context of the relevant principles of the law of the sea, bringing these two branches together and

1 These considerations also apply to other climate change-induced effects, such as desertification. However, the case of progressively submerged small island States is the most compelling.

2 IPCC, AR5 (2014).

interpreting the latter in light of the former. It is argued that the right to self-determination, which is exceptionally owned by 'people', not by 'States', can assist in interpreting existing law of the sea so as possibly to allow island States to retain control of their maritime zones, and therefore of their natural resources.

1. The effects of climate change and the right to self-determination of peoples in small island States

Chapter 29 of the Fifth Assessment Report on Climate Change deals with 'Small Islands'. It states the potential breadth and seriousness of the problem in its introduction:

> It has long been recognised that greenhouse gas emissions from small islands are negligible in relation to global emissions, but that the threats of climate change and sea-level rise to small islands are very real. Indeed, it has been suggested that the very existence of some atoll nations is threatened by rising sea levels associated with global warming. While such scenarios are not applicable to all small island nations, there is no doubt that on the whole the impacts of climate change on small islands will have serious negative effects especially on socio-economic conditions and bio-physical resources – although impacts may be reduced through effective adaptation measures.
>
> The small islands considered in this chapter are principally sovereign States and territories located within the tropics of the southern and western Pacific Ocean, central and western Indian Ocean, the Caribbean Sea, and the eastern Atlantic off the coast of west Africa, as well as in the more temperate Mediterranean Sea.[3]

The report indicates that, even if not applicable to all small island States, the problem is global rather than confined to one region. It is possible that adaptation measures will reduce the effects of climate change, but even so, the actual existence of some nations is under threat.[4] For instance, Kiribati lost an uninhabited island to sea level rise in 1998, India lost Lohachara Island, with a population of 10,000, in 2006, and the residents of the Cartaret Islands in Papua New Guinea began leaving their homes in 2007, in anticipation of complete inundation in 2015.[5]

It is not fanciful then to consider how international law might address the issue of the continued self-determination of the peoples of threatened small islands, particularly their sovereignty over natural resources.

3 Ibid, 3.

4 For an assessment of some of the economic effects of this issue see Charles de Liva and Sachiko Morita, *Maritime Rights of Coastal States and Climate Change: Should States Adapt to Submerged Boundaries?*, Law and Development Working Paper Series No 5 (World Bank, 2008) 1–10.

5 See Ann Powers and Christopher Stucko, 'Introducing the Law of the Sea and the Legal Implications of Rising Sea Levels', in Michael Gerrard and Gregory Wannier (eds), *Threatened Island Nations: Legal Implications of Rising Seas and a Changing Climate* (Cambridge University Press, 2013) 123–4.

First recognised as an aspect of the US President Wilson's Fourteen Points at the end of World War I, the right to self-determination has subsequently been considered in Article 55 of the 1945 UN Charter.[6] In 1966, Article 1 of the International Covenants on Civil and Political Rights (ICCPR)[7] and on Economic, Social and Cultural Rights (ICESCR)[8] stated in identical terms:

> 1. *All peoples* have the right of *self-determination.* By virtue of that right they freely determine their political status and freely pursue their economic, social and cultural development.
> 2. *All peoples may*, for their own ends, *freely dispose of their natural wealth and resources* without prejudice to any obligations arising out of international economic co-operation, based upon the principle of mutual benefit, and international law. *In no case may a people be deprived of its own means of subsistence.*
> 3. The States Parties to the present Covenant, including those having responsibility for the administration of Non-Self-Governing and Trust Territories, shall promote the realization of the right of self-determination, and shall respect that right, in conformity with the provisions of the Charter of the United Nations.[9]

This provision emphatically places the right to self-determination, including sovereignty over natural resources,[10] at the forefront of international human rights law.

Within this context, 'external' self-determination from colonial powers has progressively given rise to an 'internal' right to self-determination, referring to the right of peoples, that is, their constitutive groups, including indigenous peoples,[11] to freely choose and participate in the legal system of a nation-State and dispose

6 Hurst Hannum, 'Rethinking Self-Determination' (1993) 34 *Virginia Journal of International Law* 1, 1–17.

7 Opened for signature 16 December 1966, 999 UNTS 171 (entered into force 23 March 1976).

8 Opened for signature 16 December 1966, 993 UNTS 3 (entered into force 3 November 1976).

9 Emphasis added.

10 On sovereignty over natural resources within the context of the right to self-determination see UN Declaration 1803/1962, [1], recognising that 'the right of peoples and nations to permanent sovereignty over their natural wealth and resources must be exercised in the interest of their national development and of the well-being of the people of the State concerned'. Sabine Lavorel also discusses the connection between self-determination and rights to natural resources, including a right not to exploit them (Sabine Lavorel, 'Exploitations des ressources naturelles et de droit des peuples à l'autodétermination économique', in Mihaela Ailincai and Sabine Lavorel, *Exploitation des ressources naturelles et protection des droits de l'homme* (Pedone, 2013) 35, 69.

11 The right to self-determination of indigenous people is explicitly recognised in Articles 3 and 26 of the 2007 UN Declaration on the Rights of Indigenous Peoples.

of natural resources.[12] This is an *ex ante* right that could result in the future establishment of Statehood.[13]

As to the scope of the right to self-determination, in *East Timor* the ICJ recognised that it has the status of an *erga omnes* obligation.[14] Furthermore, in its advisory opinion on the *Legal Consequences of the Construction of a Wall in the Occupied Palestinian Territory*, the Court applied to a breach of the right to self-determination the consequences of a violation of *jus cogens* norms under Article 41 of the ILC's 2001 Draft Articles on State Responsibility.[15] In fact, with regard to the 'the *erga omnes* obligation to respect the right of people to self-determination' the Court held that 'all States are under an obligation not to recognise the illegal situation resulting from the construction of a wall in the Occupied Palestinian Territory' and 'not to render aid or assistance in maintaining the situation created by such construction.'[16]

This legal framework does not readily apply to the case of small island States faced with inundation, whereby the people of an already existing State have the right to self-determination. Indeed, the potential plight of small island States is relevant to internal and external self-determination beyond the dynamic of decolonisation. It is about preserving, rather than disrupting, the existing legal order in the face of a threat from the forces of nature. Sovereignty over natural resources is more relevant to the potential plight of threatened small island nations, which are already independent and thus do not seek secession, but rather the

12 See Committee on the Elimination of Racial Discrimination, General Recommendation No 21, *The Right to Self-determination* (23 August 1996) [4]. Article 29 of the 2007 UN Declaration on the Rights of Indigenous Peoples acknowledges the right of indigenous peoples to the 'conservation and protection of the environment and the productive capacity of their lands or territories and resources' so that 'States shall establish and implement assistance programmes for indigenous peoples for such conservation and protection'.

13 For a comprehensive overview of the right to self-determination see Antonio Cassese, *Self-Determination of Peoples: a Legal Reappraisal* (Cambridge University Press, 1995). See also *Accordance with International Law of the Unilateral Declaration of Independence in Respect of Kosovo (Request for Advisory Opinion)* [2010] ICJ Rep 403; Richard Falk, 'The *Kosovo* Advisory Opinion: Conflict Resolution and Precedent' (2011) 105(1) *American Journal of International Law* 50; Marko Divac Oberg, 'The Legal Effects of United Nations Resolutions in the *Kosovo* Advisory Opinion' (2011) 105(1) *American Journal of International Law* 81.

14 *Case Concerning East Timor (Portugal v Australia), (Judgment)* [1995] ICJ Rep 90.

15 See the text of the 2001 Draft Articles on Responsibility of States for Internationally Wrongful Acts in ILC, *Report to the General Assembly*, UN Doc A/56/10 (2001) 43–59,

16 *Legal Consequences of the Construction of a Wall in the Occupied Palestinian Territory (Advisory Opinion)* [2004] ICJ Rep 199–200, [155]–[159]. Along the lines of its 'inalienable' nature, also some scholars elevate the right to self-determination to the condition of a *jus cogens* norm (see Jenny Grote Stoutenberg, 'When Do States Disappear: Thresholds of Effective Statehood and the Continued Recognition of "Deterritorialised" Island States', in Michael Gerrard and Gregory Wannier (eds), *Threatened Island Nations: Legal Implications of Rising Seas and a Changing Climate* (Cambridge University Press, 2013) 57; Susannah Willcox, 'A Rising Tide: Implications of Climate Change Inundation for Human Rights and State Sovereignty' (2012) 9(1) *Essex Human Rights Review* 1, 7).

maintenance of their rights to exist and exploit natural resources in the case of progressive territorial disappearance.

Along these lines, although it was adopted in 1966 without the prospect of small islands disappearing beneath the waves, Article 1 of both the ICCPR and the ICESCR makes it clear that no people may be deprived of their own means of subsistence.[17] This is arguably relevant to small islands potentially losing their land, with particular regard to sovereignty over natural resources located in the surrounding maritime zones.

In light of this, one view is that the right to self-determination of peoples of inundated small island States is going to either be very limited, excluding sovereignty over natural resources, or simply cease to exist. Peoples of small island States would be displaced onto the territory of other States and their sovereignty over natural resources would be absorbed by the rights of all States in international waters. This nevertheless appears to be contrary to the international rule of law generally, in that it dramatically reduces the geographical area of natural resources subject to national jurisdiction.[18] It is also contrary to the spirit of the original self-determination process, as those peoples would *de facto* become subject peoples once again.

Alternatively, the right to self-determination could have a role to play in enabling peoples of small island States to argue that they should still enjoy their sovereign rights over natural resources, even if the territorial basis for asserting those rights literally disappears beneath the waves. This is ultimately due to the fact that the right to self-determination is enjoyed by 'peoples' rather than States as such.

In terms of the existing scholarly literature on the subject, there has been notable attention to the issue of island States maintaining their maritime zones under the law of the sea based on Statehood, primarily calling for new international law or a new interpretation of the existing law,[19] whereas much less attention has been paid to the more fundamental issue of self-determination.

17 See also CESCR, *Consideration of Reports Submitted by States Parties under Articles 16 and 17 of the Covenant, Madagascar*, E/C.12/MDG/CO/2 (16 December 2009); CESCR, *Consideration of Reports Submitted by States Parties under Articles 16 and 17 of the Covenant, Democratic Republic of the Congo*, E/C.12/COD/CO/4 (20 November 2009). These cases concern breaches of the right to self-determination caused by exploitation of natural resources due to foreign investment.

18 See Ann Powers and Christopher Stucko, 'Introducing the Law of the Sea and the Legal Implications of Rising Sea Levels', in Gerrard and Wannier (2013) 132–3.

19 See, for instance, Robin Warner and Clive Schofield (eds), *Climate Change and the Oceans: Gauging the Legal and Policy Currents in the Asia Pacific and Beyond* (Edward Elgar, 2013); Gerrard and Wannier (2013).

2. Interpreting the law of the sea in light of self-determination: preserving the rights of peoples of uninhabited small island States

2.1. Territory, population, government and maritime zones: with or without Statehood?

Before addressing self-determination and marine natural reources, it is first necessary to consider how the international law of the sea deals with rights to natural resources in the maritime zones of coastal States.

In particular, the exclusive economic zone (EEZ) is the maritime zone likely to be most important to small island States because, in addition to sovereignty over territory and the territorial sea, it creates exclusive rights over natural resources up to 200 nautical miles from the territorial sea baselines. For small island States reliant upon fishing, this zone is of vital economic significance.

Articles 55 and 56 of the Law of the Sea Convention (LOSC)[20] make it clear that sovereign rights over the EEZ belong to coastal States. Article 55 (Specific Legal Regime of the Exclusive Economic Zone) provides:

> The exclusive economic zone is an area beyond and adjacent to the *territorial sea*, subject to the specific legal regime established in this Part, under which the rights and jurisdiction of the coastal State and the rights and freedoms of other States are governed by the relevant provisions of this Convention.[21]

Article 56 (Rights, Jurisdiction and Duties of the Coastal State in the Exclusive Economic Zone) states:

> 1. In the exclusive economic zone, the coastal State has:
>   (a) *sovereign rights for the purpose of exploring and exploiting, conserving and managing the natural resources*, whether living or non-living, of the waters superjacent to the seabed and of the seabed and its subsoil, and with regard to other activities for the economic exploitation and exploration of the zone, such as the production of energy from the water, currents and winds [. . .][22]

This raises the question of whether uninhabitable islands continue to be States and therefore exercise their rights over the EEZ.

More specifically, Article 121 LOSC deals with the regime of islands and clearly makes the rights over the EEZ, the territorial sea and the continental shelf

20 Opened for signature 10 December 1982, 1833 UNTS 3 (entered into force 16 November 1994).
21 Emphasis added.
22 Emphasis added.

dependent on 'human habitation and economic life' as well as on the elevation of territory above the high tide:

1. An island is a naturally formed area of land, surrounded by water, which is above water at high tide.
2. Except as provided for in paragraph 3, the territorial sea, the contiguous zone, the exclusive economic zone and the continental shelf of an island are determined in accordance with the provisions of this Convention applicable to other land territory.
3. Rocks which *cannot sustain human habitation or economic life of their own* shall have no exclusive economic zone or continental shelf.[23]

As to the definition of a 'State', the best known formulation of the basic criteria for Statehood is found in Article 1 of the 1934 Montevideo Convention on the Rights and Duties of States,[24] including:

(a) a permanent population;
(b) a defined territory;
(c) government [. . .][25]

These elements further entail:

(d) capacity to enter into relations with other States.

Arguably, if a State has no habitable territory where a permanent population can live, it then fails to meet the definition of a State capable of projecting maritime zones. Along these lines, a literal interpretation of Article 121 LOSC, in conjunction with Articles 55 and 56 of the same Convention, would exclude sovereignty over natural resources of permanently displaced peoples of small island States. Indeed, Article 121 does not provide an exception for island States that subsequently become uninhabited rocks or reefs. In other words, it does not contemplate the situation of small island States becoming uninhabitable after having originally satisfied the requirements to generate an EEZ and a continental shelf. It refers to islands, reefs and rocks belonging to coastal States that have always been habitable and above water.

However, even if the Montevideo Convention, which is a regional treaty between the US and certain other Latin American States, is regarded as customary international law, there is room for the view that the definition of Statehood should evolve so as to accommodate the novel situation of uninhabitable small island

23 Emphasis added.
24 Opened for signature 26 December 1933, 165 LNTS 19 (entered into force 26 December 1934).
25 See James Crawford, *The Creation of States in International Law* (Oxford University Press, 2nd edn, 2006) 44–6.

States.[26] Thus, Burkett envisages deterritorialised States, that is, 'nations ex-situ', as a development of the law of Statehood.[27] Nonetheless, despite seeing this approach as supporting the right to self-determination, she does not argue that self-determination in itself supports the 'nation ex-situ' approach.[28] However, the right to self-determination lends weight to the idea of a deterritorialised State. Indeed, deterritorialised peoples should still have a right to exist and project maritime zones under the right to self-determination.[29]

Alternatively, it should be assumed that failure to meet the criteria for Statehood in an unsettled area of international law does not displace the right of people to self-determination, which is a peremptory claim preliminary to the idea of the State itself, particularly if it is a right that people already enjoy as an independent State and that they stand to lose through no direct fault of their own. In this respect, it is interesting to note that the CESCR invited Israel to 'recognise and respect the right of the Palestinian people to the marine resources, including the right to fish in the territorial sea and the EEZ of the Gaza Strip',[30] despite the fact that there is actually no Palestinian State. Although this situation is different from that of small island States, whose territory is disappearing, it is certainly relevant to demonstrate that it is not only possible, but also necessary, to maintain the existence of the right of peoples to self-determination, particularly sovereign rights over (marine) natural resources, outside the State.[31] This approach nevertheless

26 Ibid, 37.

27 Maxine Burkett, 'The Nation *Ex-Situ*', in Gerrard and Wannier (2013) 89, 111–14.

28 Ibid, 114. Rayfuse also discusses the idea of deterritorialised States on the basis of a development in the law of Statehood, but does not connect it to self-determination as such (Rosemary Rayfuse, 'Sea Level Rise and Maritime Zones: Preserving the Maritime Entitlements of "Disappearing" States', in Gerrard and Wannier (2013) 167, 178–80; Rosemary Rayfuse and Emily Crawford, *Climate Change, Sovereignty and Statehood* (Legal Studies Research Paper 11/59, University of Sydney, 2011).

29 In this respect, it is interesting to note that, according to Aureliu Cristescu, Special Rapporteur to the UN Sub-Commission on Prevention of Discrimination and Protection of Minorities, although the relationship of a people with a territory is an essential condition of the right to self-determination, this right exists 'even if the people in question have been wrongfully expelled from it and artificially replaced by another population' (see Aureliu Cristescu, *The Right to Self-Determination: Historical and Current Development on the Basis of United Nations Instruments* (1981) [221]).

30 *Consideration of Reports Submitted by States Parties under Articles 16 and 17 of the Covenant, Israel* E/C.12/ISR/CO/3 (16 December 2011) [12]. This is in line with UN Declaration 1803 (XVII) of 14 December 1962 on Permanent Sovereignty over Natural Resources, [2], acknowledging that 'the exploration, development and disposition of [natural] resources, as well as the import of the foreign capital required for these purposes, should be in conformity with the rules and conditions which the *peoples* and *nations* freely consider to be necessary or desirable' (emphasis added).

31 See also UN Declaration 1803/1962, Preamble and [5], regarding 'permanent sovereignty over natural resources as a basic constituent of the right to self-determination' and providing that 'the free and beneficial exercise of the sovereignty of peoples and nations over their natural resources must be furthered by the mutual respect of States based on their sovereign equality'.

still requires a progressive interpretation of international law, because also the right to self-determination is usually linked to the existence of a territory, even in the case of displaced people.[32]

2.2. Behind and beyond the State: the basic role of the right to self-determination of peoples

In light of the above, it is possible to envisage two possible scenarios: (1) the continuity of (ex situ) Statehood in the absence of a territory, including the right to self-determination of deterritorialised people; and (2) the continuity of the right to self-determination of people in the absence of a State.

The governmental element strengthens the argument for continuity of the right to self-determination of people without a territorial State. Arguably, a neighbouring State could not be compelled to host a government in exile nor to provide a base for maritime patrol vessels and aircraft, as this would impinge upon the rights of another State within its own territory. In fact, the ability to maintain a government for displaced people might have to rely upon the goodwill of any neighbouring States concerned.[33] This means that climate change threatens not only the existence of the territory of peoples of small island States, but also their government. Under Article 1(c) of the Montevideo Convention, the inability to continue a government is even more likely to lead to the loss of Statehood for uninhabitable small island States than the loss of territory. In this regard, however, it might be argued that a duty of hospitality exists as a remedy, that is, a limitation of sovereign rights as a countermeasure for GHGs emitted by States causing the disappearance of small islands.[34] The inalienable *erga omnes* nature of the right to self-determination, entailing non-derogable general obligations,[35] could provide the basis for such an argument. This view is also supported by the fact that the right to self-determination is a crucial component of the right to sustainable development.[36] However, this approach entails complex issues of causation as to the relationship between anthropogenic GHG emissions and breaches of the right to self-determination.[37]

32 According to the African Commission on Human Rights and Peoples' Rights, *Katangese Peoples' Congress v Zaire*, Communication 75/92, Eight Annual Activity Report (2000) [4], the right of peoples to self-determination 'is fully cognizant of other recognised principles such as sovereignty and territorial integrity'. See also Cristescu (1981) [221].

33 Stoutenberg (2013) 68–70.

34 In this respect, for instance, Article 10 of the 2007 UN Declaration on the Rights of Indigenous Peoples provides:

> Indigenous peoples shall not be forcibly removed from their lands or territories. No relocation shall take place without the free, prior and informed consent of the indigenous peoples concerned and after agreement on just and fair compensation and, where possible, with the option of return.

35 On responsibility for breaches of *erga omnes* obligations, see Christian Dominicé, 'The International Responsibility of States for Breach of Multilateral Obligations' (1999) 10 *EJIL* 353 ff.

36 See Varayudej in this volume.

37 See the contribution by Quirico, Bröhmer and Szabó in this volume.

In any case, a loss of government should not entail a loss of the right to self-determination. In fact, the claim to self-determination does not necessarily require the existence of a government, but provides the basis upon which to establish a new one.[38] Since it is possible that in some cases peoples will be granted hospitality abroad, allowing the continuity of government, and in other cases not, in some circumstances sovereign rights over natural resources could continue based on Statehood, underpinned by the right to self-determination; otherwise, the right to self-determination alone could be relied upon as *extrema ratio* to justify such sovereign rights.[39]

Among legal scholars, Willcox still ultimately relies upon the continuity of the State, but claims that this needs to be reconceptualised in conjunction with those most affected. She argues that human rights law needs to develop so as to address the changing circumstances which threaten small island States and puts forward that 'the State can no longer be the sole – nor primary – framework for the recognition of human rights and self-determination'.[40] In other words, Willcox does not 'seek to dictate the terms of the proposed reconceptualisation', but does assume that current arguments for new international law 'presuppose the continuation of a State-centric international framework and thus fail to address the challenge posed by climate change to Statehood and self-determination more broadly'.[41] It is hard to disagree with this position, which nevertheless does not take matters very far, since Willcox only seeks to 'encourage broader dialogue' about possible alternatives to Statehood and suggests that they are 'likely to be less radical than one might expect'.[42]

Stoutenberg goes further and connects the right to natural resources and the right to self-determination.[43] She refers to the arguments of Nauru in its action in the ICJ against Australia to the effect that excessive depletion of phosphate mines during Australia's mandate and trusteeship over Nauru affected the latter's

38 See *Katangese Peoples' Congress* (2000) [4]: 'The [African] Commission believes that self-determination may be exercised in any of the following ways – independence, self-government, local government, federalism, confederalism, unitarism or any other form of relations that accords with the wishes of the people.' See also Jane Wright, 'Minority Groups, Autonomy, and Self-Determination' (1999) 19(4) *Oxford Journal of Legal Studies* 605. In this respect, it must be noted that international and regional bodies tend to provide a broad interpretation of the notion of 'people', in line with the open formulation of Article 1 of the ICCPR and the ICESCR (see *Centre for Minority Rights Development (Kenya) and Minority Rights Group (on behalf of Endorois Welfare Council) v Kenya*, 276/03 (AComHPR, 25 November 2009); *Kevin Mgwanga Gunme et al v Cameroon*, 266/03 (AComHPR, 27 May 2009); *Katangese Peoples' Congress* (2000)). According to Aureliu Cristescu, besides territory, a 'clear identity' is the main feature of a 'people' (see Cristescu (1981) [221]).

39 This could realistically only apply to some sort of international judicial proceedings, as the enforcement of sovereign rights at sea could only be allowed to officially recognised government vessels and aircraft. This would be necessary to avoid a potentially dangerous anarchist approach to the protection of maritime resources, particularly in light of the broad notion of 'people'.

40 Willcox (2012) 18.

41 Ibid, 17.

42 Ibid.

43 Stoutenberg (2013) 57.

permanent sovereignty over its natural resources as part of Nauru's right to self-determination.[44] Even though the case did not proceed to the merits stage, it does illustrate the argument for permanent sovereignty over natural resources within the right to self-determination, drawing upon the provision that 'in no case may a people be deprived of its own means of subsistence' embedded in common Article 1 of the 1966 Human Rights Covenants. Although she does not address the details of the relevant LOSC provisions, Stoutenberg links this argument with the loss of maritime zones and argues that such a loss 'could be prevented only by amending the law of the sea to provide for stable maritime zones'.[45] This approach still presupposes the critical function of Statehood.[46]

Fundamentally, it is not strictly necessary (and possibly not realistic) to amend the LOSC, but it is suitable to interpret the Convention based on the right to self-determination, with particular regard to sovereignty over natural resources, in order to retain 'island' as opposed to 'rock' status, maritime zones and further fix baselines. The argument for sovereignty over natural resources being permanent as part of the right to self-determination could thus be relied upon to support an extensive interpretation of the LOSC, possibly beyond Statehood. A right with such prominence in international human rights law as the claim to self-determination should not be readily displaced by Articles 55, 56 and 121 LOSC, which do not take away rights, but are rather concerned with establishing them in the first place. Thus, it can be argued that, if the LOSC is silent on the point of small island States threatened by climate change, it is then subject to interpretation, in the sense that the regime of islands is not only designed to establish rights, but is also concerned with maintaining them once established. In light of this, Articles 55, 56 and 121 should not justify taking away rights previously enjoyed under the LOSC, particularly the right to self-determination, including 'inalienable' sovereign rights over natural resources. The inalienable nature of the right in question, which has been pointed out by the CESCR with regard to indigenous people living in the Indonesian archipelago,[47] can be a strong argument in favour of an evolving interpretation of Articles 55, 56 and 121 LOSC, so as to preserve sovereign rights over natural resources in the case of progressively submerged and uninhabitable States. This stance is also reinforced by the fact that the right to self-determination is an essential component of the right to development, which would be thoroughly impaired by the loss of sovereignty over natural resources. Invoking self-determination to provide an extensive

44 *Case Concerning Certain Phosphate Lands in Nauru (Nauru v Australia) (Preliminary Objections), Judgment* [1992] ICJ Rep 240, 243.

45 Stoutenberg (2013) 70.

46 See also Clive Schofield and David Freestone, 'Options to Protect Coastlines and Secure Maritime Jurisdictional Claims in the Face of Global Sea Level Rise', in Gerrard and Wannier (2013) 160–1.

47 The CESCR mentioned the 'inalienable right to own, develop, control and use their [indigenous peoples'] customary lands and resources' (CESCR, *Concluding Observations on the Initial Report of Indonesia*, E/CN.12/IDN/CO/1 (19 June 2014) [37]).

interpretation of the existing law of maritime zones, and only for the purpose of maintaining existing rights to natural resources, could be a possible means to reconceptualise international law as to the situation of small island States. This hardly seems a radical step and does go some way to broadening the dialogue on Statehood.

2.3. Fixing baselines

The issue of fixed baselines exemplifies how the law of sea can be interpreted in light of the right to self-determination. Indeed, a particular impediment to small island States maintaining their maritime zones concerns the point from which such zones are measured. The territorial sea baseline is the point from which all other maritime zones are generated and Article 5 LOSC (Normal Baseline) defines it as being the naturally occurring low water line in the following terms:

> Except where otherwise provided in this Convention, the normal baseline for measuring the breadth of the territorial sea is the low-water line along the coast as marked on large-scale charts officially recognised by the coastal State.

On the face of it, a small island State with a retreating or disappearing coastline might have difficulties arguing for a fixed point from which to generate maritime zones.[48]

Fortunately, perhaps, albeit not addressing small island States disappearing beneath the waves, Article 7 LOSC (Straight Baselines) does provide for fixed baselines in certain circumstances as follows:

> 1. In localities where the coastline is deeply indented and cut into, or if there is a fringe of islands along the coast in its immediate vicinity, the method of straight baselines joining appropriate points may be employed in drawing the baseline from which the breadth of the territorial sea is measured.
> 2. Where because of the presence of a delta and *other natural conditions* the coastline is highly unstable, the appropriate points may be selected along the furthest seaward extent of the low-water line and, notwithstanding subsequent regression of the low-water line, the straight baselines shall remain effective until changed by the coastal State in accordance with this Convention.[49]

This provision recognises that coastlines are not always permanent. Schofield and Freestone discuss this in some detail and argue that fixing baselines is becoming

48 Powers and Stucko (2013) 132–3, discussing the baseline as being ‘ambulatory’. See also Clive Schofield and Andi Arsana, ‘Climate Change and the Limits of Maritime Jurisdiction’, in Warner and Schofield (2013) 134–6; Schofield and Freestone (2013) 144–8.

49 Emphasis added.

widely accepted in respect of sea level rise.[50] They do argue that customary international law will develop to accommodate the threat posed by climate change to small island States.

Self-determination should assist in interpreting the phrase 'other natural conditions' and play a role in this development. Although this would require disregarding the reference to a delta, a purposive interpretation of sub-section 2 of Article 7 LOSC can encompass the assumption that the people of a threatened small island State should maintain their right to self-determination, enabling retention of existing maritime jurisdiction over natural resources.

As Rayfuse notes, small island States without straight baselines would need to put them in place, adding that many small island States are actually archipelagos within the terms of Articles 46 and 47 LOSC.[51] These are therefore entitled to draw archipelagic straight baselines around groups of islands that meet the required ratios of land and water.[52] Such baselines effectively act in the place of territorial sea baselines. Where such baselines exist, it would be necessary to permit them to remain in place despite the alteration of the water to land ratio as islands become progressively submerged, in order to preserve the sovereignty of peoples over natural resources, particularly in the EEZ.[53]

Actually, interpreting Article 121 LOSC broadly could facilitate the maintenance of jurisdictional maritime zones,[54] particularly via habitable or uninhabitable artificial structures in derogation from Article 60(8) LOSC.[55] Along these lines, the Maldives, a leader in complex engineering to fight sea level rise, went so far as to complete a project for the artificial island 'Hulhumalé' within waters under its sovereignty. This structure would be able to host people from Maldives' atolls, thus also sustaining life, if these were to be submerged by rising waters.[56]

50 Schofield and Freestone (2013) 157–9. See also Schofield and Arsana (2013) 142–9; Rayfuse (2013) 167, 181–2, 187–91.

51 Ibid, 182–5.

52 LOSC, Art 47.

53 These arguments are also relevant to continental States with coastlines affected by rising sea levels (see Schofield and Arsana (2013) 135).

54 It must be noted that practice with respect to Article 121(3) is largely inconsistent and proposals have already been put forward to withdraw this provision from the LOSC (Yann-huei Song, 'The Application of Article 121 of the Law of the Sea Convention to the Selected Geographical Features Situated in the Pacific Ocean' (2010) 9 *Chinese Journal of International Law* 663 ff).

55 Article 60(8) (Artificial Islands, Installations and Structures in the Exclusive Economic Zone): 'Artificial islands, installations and structures do not possess the status of islands. They have no territorial sea of their own, and their presence does not affect the delimitation of the territorial sea, the exclusive economic zone or the continental shelf.' See also Grigoris Tsaltas *et al*, *Artificial Islands and Structures as a Means of Safeguarding State Sovereignty against Sea Level Rise: a Law of the Sea Perspective* (UNSW) 3 ff, <www.gmat.unsw.edu.au/ablos/ABLOS10Folder/S2P3-P.pdf>.

56 See Michael Gagain, 'Climate Change, Sea Level Rise, and Artificial Islands: Saving the Maldives' Statehood and Maritime Claims through the "Constitution of the Oceans"' (2012) 23 *Colo J Int'l Envtl L & Pol'y* 77, 81–2.

Conclusion

There has been quite a body of scholarship on the issue of the maritime zones of island States threatened by disappearance owing to climate change. This primarily concerns the interpretation of the law of the sea based on Statehood, without addressing the right to self-determination, specifically with regard to natural resources. In particular, an explicit connection has not been established between the right to self-determination and the LOSC provisions relating to the maritime zones of small island nations, specifically the EEZ.

This chapter argues that the right to self-determination, which is not State-centric, but based on the concept of 'people', should prompt an extensive interpretation of the law of the sea, with specific regard to sovereign rights over natural resources. Self-determination buttresses arguments for fixing the baselines of threatened small island States and maintaining their status as islands rather than rocks, so as to preserve existing maritime zones in circumstances originally not envisaged by the drafters of the LOSC. While this is a novel application of both the right to self-determination and the law of the sea, it is entirely consistent with their formulation and a systemic interpretation under Article 31 of the 1969 Vienna Convention on the Law of Treaties.[57]

57 ILC, *Fragmentation of International Law: Difficulties Arising from the Diversification and Expansion of International Law* (UN Doc A/CN.4/L.682, 2006) 25 ff, [37] ff.

7 Two-pronged right to development and climate change

Reciprocal implications

Same Varayudej

Introduction

As a result of the overwhelming weight of scientific evidence, it is now reasonable to maintain that peoples of the globe, especially those living in developing countries, are facing high risks posed by anthropogenic climate change.[1]

A recent report commissioned by the World Bank provides a forecast of the frightening impacts of 2°C warming on three tropical regions: Sub-Saharan Africa, South Asia and South East Asia.[2] The report highlights the key findings that have potentially devastating implications for development, such as unprecedented and more frequent heat extremes covering greater land areas both globally and in the three regions examined. For example, heat extremes in South East Asia are projected to increase substantially in the near term, and will have significant adverse effects on humans and ecosystems with as little as 2°C rise in temperature.[3]

There is therefore no doubt that unmitigated climate change is incompatible with sustainable development. If unregulated, global warming will reverse progress and prevent peoples in developing countries from realising their rights: climate change and development are inextricably linked. Nonetheless, the legal status of the right to development is unclear, since it is basically a general concept encompassing a variety of collective and individual claims. Furthermore, its binding force is still uncertain. In light of such a framework, this chapter explores the extent to which the right to development compels States to act on climate change. Reciprocally, some conclusions are made on the impact of climate change on the evolution of the right to development.

1 John Cook *et al*, 'Quantifying the Consensus on Anthropogenic Global Warming in the Scientific Literature' (2013) 8(2) *Environmental Research Letters* 1 ff; James S Risbey *et al*, 'Well-estimated Global Surface Warming in Climate Projections Selected for ENSO Phase' (2014) *Nature Climate Change* 835.

2 World Bank, *Turn Down the Heat: Climate Extremes, Regional Impacts, and the Case for Resilience* (2013).

3 Ibid, XV ff.

1. The right to development as a subjectively and objectively 'hybrid' claim and as a two-pronged claim: first and second generation human rights (first limb) and environmental sustainability (second limb)

The origin of the right to development can be traced back to the goals of the formation of the United Nations in 1945, promoting 'higher standards of living, full employment, and conditions of economic and social progress and development'.[4]

Senegalese jurist Keba M'Baye is generally credited with the first articulation of the right to development in 1972.[5] Subsequently, the right to development was popularised by Karel Vasak as one of several peoples' rights belonging to a 'third generation' of solidarity claims, together with the right to environment.[6]

Since 1977, there has been a new trend, led by the then UN Commission on Human Rights, to identify and attempt to remove structural obstacles to the full realisation of fundamental rights, particularly the right to development as a human right in developing countries.[7] However, such a claim was not formally recognised as a human right until the 1980s.

In 1986 the UN General Assembly overwhelmingly adopted the Declaration on the Right to Development (UNDRD),[8] which provides:

> The right to development is an inalienable human right by virtue of which *every human person* and all *peoples* are entitled to participate in, contribute to, and enjoy economic, social, cultural and political development, in which *all human rights and fundamental freedoms can be fully realised.*[9]

4 See UN Charter, Preamble (declaring the formation of the United Nations and outlining the goals of the organisation) and Art 55.

5 Keba M'Baye, 'Le droit au développement comme un droit de l'homme' (1972) 5 *Revue des droits de l'homme* 503.

6 Kamrul Hossain, 'The Realisation of the Right to Environment and the Right to Development in respect to the Arctic Indigenous Peoples' (2011) 3 *The Yearbook of Polar Law* 129, 131 ff; Henry J Steiner *et al*, *International Human Rights in Context: Law, Politics, Morals* (2007) 1442; Carl Wellman, 'Solidarity, the Individual, and Human Rights' (2000) 22 *Human Rights Quarterly* 639; Karel Vasak, 'Pour une troisième génération des droits de l'homme', in Christophe Swinarski (ed), *Studies and Essays on International Humanitarian Law and Red Cross Principles* (Martinus Nijhoff, 1984) 837, 839; Stephen Marks, 'Emerging Human Rights: a New Generation for the 1980's' (1981) 33 *Rutgers Law Review* 435; Karel Vasak, 'A Thirty Year Struggle: the Sustained Efforts to give Force of Law to the Universal Declaration of Human Rights', *UNESCO Courier* (November 1977) 29.

7 UN Commission on Human Rights, *Report of the Thirty-third Session*, E/CN.4/1257 (1977) Resolution 4 (XXXIII) of 21 February 1977. See also Roland Rich, 'The Right to Development: a Right of Peoples?', in James Crawford (ed), *The Rights of Peoples* (Clarendon Press, 1988) 39, 41.

8 See UN GA, *Declaration on the Right to Development*, Res 41/128 of 4 December 1986. The Resolution was adopted by 146 votes. The United States cast the only negative vote and eight other countries abstained (Denmark, Finland, Federal Republic of Germany, Iceland, Israel, Japan, Sweden and the United Kingdom).

9 Ibid, Art 1(1), emphasis added.

Objectively, the right to development thus emerges as the 'hybrid' sum of all other human rights, encompassing both individual human rights and collective rights. This connection is relevant to all individuals, peoples and States, but specifically to indigenous peoples, particularly those living in the Arctic region, who mostly have a subsistence economy and a strict cultural relationship to natural resources.[10]

Given the amorphous nature and imprecise content of the right to development, an extensive definition virtually becomes an amalgamation of a variety, if not all, of existing (first and second generation) individual and collective human rights.[11] Therefore, the debate is now focusing on the search for a narrower definition and a more precise content.[12]

Correspondingly, Article 3(1) UNDRD recognises that '*States* have the primary responsibility for the creation of national and international conditions favourable to the realisation of the right to development.'[13] Furthermore, 'States have the duty to co-operate with each other in ensuring development and eliminating obstacles to development' (Article 3(3)) and 'have the duty to take steps, individually and collectively, to formulate international development policies with a view to facilitating the full realisation of the right to development' (Article 4(1)). However, for States this is not only a duty, but also a right, given that under Article 2(3) '*States* have the *right and the duty* to formulate appropriate national development policies that aim at the constant improvement of the *well-being of the entire population and of all individuals*'.[14] The latter provision particularly stresses the all-encompassing character of the duties correlative to the right to development.

Subjectively, it must be assumed that the right to development is also hybrid, since it incorporates State–State, State–people and State–individual duty–right relationships, correlative to an individual right to be enjoyed by every person and a collective right belonging to all peoples and States.[15] According to some scholars,

10 Hossain (2011) 139 ff.

11 For this reason, developed countries are rather reluctant to accept the right to development as a fundamental claim, fearing that it may become 'a right to everything' or a right to 'development assistance', allowing a poor developing State (as a right holder) to sue a wealthy country (as a duty bearer) for the latter's failure to provide adequate financial assistance necessary for the fulfilment of the former's right (Yashi Gai, *Whose Human Right to Development?* (Commonwealth Secretariat Series of Occasional Papers on the Right to Development, 1989); Felix Kirchmeier, *The Right to Development – Where Do We Stand?* (Friedrich Ebert Stiftung, Occasional Papers, No 23, July 2006) 4, 12; Donnelly (1985) 481).

12 See Working Group on the Right to Development, *Report on Its Fifteenth Session*, A/HRC/27/45 (2014) 22, Recommendations.

13 Emphasis added.

14 Emphasis added. This has been ultimately interpreted as an expression of peoples' rights (Hossain (2011) 138). See also Marcos Orellana, *Climate Change and the Right to Development: International Cooperation, Financial Arrangements, and the Clean Development Mechanism*, A/HRC/15/WG.2/TF/CRP.3/Rev.139 (10 February 2010) [22].

15 This subjective complexity, which is proper to third generation human rights, has nonetheless been criticised as a source of confusion (Noel G Villaroman, 'The Right to Development: Exploring the Legal Basis of a Supernorm' (2010) 22 *Florida Journal of International Law* 299, 302).

the subjectively complex nature of the right to development would also entail an obligation to respect and fulfil it on the part of the international community as a whole.[16]

From the standpoint of hard law, the ACHPR,[17] which entered into force in 1986 at almost the same time as the adoption of the UNDRD, recognises the right to development from a more collective perspective. Indeed, Article 22 provides that 'all *peoples* shall have the right to their economic, social and cultural development with due regard to their freedom and identity and in the equal enjoyment of the common heritage of mankind.'[18] States have a correlative individual and collective duty to 'ensure the exercise of the right to development.'[19]

The ACHPR also links the right to development to the rights of peoples to equality,[20] existence,[21] and free disposal of natural resources, including food and water.[22] Along these lines, scholars have highlighted a link between the right to development and the right to (internal) self-determination,[23] particularly with regard to indigenous peoples, not only substantively, but also as to participation in decision-making processes.[24]

Most importantly, Article 24 ACHPR takes a step further with respect to the contemporaneous UNDRD, which does not mention 'environment' and 'sustainability'. Indeed, the ACHPR envisages a right to environment functional to the right to development and preliminary to all other human rights: 'All peoples shall have the right to a general satisfactory environment favourable to their development.' Such a provision follows the recognition of a link between environment and development by the World Bank in the 1970s.[25] This link is particularly important because the right to environment emerges as a correlated but separate category with respect to the right to development, unlike first and second generation human rights, which are indistinctly encompassed by the right to development.

Along the lines of the ACHPR, the 1987 Brundtland Report clearly brought together the notions of 'development' and 'sustainability', under the concept of

16 Hossain (2011) 137.

17 *African Charter on Human and Peoples' Rights*, signed 27 June 1981, OAU Doc CAB/LEG/67/3 Rev 5 (entered into force 21 October 1986).

18 Ibid, Art 22(1) (emphasis added).

19 Ibid, Art 22(2).

20 Ibid, Art 19.

21 Ibid, Art 20.

22 Ibid, Art 21.

23 Alan Boyle, 'Environment and Development: Accountability through International Law' (1993) 12 *Third World Legal Studies* 95 ff; Robert E Lutz *et al*, 'Environment, Economic Development and Human Rights: a Triangular Relationship? Remarks' (1988) 82 *Am Soc'y Int'l L Proc* 40.

24 Hossain (2011) 152.

25 'Environment, Economic Development and Human Rights: a Triangular Relationship? Remarks by Ibrahim Shihata, David Wirth and Philip Alston' (1988) 44–55.

'sustainable development', whereby 'sustainability' refers to the idea of maintaining well-being '*over time*':[26]

> Sustainable development is development that meets the needs of the present without compromising the ability of future generations to meet their own needs. It contains within it two key concepts:
>
> - the concept of 'needs', in particular the essential needs of the world's poor, to which overriding priority should be given; and
> - the idea of limitations imposed by the state of technology and social organisation on the environment's ability to meet *present and future needs.*[27]

Similarly, the right to environment tends to be associated with the right to development in subsequent regulatory instruments. Thus, the 1990 UN Declaration on International Economic Cooperation states that 'economic development must be environmentally sound and sustainable'.[28] UN GA Resolution 48/141 recognises the concept of 'sustainable development' in establishing the mandate of the High Commissioner for Human Rights.[29]

Crucially, Principle 1 of the 1992 Rio Declaration on Environment and Development provides that 'human beings are at the centre of concerns for sustainable development' and are 'entitled to a healthy and productive life in harmony with nature.'[30] Principle 27 of the Rio Declaration is also particularly relevant in this respect, since it provides that 'the *right to development* must be *fulfilled* so as to *equitably* meet *developmental* and *environmental* needs of present and future generations.'[31] The environmental need thus emerges as a separate branch with respect to the necessity of eliminating poverty.

Even more explicitly, the 1995 IUCN Draft International Covenant on Environment and Development (IUCN DICED), as re-edited in 2014, points out the need to integrate environment and development so as to outline a 'comprehensive legal framework with the aim of achieving environmental conservation', which is considered an 'indispensable foundation for sustainable

26 Ana Markulev and Anthea Long, *On Sustainability: an Economic Approach* (Commonwealth of Australia, 2013) 3, emphasis added.

27 World Commission on Environment and Development, *Our Common Future* (Brundtland Report, 1987) [2.1], emphasis added. A right to environment is consequently provided for in Annex 1 to the Brundtland Report, Principle 1: '*All human beings* have the fundamental right to an environment adequate for their health and well-being' (*Summary of Proposed Legal Principles for Environmental Protection and Sustainable Development Adopted by the WCED Experts Group on Environmental Law* (1987), emphasis added).

28 UN GA Res A/RES/S-18/3, *Declaration on International Economic Cooperation, in particular the Revitalisation of Economic Growth and Development of the Developing Countries* (1 May 1990) [16].

29 GA Res 48/141, UN Doc A/48/141 (20 December 1993) 3(c): '[The High Commissioner for Human Rights shall] recognise the importance of promoting a balanced and *sustainable development* for all people and of ensuring realisation of the right to development' (emphasis added).

30 UN GA, *Report of the United Nations Conference on Environment and Development*, UN Doc A/CONF.151/26 (12 August 1992) Annex I.

31 Emphasis added.

development'.[32] More specifically, the right to development 'entails the obligation to meet environmental, as well as social and economic needs of humanity in a sustainable and equitable manner.'[33]

Also the 1993 Vienna Declaration and Programme of Action highlights both the economic and environmental components of the right to development.[34] A similar stance has been taken by the UN Millennium Declaration,[35] the 2002 Monterrey Consensus[36], the 2005 World Summit Outcome Document,[37] and even the 2007 Declaration on the Rights of Indigenous Peoples.[38]

Finally, the HRC High-level Task Force on the Implementation of the Right to Development and legal scholars have recognised the privileged link between the right to development and the right to environment, standing on the same footing.[39]

2. Climate change and right to development (first limb): balancing economic inequalities

Peoples living in poverty in developing countries have been disproportionately affected by the impacts of climate change, as their culture is dependent on natural resources like land, water, fisheries or forestry, which are profoundly affected by global warming.[40]

South Asian populations are likely to be increasingly vulnerable to the greater variability of rainfall changes that could put water and food resources at severe

32 IUCN DICED (4th edn, 2014), Articles 1 and 16. The commentary to Article 1 of the 3rd edition of the Draft Covenant (2004) specifies that 'environment and development are indivisible'.

33 Ibid, Article 8.

34 *Vienna Declaration and Programme of Action*, Note by the Secretariat, World Conference on Human Rights, UN Doc A/CONF.157/23 (25 June 1993) pt I, [10]: 'Lasting progress towards the implementation of the right to development requires effective development policies at the national level, as well as equitable economic relations and a favourable economic environment at the international level'; [11]: 'The right to development should be fulfilled so as to meet equitably the developmental and environmental needs of present and future generations.' The Vienna Declaration and Programme of Action was unanimously adopted by 171 Member States.

35 GA Res 55/2, *United Nations Millennium Declaration* (18 September 2000).

36 *Report of the International Conference on Financing for Development*, Monterrey, Mexico, A/CONF.198/11 (2002).

37 GA Res 60/1, *World Summit Outcome* (24 October 2005).

38 GA Res 61/295, *UN Declaration on the Rights of Indigenous Peoples* (27 October 2007), Preamble, mentioning the 'sustainable and equitable development and proper management of the environment'.

39 High-level Task Force on the Implementation of the Right to Development, *Report on Its Sixth Session*, Doc A/HRC/15/WG.2/TF/2/Add.2 (2010) Annex, 1(h), Environmental Sustainability and Sustainable Use of Natural Resources; Hossain (2011) 137; Fatma Zohra Ksentini, *Human Rights and the Environment, Review of Further Developments in Fields with which the Sub-Commission Has Been Concerned*, Final Report of the UN Sub-Commission on Prevention of Discrimination and Protection of Minorities, ECOSOC, E/CN.4/Sub.2/1994/9 (6 July 1994) [7], [63]–[64]).

40 Climate and Development Knowledge Network, *Integrating Climate Change in the Post-2015 Development Agenda* (March 2014) 3. See also the contribution by Faustin Ntoubandi and Roland Adjovi in this volume.

risk, since shifting rain patterns will leave some areas under water and others without enough water for power generation, agriculture or drinking. Events like the devastating Pakistan floods of 2010, which affected more than 20 million people, could become commonplace. Declines of up to 20 per cent in water availability are projected for many regions below a 2°C temperature-rise and up to 50 per cent for some regions below a 4°C temperature-rise.[41] With declining water availability, crop production systems are under increasing pressure to meet growing global demand in the future. Significant crop yield impacts are already being felt at 0.8°C warming. These impacts will have strong repercussions on food security and are likely to negatively influence economic growth and poverty reduction in affected regions.[42] Furthermore, farmers, coastal communities, and large urban centres are facing mounting pressure as sea level rises, tropical cyclones increase in intensity, fish catches decrease, and coastal zones lose the protection of coral reefs.[43]

Similarly, in Sub-Saharan Africa, a rapidly developing region of over 800 million people with 49 countries, food shortages are becoming more common. Along these lines, drought and heat as a result of a 1.5°C warming by 2030 will leave 40 per cent of the land now growing maize unable to support that crop. Rising temperatures cause major loss of savannah grasslands, threatening pastoral livelihoods. More extreme droughts could stall power generation and turn rural fields barren, leading to lost income for farmers and widespread food shortages. For instance, a warming of less than 2°C by 2050 could lead to a 10 per cent reduction in total crop production. As agriculture falters, and food and water become scarce, the pace of people migrating into informal urban settlements accelerates. Life in these slums is no less difficult than the life people fled. Housing made from scraps traps heat yet provides little protection from storms, landslides or floods. Diseases like malaria, dengue and cholera become easier to contract and tension over access to dwindling resources may generate conflicts.[44]

Overall, the OHCHR has highlighted that poorer regions and countries are disproportionately impacted by the 'catastrophic consequences' of climate change and also suffer from a low adaptation capacity.[45] To sum up, climate change is intrinsically linked to the economic implications of the right to development, as the former has already undermined progress and prospects of vulnerable peoples, particularly in developing countries, in achieving the Millennium

41 Climate and Development Knowledge Network (2014) 3.

42 Ibid.

43 Jim Yong Kim, President of the World Bank Group, *Op-Ed: Ending Poverty Includes Tackling Climate Change* (10 July 2013).

44 Ibid.

45 OHCHR, *Report on the Relationship between Climate Change and Human Rights*, UN Doc A/HRC/10/61 (15 January 2009) [26], [42], [77] and [81]; Human Rights Council Panel Discussion on the Relationship between Human Rights and Climate Change, Kyung-wha Kang, Deputy High Commissioner for Human Rights, *Opening Remarks* (15 June 2009) 2. See also Marc Limon, 'Human Rights Obligations and Accountability in the Face of Climate Change' (2010) *Georgia Journal of International and Comparative Law* 543, 551.

Development Goals (MDGs), which include the commitment 'to making the right to development a reality for everyone and to freeing the entire human race from want'.[46] Unmitigated climate change will continue to deepen vulnerabilities and erode hard-won gains in development and eradication of poverty that are essential for realising other human rights. This raises an issue of justice, concerning inequity and inequalities relevant to the prohibition of discrimination under Article 26 of the ICCPR.[47]

Mary Robinson, former UN High Commissioner for Human Rights, rightly pointed out:

> It is poor communities who are suffering most from the effects of climate change, and it is rich countries that are contributing most to the problem. The human rights approach, emphasising the equality of all people, is a direct challenge to the power imbalances that allow the perpetrators of climate change to continue unchecked. The human rights framework gives us the legal and normative grounds for empowering the poor to seek redress.[48]

Normatively, collective and individual first and second generation human rights obligations encompassed by the economic limb of the right to development compel developed States to a more equitable and equal sharing of the planet's resources.[49] At the same time, developing States are also compelled to take GHG mitigation and adaptation policies in light of classical fundamental rights of their own people. This framework requires a proactive approach by developed and developing States under the obligation to fulfil the kaleidoscope of fundamental rights held especially by individuals and peoples in accordance with the 'right to development'.[50]

In particular, the effects of GHG emissions impinge upon natural resources and thus violate natural resources-based first and second generation fundamental rights encompassed by the right to development, such as the basic claims to food, water, health and self-determination, as highlighted universally, *de lege lata*, in Article 8 UNDRD and regionally in Articles 20–4 ACHPR.[51] A broad interpretation of the right to self-determination can also be maintained in light of

46 UN GA Res 55/2 (2000) [11].

47 See Olivier De Schutter, *International Human Rights Law* (Cambridge University Press, 2nd edn, 2014) 647.

48 Mary Robinson, 'Climate Change and Justice' (2007) 36 *Scottish Human Rights Journal* <www.scottishhumanrights.com>.

49 Natalia Cardona, *Climate Justice for a Changing Planet: Placing the Right to Development and Justice at the Heart of the Response to Climate Change* (UN NGLS, 2009); Marcos A Orellana, *Climate Change and the Millenium Development Goals: the Right to Development, International Cooperation and the Clean Development Mechanism*, [2.1], <www.surjournal.org>.

50 Noel G Villaroman, 'Rescuing a Troubled Concept: an Alternative View of the Right to Development' (2011) 29 *Netherlands Quarterly of Human Rights* 13, 49 ff.

51 See Marcos Orellana, 'A Rights-Based Approach to Climate Change Mitigation', in Thomas Greiber (ed), *Conservation with Justice: a Rights-based Approach* (International Union for Conservation of Nature, 2009) 37.

Article 21 of the Stockholm Declaration, so as to encompass environmental protection under the no-harm rule.[52]

Significantly, under the UNDRD 'States have the right and the duty to formulate appropriate national development policies that aim at the constant improvement of the well-being of the *entire population* and of *all individuals*.'[53] Furthermore, Resolution 41/181 provides that 'States should realise their rights and fulfil their duties in such a manner as to promote a new international economic order based on sovereign equality, interdependence, mutual interest and co-operation among all States, as well as to encourage the *observance and realisation of human rights*.'[54] The UNDRD also envisages that States should take steps 'to ensure the *full exercise* and *progressive enhancement* of the right to development, including the formulation, adoption and implementation of policy, legislative and other measures at the national and international levels'.[55] A cooperative framework for the implementation of these obligations is outlined by the necessity to take '[s]ustained action . . . to promote more rapid development of developing countries', by 'providing these countries with appropriate means and facilities to foster their comprehensive development'.[56]

3. Climate change and right to development (second limb): achieving environmental sustainability

As Marcos Orellana rightly pointed out, 'if [GHG] emissions cannot continue to increase without causing severe global environmental and social harm, then by necessity development must follow a *sustainable, non-carbon path*.'[57]

The second limb of the right to development, that is, its environmental component, provides a crucial link between climate change and consequent economic inequalities. In light of the hybrid subjective nature of the right to development, this limb imposes on developed and developing countries two main duties.[58]

52 Article 21 of the Stockholm Declaration provides:

> States have, in accordance with the Charter of the United Nations and the principles of international law, the sovereign right to exploit their own resources pursuant to their own environmental policies, and the responsibility to ensure that activities within their jurisdiction or control do not cause damage to the environment of other States or of areas beyond the limits of national jurisdiction.

53 Resolution 41/128, Art 2(3), emphasis added.

54 Ibid, Art 3(3), emphasis added.

55 Ibid, Arts 4(1) and 10, emphasis added.

56 Ibid, Art 4(2).

57 Marcos Orellana, 'Climate Change, Sustainable Development and the Clean Development Mechanism', in *Realising the Right to Development: Essays in Commemoration of 25 Years of the United Nations Declaration on the Right to Development* (OHCHR, 2013) 321, emphasis added.

58 See John Knox, UN Independent Expert on Human Rights and the Environment, *Mapping Human Rights Obligations Relating to the Enjoyment of a Safe, Clean, Healthy and Sustainable Environment: Focus Report on Human Rights and Climate Change* (June 2014) Ch VI, [VII].

First, developed and developing States should take mitigation and adaptation measures to 'respect' and 'protect' the right to 'sustainable development', that is, the right to environment of States, their people and individuals.[59] This duty is implied by the fact that development itself must be 'sustainable'.[60] In this regard, the 1990 UN Declaration on International Economic Cooperation significantly provides that environmental deterioration is a cause of 'common concern for *all*', so that '*all countries* should take effective actions for the protection and enhancement of the environment in accordance with their respective capacities and responsibilities, and taking into account the specific needs of developing countries'.[61] The IUCN DICED provides that States have a right and a duty as to compliance with the no-harm rule.[62] More specifically, the Draft Covenant, which is supposed to complement the UNFCCC,[63] explicitly provides that States 'shall take precautionary measures to protect the Earth's climate system and mitigate the adverse effects of climate change' within the context of the integration of environment and development.[64]

Second, developed States should provide financial and technological assistance to developing States, so as to enable them to achieve environmentally sound development, thus 'fulfilling' the right to development.[65] This is implicit in the principle of 'common but differentiated responsibility' (CBDR). In this regard, the IUCN DICED provides that '[p]arties undertake to achieve progressively the full realisation of the right of all persons to live in an ecologically sound environment adequate for their development, health, well-being and dignity'.[66] Such a duty is underpinned by an intrinsic obligation to cooperate,[67] which perfectly matches the collaborative framework established under the UNFCCC,[68] compelling industrialised States to provide financial and technological support to poor countries.[69] Within this context, Article 3 UNFCCC emerges as a key provision.

59 On the 'dual' nature of the right to development as a right and a duty for developing countries see Hossain (2011) 138.

60 *Centre for Minority Rights Development (Kenya) and Minority Rights Group International on behalf of Endorois Welfare Council v Kenya*, Communication 276/2003 (AComHPR, 2009) [22], [298]. In the *Endorois* case, a State was held responsible for violating the right to development under Article 22 ACHPR.

61 UN GA Res S-18/3, *Declaration on International Economic Cooperation, in particular the Revitalisation of Economic Growth and Development of the Developing Countries* (1 May 1990) [29], emphasis added.

62 IUCN DICED, Article 13.

63 Ibid, Comment to Article 20.

64 Ibid, Article 20.

65 On positive and negative State duties associated with the right to development see Orellana (2010) 9, [23].

66 IUCN DICED, Article 14.

67 Orellana (2010) 9, [24] ff.

68 See, for instance, UNFCCC, Preamble and Article 3(3).

69 David Hunter, James Salzman and Durwood Zaelke, *International Environmental Law and Policy* (Foundation Press, 2nd edn, 2002) 495; Duncan French, 'Developing States and International Environmental Law: the Importance of Differentiated Responsibilities' (2000) 49 *International & Comparative Law Quarterly* 35.

This rule, which is also known as the 'equity article', seeks to strike a balance between developed and developing countries as to climate change via the principle of 'equity', specifically via CBDR. Along these lines, developed countries are supposed to take the lead in combating climate change and its adverse effects. Furthermore, the needs of developing countries, which are particularly vulnerable to climate change and bear a disproportionate burden under the UNFCCC, must be taken into account.[70] This entails a proactive approach to environmental sustainability, in accordance with the obligation to fulfil human rights.[71]

In line with the principle of equity and cooperation embodied in the UNFCCC, at the intersection between the responsibility of developed and developing countries, the Kyoto Protocol provides GHG mitigation via joint implementation of reduction commitments, an international emission trading system, and clean development mechanisms (CDMs).[72] These processes can be seen through the lens of the right to environmental sustainability for both developed and developing countries.[73] For instance, according to CDMs established under Articles 10 and 12 of the Kyoto Protocol, developed States are to assist developing countries in achieving sustainable development. More specifically, CDMs aim to stimulate high emitter Annex I countries to invest in projects that remove or reduce GHGs in developing countries, so as to meet their reduction targets under Annex B, and for which both Annex I and non-Annex I countries obtain certified emissions reductions (CERs).[74]

70 UNFCCC, Article 3:

> In their actions to achieve the objective of the Convention and to implement its provisions, the Parties shall be guided, inter alia, by the following: 1. The Parties should protect the climate system for the benefit of present and future generations of humankind, on the basis of equity and in accordance with their common but differentiated responsibilities and respective capabilities. Accordingly, the developed country Parties should take the lead in combating climate change and the adverse effects thereof; 2. The specific needs and special circumstances of developing country Parties, especially those that are particularly vulnerable to the adverse effects of climate change, and of those Parties, especially developing country Parties, that would have to bear a disproportionate or abnormal burden under the Convention, should be given full consideration.'

See also OHCHR (2009) [11]. Derek Bell argues that the principle of environmental equity and justice is a necessary condition for securing the path to sustainability (Derek Bell, 'Climate Change and Human Rights' (2013) 4(3) *Wiley Interdisciplinary Reviews: Climate Change* 159).

71 HRC, Res 26/L.33, *Human Rights and Climate Change* (23 June 2014) [3], [5], [6].

72 1997 *Kyoto Protocol to the UNFCCC.*

73 See, in more detail, Orellana (2010) 13, [39] ff.

74 The Kyoto Protocol was adopted in 1997 under the UNFCCC as a legally binding international agreement to reduce GHG emissions and entered into force in 2005, with emissions reduction targets for 37 industrialised and transition countries, and the European Community. These are known as Annex I countries, for which specific targets are set out in Annex B. However, compared to Annex B countries, Annex I countries include two additional States, that is, Belarus and Turkey. Furthermore, although the US is listed on Annex B, it has not signed up to the Kyoto Protocol.

The Kyoto Protocol's CDM provides a mode of cooperation between industrialised and developing countries. CDMs are designed to assist developing countries in the move to green technology, so as to enhance their prospects for sustainable development while helping industrialised countries to achieve their legally binding targets under the Protocol.[75] CDMs are thus a means for developed States to fulfil the duty to provide environmental protection within the context of the CBDR principle. Since rich countries have contributed most to carbon emissions, they have an obligation to provide financial and technological support to poor countries and enable them to develop in an ecologically sound way.[76] In recognition of this, the High-level Task Force of the Working Group on the Right to Development has described the CDM as 'an arrangement of value to the *climate change dimension of the right to development* insofar as the transfer of green technology can enhance the prospects for *sustainable development* in *developing countries*.'[77] Collateral projects, such as the 'debt-for-nature swap', can also be framed as a means for developed and developing States to respect and fulfil environmental sustainability.[78]

4. Soft law or hard law? Disentangling substance and procedure

Twenty-eight years have elapsed since the right to development was recognised by the General Assembly in the UNDRD and the ACHPR. Nonetheless, the implementation and consolidation of the right to (sustainable) development in a legally binding document has made slow progress owing to the fact that the process of building consensus on its meaning and practical implementation is highly politicised.

The right to development is viewed by industrialised nations as an integral part of the controversial demands made by less developed countries for establishing a *New International Economic Order* (NIEO), challenging the existing international economic regime established in the 1970s.[79] One of the most contentious issues arising from the debate on the right to development is thus whether this right has already crystallised into the body of customary international law (*lex lata*) or is only

75 Marcos Orellana maintains that CDMs establish 'an international partnership between the global South and the industrialised North to achieve the twin objectives of promoting sustainable development and mitigating climate change' (Orellana (2010) [3]).

76 Barbara Adams and Gretchen Luchsinger, *Climate Justice for a Changing Planet: a Primer for Policy Makers and NGOs* (UN Non-Governmental Liaison Service, 2009). Similarly, Marcos Orellana argues that 'given historical responsibility and current capabilities, the industrialised countries bear the obligation to provide the financial and technological support to make this leap [development] possible' (Orellana (2013) 321).

77 Orellana (2010) 13, [40] (emphasis added).

78 'Environment, Economic Development and Human Rights: a Triangular Relationship?, Remarks by John Porter' (1988) 57 ff.

79 See UN GA Res 3201 (S-VI), *Declaration on the Establishment of a New International Economic Order* (1 May 1974).

an emerging human right (*lex ferenda*). Some scholars argue that at least certain obligations under the right to development have gradually attained the status of customary international law and should be distributed according to CBDR.[80] In fact, although it has not yet achieved the status of customary international law,[81] the CBDR principle is emerging as such, since it is increasingly acknowledged by international environmental instruments.[82] In practice, however, substantively the right to development has not been generally accepted by States, particularly those specially interested in its implementation. In fact, a number of major developed countries still refuse to accept the right to development as part of international human rights law. The US, for instance, is 'willing to talk about an individual's right to development but not a nation's right to development . . . for the simple reason that nations do not have human rights.'[83] Instead, Germany fully supports the content of the right to development as 'growth with equity' in terms of both development outcomes and development processes, and is prepared to acknowledge the bindingness of this principle.[84] The UK is committed to supporting partner governments to fulfil their human rights obligations and is willing to discuss with these governments how to assess progress in this area but, like other countries in the EU, prefers to keep its commitments voluntary.[85] Specifically, the environmental component of the right to development cannot be considered a clearly established customary international law duty,[86] unless it is regarded as a general principle of law inferred from the majority of the legal systems of the world.[87]

In light of the fact that the two-pronged right to development has not yet achieved the status of hard law after its twenty-fifth anniversary, some scholars have finally taken the pessimistic view that States are unlikely to ever agree on a generally binding instrument on the right to development. Vandenbogaerde thus

80 See Nienke van der Have, *The Right to Development and State Responsibility: Can States Be Held to Account?* (SHARES Research Paper 23, 2013) 17–18.

81 Christopher D Stone, 'Common but Differentiated Responsibilities in International Law' (2004) 98 *American Journal of International Law* 276, 299.

82 The CBDR principle is expressly recognised in the Rio Declaration on Environment and Development and has been affirmed in various international environmental instruments, including the Montreal Protocol on Substances that Deplete the Ozone Layer, the UNFCCC, the Kyoto Protocol, and the Stockholm Convention on Persistent Organic Pollutants. UNFCCC COP20 restated that CBDR is a basis for taking mitigation and adaptation measures to protect the climate system for the benefit of present and future generations (*Lima Call for Climate Action*, Decision -/CP.20 (December 2014) [3]). See also Orellana (2010) [92].

83 Statement by Lino J Piedra, Public Member of the US Delegation (22 March 2005), cited in Kirchmeier (2006) 14. This is nevertheless criticisable, particularly in light of the right of peoples to self-determination.

84 Ibid, 13–14.

85 Ibid, 14–15.

86 Hossain (2011) 133.

87 Knox, Preliminary Report, UN Doc A/HRC/22/43 (2012) [12] and [14]; David R Boyd, *The Constitutional Right to a Healthy Environment* (2013) <www.lawnow.org/right-to-healthy-environment>.

suggests that the right to development should be dissolved, as it duplicates and detracts from other human rights, particularly economic, social and cultural rights.[88] This stance is supported by the assumption that, even if substantively the right to development were to provide a basis for climate change action for both developed and developing countries, procedurally it is unclear what avenues might be exploited to enforce this right.[89] Therefore, the right to development should be regarded as nothing more than soft law.[90] Nonetheless, the enforcement argument is weak. *De lege ferenda*, it is possible to envisage State action before international adjudicatory bodies, grounded in a fully established substantive right to development, subject to the principle of consensual jurisdiction.[91] After all, this has already taken place in *Trail Smelter*, *Lac Lanoux*, and the *Nuclear Tests* cases, based on the no-harm rule, given that environmental protection is part of the right to development.[92] Individual and group action is also possible both domestically and internationally within the framework of administrative, civil, criminal and human rights protection mechanisms, as the *Endorois* case proves.[93] *De lege lata*, the procedural implications of the right to development in terms of access to information, justice and participation in decision-making are already well-established in international law, with respect to both its economic and environmental branches. With regard to the economic component, classical first and second generation human rights entail procedural guarantees since at least the time of the UDHR.[94] The environmental element has been decisive in outlining these procedural rights, which are nowadays firmly established under the Aarhus Convention.

Therefore, *de lege lata*, it can already be assumed that the right to development, bringing together first, second and third generation human rights, strengthens the rights to information, participation and access to justice in climate change

88 Arne Vandenbogaerde, 'The Right to Development in International Human Rights Law: a Call for Its Dissolution' (2013) 31 *Netherlands Quarterly of Human Rights* 187, 208; Villaroman (2011) 39.

89 See, for instance, Anna Olga Dabrowska, *Legal Status of the Right to Development* (Unversitas Islandiae, 2010) 5–7; Philip Alston, 'Making Space for New Human Rights: the Case of the Right to Development' (1988) 1 *Harvard Human Rights Year Book* 3, 20; Jack Donnelly, 'In Search of the Unicorn: the Jurisprudence and Politics of the Right to Development' (1985) 15 *Cal W Int'l LJ* 473.

90 On the concept of soft law, see Prosper Weil, 'Towards Relative Normativity in International Law' (1983) 77 *American Journal of International Law* 413, 421; Christine M Chinkin, 'The Challenge of Soft Law: Development and Change in International Law' (1989) 38 *International and Comparative Law Quarterly* 850.

91 See IUCN DICED, Articles 68–69 and related comments. See also Boyle (1993) 98 ff.

92 *Trail Smelter (USA v Canada)*, Award of 16 April 1938 and 11 March 1941 (2006) III *RIAA* 1905; *Lake Lanoux Arbitration, France v Spain* (1957) 12 *RIAA* 281; *Nuclear Tests, Australia v France* [1974] ICJ Rep 253; *Nuclear Tests, New Zealand v France* [1974] ICJ Rep 457. See also Villaroman (2011) 47–8. On the degree of responsibility with respect to the no-harm rule see the contributions in Part I of this volume.

93 *Centre for Minority Rights Development* (2009).

94 See, for instance, UDHR, Articles 8 and 10.

matters.[95] In this regard, Article 2(3) UNDRD emphasises the importance of active, free and meaningful participation in the process of development.[96] Along these lines, the Inuit claim for involvement in climate change policy-making, a matter that is essential to ensure their sustainable development, which has been convincingly defined as 'self-determined development'.[97] Similarly, the High-level Task Force on the Implementation of the Right to Development has proposed to improve participation in decision-making concerning the implementation of CDMs.[98]

Conclusion

The right to (sustainable) development is two-pronged and entails an economic and an environmental component. From a legal standpoint, the former is the result of a sum of first and second generation human rights, whereas the latter is a third generation claim. Both elements, particularly the preliminary environmental duty, are relevant to climate change. They should be interpreted as imposing on developed and developing States an obligation to take GHG mitigation and adaptation measures, according to the principle of CBDR.

In practice, although the international legal status of the economic and environmental components of the right to development is substantively debatable, an analytical investigation demonstrates that their procedural implications are firmly established in international law. Therefore, *de lege lata* the right to development is relevant to facilitate participation in decision-making and access to information and justice in climate change-related matters, which is particularly important for developing countries. Whether or not this exercise of procedural rights can lead to a fully established substantive right to development is a matter that the current Working Group on the Right to Development will be hopefully able to clarify in the years to come.

95 See Orellana (2010) 7, [13] See also International Bar Association, *Achieving Justice and Human Rights in an Era of Climate Disruption* (2014) 122; Hossain (2011) 138.

96 *Centre for Minority Rights Development* (2009) [297]. See also IUCN DICED, Article 14; Ksentini (1994) [67] ff.

97 See Inuit, *Petition to the Inter-American Commission on Human Rights, Seeking Relief from Violations Resulting from Global Warming Caused by Acts and Omissions of the United States* (7 December 2005) 1, 118, Request for Relief; Hossain (2011) 141, 152. See also Arctic Athabaskan Council on Behalf of All Arctic Athabaskan Peoples of the Arctic Regions of Canada and the United States, *Petition to the Inter-American Commission on Human Rights Seeking Relief from Violations of the Rights of Arctic Athabaskan Peoples Resulting from Rapid Arctic Warming and Melting Caused by Emissions of Black Carbon by Canada* (2013) 87.

98 Orellana (2010) [101] ff.

8 Untying the Gordian Knot

Towards the human right to a climatically sustainable environment?

Francesco Francioni and Ottavio Quirico

Introduction

The Declaration adopted at the 1972 UN Conference on the Human Environment held in Stockholm, Sweden, provides that the concept of 'environment' includes the 'natural resources of the earth, including the air, water, land, flora and fauna'.[1]

The most important reason for focusing today on a human rights approach to environmental protection lies in the existential challenge posed to humanity by the impending threat of anthropogenic climate change.[2] However, is this situation sufficient in itself to justify such an approach, particularly based on a right to a (climatically) sustainable,[3] satisfactory, safe, clean, healthy[4] environment(al)[5] protection[6]?

This chapter traces the evolution from obligations related to the enjoyment of a sustainable environment inferred from first and second generation human rights

1 Principle 2. See also *Legality of the Threat or Use of Nuclear Weapons (Advisory Opinion)* [1996] ICJ Rep 226, 241 [29].

2 IPCC, Fifth Assessment Report, *Climate Change 2014: Impacts, Adaptation, and Vulnerability*, Yokohama, Japan (29 March 2014); OHCHR, *Report on the Relationship between Climate Change and Human Rights*, UN Doc A/HRC10/6 (15 January 2009).

3 Fatma Zohra Ksentini, *Human Rights and the Environment, Review of Further Developments in Fields with which the Sub-Commission Has Been Concerned*, Final Report of the UN Sub-Commission on Prevention of Discrimination and Protection of Minorities, United Nations Economic and Social Council, UN Doc E.CN.4/Sub.2/1994/9 (6 July 1994) [119].

4 John Knox, *Mapping Report of the Independent Expert on the Issue of Human Rights Obligations Relating to the Enjoyment of a Safe, Clean, Healthy and Sustainable Environment*, UN Doc A/HRC/25/53 (30 December 2013); Id, *Preliminary Report of the Independent Expert on the Issue of Human Rights Obligations Relating to the Enjoyment of a Safe, Clean, Healthy and Sustainable Environment*, UN Doc A/HRC/22/43 (24 December 2012); Ksentini (1994) [242]. See also Burns H Weston and David Bollier, *Green Governance: Ecological Survival, Human Rights, and the Law of the Commons* (CUP, 2013) Addendum, The International Legal Status of the Human Right to a Clean and Healthy Environment.

5 See, for instance, UNESCO, *Declaration of Bizkaia on the Right to Environment*, 30 C/INF.11 (24 September 1999).

6 *Bizkaia Declaration*, Article 2; *Draft American Declaration on the Rights of Indigenous Peoples* (1997) Article XIII.

provisions to the status of an independent third generation human right to a (climatically) sustainable environment under the pressure of various forms of environmental degradation, and more specifically under the adverse impact of climate change on humanity.

The chapter takes a systemic approach to the right in issue from three complementary perspectives: (1) law making; (2) substantive law; and (3) adjudication and enforcement. In light of the existing terminological uncertainty, we refer to the comprehensive expression 'right to a sustainable environment', along the lines of the concept of 'sustainable development' promoted by the Brundtland Report.[7]

1. The environmental dimension of human rights

1.1. Environmental interpretation of substantive first and second generation human rights

A collateral environmental dimension of a few existing human rights has evolved over time. By way of judicial interpretation of relevant treaty provisions, a considerable degree of progress has been made towards the integration of environmental values into existing human rights norms. This integration has occurred both in a *positive* sense, that is, for the purpose of affording enhanced protection against environmental harm, and in a *negative* sense, by referring to environmental considerations as a legitimate aim to limit to the enjoyment of human rights.[8]

The most common provisions relied upon in the development of the environmental dimension of human rights are those concerning the rights to life, private and family life, property and information.[9] However, the trend is not limited to traditional civil and political rights, but extends to economic, social and cultural rights, although the interface between the latter and climate change has been underplayed.

In interpreting and implementing the ECHR, the ECtHR has recognised: (1) the positive obligation to regulate activities of an industrial or technological nature, which are likely to produce adverse effects on the sphere of protected human rights, particularly the right to private and family life;[10] and (2) the positive obligation to effectively enforce legal, administrative, and judicial measures designed to prevent or remedy unlawful interference with protected human rights.[11]

7 World Commission on Environment and Development, *Our Common Future* (Brundtland Report, 1987).

8 See, for instance, *Salvador Chiriboga v Ecuador*, Preliminary Objection and Merits, Series C No 179 (IACtHR, 6 May 2008) [76].

9 For a recent in-depth analysis, see Riccardo Pavoni, *Interesse pubblico e diritti individuali nella giurisprudenza della Corte europea dei diritti umani* (Editorial Scientifica, 2014).

10 Article 8.

11 *Lopez Ostra v Spain*, Appl No 16798/90 (ECtHR, 9 December 1994); *Fadeyeva v Russia*, Appl No 55723/00 (ECtHR, 9 June 2005). See further the contribution by Quirico in this volume.

A similar approach to the interpretation of human rights law in environmental disputes can be found in a number of well-known cases decided under the ACHR. These cases mostly concern environmental impacts caused by extraction, logging or infrastructure projects in ancestral lands of indigenous peoples, which have been held to be in violation of the right to property under Article 21, extensively interpreted so as to include communal customary property of the indigenous peoples, and the right to life under Article 4 of the Convention.[12]

Even more explicitly, the 2004 Arab Charter on Human Rights states that 'every person has the right to an adequate standard of living for himself and his family, which ensures their well-being and a decent life, including food, clothing, housing, services and the right to a healthy environment'.[13] Along the same lines, the 2012 ASEAN Declaration on Human Rights recognises the 'right to a safe, clean and sustainable environment' as part of the 'right to an adequate standard of living'.[14]

At the universal level, the ICESCR provides for 'the improvement of all aspects of environmental and industrial hygiene'.[15] Nonetheless, the development of an environmental dimension of human rights provisions via the 1966 UN Covenants has been modest.[16] So far, there have been a very limited number of cases involving environmental claims, and most of them have been brought before the HRCte under the minority clause of Article 27 ICCPR.[17]

These principles are certainly very important for 'greening' traditional human rights instruments,[18] but they fall short of recognising or accrediting the existence of a true substantive right to environment. This is also the stance taken by the HRC in the Resolutions so far adopted on climate change and human rights. Indeed, Resolution 7/23 of 28 March 2008 refers to 'the link between the enjoyment of human rights and the protection of the environment'.[19] Along the same lines, Resolution 10/4 of 25 March 2009 notes that 'climate change-related impacts have a range of implications, both direct and indirect, for the effective

12 These cases include *Kichwa Peoples of Sarayaku v Ecuador*, Series C No 245 (IACtHR, 27 June 2012); *Maya Indigenous Communities of the Toledo District (Belize)*, Case 12.053, Report No 40/04 (IAComHR, 12 October 2004); *Mayagna (Sumo) Awas Tingni Community v Nicaragua*, Series C No 79 (IACtHR, 31 August 2001); *Indians Yanomami (Brazil)*, Case 7615, Resolution No 12/85 (IAComHR, 5 March 1985). See further Boumghar in this volume.

13 Article 38.

14 Principle 28(f).

15 Article 12(2).

16 For a survey of international instruments highlighting environmental aspects of substantive human rights see Ksentini (1994) [39] ff. More specifically, with regard to climate change aspects of human rights see John Knox, *Mapping Human Rights Obligations Relating to the Enjoyment of a Safe, Clean, Healthy and Sustainable Environment: Focus Report on Human Rights and Climate Change* (June 2014).

17 For a review of these cases see Francesco Francioni, 'International Human Rights in an Environmental Horizon' (2010) 21 *EJIL* 41.

18 See Council of Europe, *Manual on Human Rights and the Environment – Principles Emerging from the Case Law of the European Court of Human Rights* (2006).

19 Preamble.

enjoyment of human rights',[20] and Resolution 18/22 of 30 September 2011 takes a similar stand. Analogous approach is adopted by UNFCCC institutions, since the main reference to the relationship between climate change and human rights is embedded in the Cancun Agreements and provides that 'the Parties [to the UNFCCC] should, in all climate change-related actions, fully respect human rights'.[21]

1.2. Environmental 'proceduralisation' of substantive human rights

Contemporary developments at the level of treaty law have tended to fashion the interest of individuals and groups to take part in environmental decisions affecting their lives, and to have access to environmental justice in terms of 'human rights'. This phenomenon is currently referred to as the 'proceduralisation of environmental rights' and is crucial to climate change.[22] It refers to an individual and social empowerment to participate in the deliberative process leading to important environmental decisions and to activate legal remedies against environmental harm.[23]

This stance is based on the fact that procedural rights to information, participation and access to justice are commonly recognised in both environmental and human rights instruments, for instance, Article 6 UNFCCC and Article 8 UDHR.[24] Such 'approximation' is based on classical first and second generation human rights, so much so that scholars speak of 'procedural environmental rights' in the plural,[25] and is summarised in Principle 10 of the 1992 Rio Declaration.[26]

20 Ibid.

21 UNFCCC COP, Decision 1/CP.16, *The Cancun Agreements: Outcome of the Work of the Ad Hoc Working Group on Long-term Cooperative Action under the Convention*, UN Doc FCCC/CP/2010/7/Add.1 (15 March 2011) 4, [8]. See, more extensively, the contribution by Aktypis, Decaux and Leroy in this volume.

22 See Svitlana Kravchenko, 'Procedural Rights as a Crucial Tool to Combat Climate Change' (2010) 38 *Georgia J Int'l & Comp L* 613 ff.

23 Alan Boyle, 'Human Rights and the Environment: Where Next?' (2012) 23 *EJIL* 613, 616; Catherine Redgwell, 'Access to Environmental Justice', in Francesco Francioni (ed), *Access to Justice as a Human Right* (OUP, 2007) 153 ff; James Cameron and Ruth Mackenzie, 'Access to Environmental Justice and Procedural Rights in International Institutions', in Alan Boyle and Michael R Anderson (eds), *Human Rights Approaches to Environmental Protection* (OUP, 1996) 129 ff.

24 For further analysis of relevant international regulatory instruments see Knox (2013) [29] ff.

25 Weston and Bollier (2013) 328.

26 Principle 10:

> Environmental issues are best handled with participation of all concerned citizens, at the relevant level. At the national level, each individual shall have appropriate access to information concerning the environment that is held by public authorities, including information on hazardous materials and activities in their communities, and the opportunity to participate in decision-making processes. States shall facilitate and encourage public awareness and participation by making information widely available. Effective access to judicial and administrative proceedings, including redress and remedy, shall be provided.

The trend is all the more significant when considering that the main weakness of the current system of environmental protection is the lack of compulsory dispute settlement mechanisms and enforcement institutions, which reduces its effectiveness as compared to the international economic law regime, including compulsory investment arbitration and binding WTO dispute settlement procedures.

Along these lines, in *Taşkin v Turkey*, after considering the status of the 'right to a healthy environment',[27] the ECtHR found that the respondent State had violated the right to private and family life under Article 8 ECHR by failing to enforce protective measures requested by the applicant to stop the environmentally noxious operation of a mine.[28] In so doing, the Court provided an extensive interpretation, holding that 'while Article 8 contains no explicit procedural requirement, the decision making process leading to measures of interference must be fair and afford due respect to the interests of the individual as safeguarded by Article 8.'[29]

The jurisprudence of the Strasbourg Court is nevertheless limited, to the extent that it takes a merely individualistic approach to the determination of environmentally entangled human rights, particularly the right to private and family life.[30] On this basis, it is not possible to maintain the existence of a specific 'human right' to information, participation and access to justice in environmental matters. This also entails the risk that unlawful environmental impacts may fall below the radar of the human rights review, if they do not entail a direct interference with the sphere of the individual applicant's human rights.

In *Maya Indigenous Communities of the Toledo District (Belize Maya)*, the IAComHR adjudicated upon the claims that the State of Belize breached rights over lands traditionally used and occupied by the Maya people, not only by granting logging and oil concessions and failing to adequately protect those lands and the territorial rights of the Maya people, but also by failing to afford the Maya people adequate judicial safeguard. The Commission held that Belize's conduct was in breach of the right to property under Article XXIII ADRDM and the right to judicial protection under Article XVIII ADRDM. It consequently condemned Belize to restore the right to property through fully informed consultations with the Maya people and, importantly, to repair environmental damage.[31]

27 *Taşkin et al v Turkey*, Appl No 46117/99 (10 November 2004) [98]–[100].

28 Ibid, [125]–[126].

29 Ibid, [118]. See also *Guerra v Italy*, Appl No 116/1996/735/932 (19 February 1998).

30 In *Kyrtatos v Greece*, the applicants lodged a complaint against the draining of a wetland in the vicinity of their home. Although the drainage and consequent destruction of the wetland resulted in unlawful conduct, the Court stated that 'neither Article 8 nor any other Articles of the Convention [ECHR] are specifically designed to provide general protection of the environment as such' (*Kyrtatos v Greece*, Appl No 41665/98 (EctHR, 22 May 2003) [52]. For a critique of this approach, see Francioni (2010).

31 *Maya Indigenous Communities* (2004) [193]–[197]. As to the case law of the IACtHR see, for instance, *Saramaka People v Suriname*, Preliminary Objections, Merits, Reparations, Costs, Series C No 172 (IACtHR, 28 November 2007).

In *Centre for Minority Rights Development on behalf of Endorois Community v Kenya*, the AComHPR upheld the complaint that the Government of Kenya, in establishing a game reserve, forcibly removed the Endorois from their ancestral lands. Indeed, Kenya proceeded without proper prior consultations and adequate and effective compensation, in violation of provisions of the ACHPR protecting fundamental rights such as property and culture. Relying on the stance of the UN Special Rapporteur on the Situation of Human Rights and Fundamental Freedoms of Indigenous People, the Commission also referred to the fundamental duty not to pollute traditional environments.[32]

Most significantly, in *Ogoniland*, a seminal case concerning environmental devastation caused by oil extraction in Nigeria, the AComHPR held:

> Government compliance with the spirit of Articles 16 [of the Banjul Charter: Right to enjoy the best attainable state of physical and mental health] and 24 [Right to a general satisfactory environment] must also include ordering or at least permitting independent scientific monitoring of threatened environments, requiring and publicising environmental and social impact studies prior to any major industrial development, undertaking appropriate monitoring and providing information to those communities exposed to hazardous materials and activities and providing meaningful opportunities for individuals to be heard and to participate in the development decisions affecting their communities.[33]

Interestingly, unlike the ECtHR, the Inter-American and African Commissions tend to recognise a 'human right' to information, participation and access to justice in environmental matters not only based on classical first and second generation human rights, but also in light of a human right to environmental protection *per se*, thus outlining independent fundamental procedural rights. This trend is also recognised in domestic constitutions,[34] and is summarised by the 1998 Aarhus Convention on Access to Information, Public Participation in Decision-making and Access to Justice in Environmental Matters,[35] which 'codifies' procedural environmental claims as self-standing human rights. In fact, according to the Convention, the 'rights of access to information, public participation in decision-making, and access to justice in environmental matters' are set out in order to 'contribute to the protection of the *right of every person of present and future generations to live in an environment adequate to his or her health and well-being*'.[36] The fact that the Convention has only been ratified by around fifty States, most of which are

32 *Centre for Minority Rights Development on behalf of Endorois Community v Kenya*, Comm No 276/2003 (2010) [293].

33 *The Social and Economic Rights Action Center and the Center for Economic and Social Rights v Nigeria*, Comm No 155/96 (AComHPR, 2001) [53].

34 Knox (2012) [12].

35 Adopted 25 June 1998, entered into force on 30 October 2001, 2161 UNTS 447.

36 Article 1 (emphasis added).

European, may nonetheless prevent the universal recognition of 'human' environmental procedural rights.[37]

2. An independent substantive human right to a sustainable environment?

2.1. Non-governmental initiatives

The recognition of a right to a sustainable environment has been dealt with as a fundamental matter in different projects of non-governmental organisations, and also based on climate change. For instance, the 2000 Earth Charter promotes 'action to avoid the possibility of serious or irreversible environmental harm even when scientific knowledge is incomplete or inconclusive'.[38] However, the Charter establishes a set of moral principles and thus 'only' provides a general background for a possible legal right to a sustainable environment.

From a strictly legal viewpoint, the Siena Forum on International Law of the Environment organised by the government of Italy in preparation for the 1992 G7 and Rio Summit highlighted the necessity of regulating climate change and included a whole section on the right to a sustainable environment as a fundamental right.[39] Along these lines, a Draft Statute for the International Environmental Agency and the International Court of the Environment was presented by the ICE Coalition at the 1992 UNCED Conference in Rio de Janeiro.[40] The Draft Statute provides a definition of the 'right to the environment', based, inter alia, on the necessity of capping GHG emissions.[41] Although it is not binding, this initiative is interesting, because it seeks to tackle the complex substantive and procedural problems raised by the right to environment head-on. The Draft was proposed more than 20 years ago and partially renewed via a petition submitted to the Rio+20 Conference.[42]

Articles 1–8 of the ICE Draft Statute provide a definition of the substantive 'right to the environment', from the viewpoint of both active claims and passive duties. Article 1 posits the basic principle that 'everyone has a *fundamental right to the environment* and an *absolute duty to preserve life* on earth for the benefit of *present and future generations*'.[43] This simple provision entails that the right to environment

37 The Convention has been adopted within the framework of the UN Economic Council for Europe. See further the contribution by Quirico in this volume.

38 Earth Charter Initiative, *The Earth Charter* (2000) Principle 6, <www.earthcharterinaction.org/content>.

39 Published in Presidenza del Consiglio dei Ministri, *Vita Italiana*, Suppl No 1 (1990) and *Yb Int'l Env L* (1990) 696.

40 ICE Coalition, *Draft Statute of the International Environmental Agency and the International Court of the Environment* (June 1992) <www.icef-court.org>.

41 Ibid, Preamble.

42 ICE Coalition, *Call for Action from Lawyers and Environmental Law Organizations* (2012) <www.petitions24.com/signatures/rio20_call_from_lawyers_and_organizations>.

43 Emphasis added.

is a fundamental human claim indefinitely extending in time. Such a right is also correlative to an obligation to protect life, which can probably be read as a general reference to biodiversity, although this concept does not fully match the notion of 'environment'.[44] On this basis, each legal person is considered bound by a duty to rationally use natural resources, which is another obligation correlative to the right to environment.[45] Along the lines of the duty to respect human rights, State responsibility is conceived of as *erga omnes*, peremptory and extraterritorial.[46] State policies are framed so as to prevent and repair environmental damage,[47] within the context of a general obligation to fulfil the right to environment.[48]

These substantive provisions are matched by corresponding procedural rights on participation in decision-making and access to information and judicial remedies.[49] As to decision-making, the Statute provides for the establishment of an International Environmental Agency, vested with the power to 'control and

44 See further Lenzerini and Piergentili in this volume.

45 Article 5:

> Everyone is under a duty to utilise natural resources with equity and care, by ensuring the maximum saving of energy . . . [and] minimum consumption of resources.

46 Article 7:

> States are legally responsible to the *entire* International Community for acts that cause substantial damage to the environment in their own territory, in that of other States or in areas *beyond the limits of national jurisdiction* [and] shall adopt all measures to prevent such damage [emphasis added].

47 Article 8:

> States, in particular, shall . . . prohibit all activities that may cause irreversible damage to the basic natural processes of the biosphere and, as a precautionary measure, suspend those activities whose effects cannot be determined until all such uncertainty has been removed; e) take action to restore degraded ecosystems; f) prevent the transfer of environmental harm and risks to other parts of the world.

48 Article 6:

> States shall recognise and guarantee the human right to the environment, and foster conditions that make this right effective.

49 Article 2:

> Everyone has the right of access to environmental information and the duty to provide any environmental information in his possession/her possession.

Article 3:

> Everyone has the right to participate in procedures that may involve the environment, subject to the fact that the public authorities are deemed to have final responsibility with regard to the environmental decision-making process.

Article 4:

> Everyone, whether an individual or an association, has the right to take legal action to prevent activities that are harmful to the environment and to seek compensation for any environmental damage.

monitor the state of the environment on the planet'.[50] With regard to adjudication, the Statute envisages the creation of a permanent International Court of the Environment, having jurisdiction over both State and non-State responsibility arising from the breach of the fundamental right to environment.[51] Rules on sanctions are nonetheless quite general, since they provide for the adoption of 'any measures considered necessary for remedying the violated right [to environment]'.[52] The Draft Statute also falls short of considering enforcement mechanisms, which are thus left to the problematic system of bilateral and multilateral countermeasures.[53]

This pioneering work is significant, to the extent that it seeks to afford thorough legal protection for the right to a sustainable environment, outlining a substantive claim and its procedural implications.[54] In accordance with this approach, in preparation for the 2005 World Summit, the French government commissioned the European University Institute to give an expert opinion on possible options for reforming UN environmental institutions. Professors Dupuy and Francioni drafted the opinion, assessing different environmental governance mechanisms and prioritising the creation of the UN Environment Organisation (UNEO) as a new specialised agency complemented by adequate enforcement mechanisms. The opinion was later endorsed by the EU, but no meaningful reform ensued for lack of political will.[55]

50 Article 9.

51 Literally, the Court is charged to 'protect the environment as a fundamental human right in the name of the International Community' (Article 10, Functions (a)) and adjudicates upon 'any disputes concerning environmental damage, caused by private or public parties, including the State, where it is presumed that, due to its size, characteristics and kind, this damage affects interests that are fundamental for safeguarding and protecting the human environment on earth' (Article 10(b)).

52 Article 10 (Sanctions):

> Whenever the Court finds in favour of an individual or association, it shall adopt any measures considered necessary for remedying the violated right, by ordering, in accordance with the circumstances, whatever the party, or even the State, guilty of the alleged violation is or is not required to do.

53 See the ILC's 2001 Draft Articles on Responsibility of States for Internationally Wrongful Acts, in *ILC Report to the General Assembly*, UN Doc A/56/10 (2001) 43 ff, Articles 42 ff.

54 Ole W Pedersen, 'An International Environmental Court and International Legalism' (2012) 24 *Journal of Environmental Law* 547. On the compatibility of the Court with UNFCCC-based enforcement mechanisms see Stuart A Bruce, *An International Court for the Environment and Climate Change Law Compliance and Enforcement* (AID, 2012).

55 Francesco Francioni and Pierre-Marie Dupuy, *Options and Modalities for the Improvement of International Environmental Governance through the Establishment of a UN Environmental Organisation*, Study Commissioned by the French Government (2005) <www.diplomatie.Gouv.fr/IMG/pdf/Etudes_iddri_juridique_EN.pdf>.

2.2. Soft law

Article 28 of the 1948 UDHR provides that 'everyone is entitled to a social and international order in which the rights and freedoms set forth in this Declaration can be fully realised.' The notion of 'order' has been interpreted so as to encompass 'environmental concerns'.[56] Only in 1972, however, did the Stockholm Declaration on the Human Environment clearly recognise a 'fundamental' right to an 'environment of quality'.[57] In its simplicity, this statement contains all the elements for a legal combination of an ecological and human rights approach to the question of environmental protection. It was couched in terms of a solemn *erga omnes* commitment towards the international community as a whole and for the protection of a global common good, rather than as a bilateral or multilateral obligation between States, thus echoing the language of human rights treaties,[58] including the concept of inter-generational responsibility.[59]

The legacy of the Stockholm Declaration has not effectively spurred the development of a human right to a sustainable environment. Twenty years after Stockholm, the 1992 Rio Conference on Environment and Development produced a Declaration that, unfortunately, departs in a significant manner from the original idea of a close connection between human rights and environmental protection. As pointed out elsewhere,[60] Principle 1 of the Rio Declaration only proclaims that human beings are at the centre of concerns for sustainable development and are 'entitled to a healthy and productive life in harmony with nature'. This is hardly human rights language. The main concern of the Declaration is the conjugation of environmental protection with economic development, not the safeguarding of human rights through enhanced protection of the environment, let alone the recognition of a self-standing right to a sustainable environment. Neither did the 2002 World Summit on Sustainable Development

56 Ksentini (1994) [34].

57 Principle 1:

> Man has the fundamental right to freedom, equality and adequate conditions of life, in an environment of a quality that permits a life of dignity and well-being, and he bears a solemn responsibility to protect and improve the environment for present and future generations.

It is doubtful, however, whether the adjective 'fundamental' refers to the 'environment' or rather to the 'right to freedom, equality and adequate conditions of life'.

The first preambular paragraph, instead, takes a stance on environmental protection that is based on first and second generation human rights:

> Man is both creature and molder of the environment . . . Both aspects of man's environment, the natural and the man-made, are essential to his well-being and to the enjoyment of basic human rights – even the right to life itself.

58 Note that six years before the Stockholm Declaration the UN had adopted the two most important human rights treaties under the name of 'Covenants', thus underscoring at the political level their character as solemn *erga omnes* commitments rather than mere contractual instruments.

59 Principle 6.

60 See Francesco Francioni, 'Comments on the Preamble and on Principle 1', in Jorge E Viñuales (ed), *The Rio Declaration on Environment and Development: a Commentary* (CUP, 2014).

and the Rio+20 Conference endorse the idea of a human right to a sustainable environment, despite several references to first and second generation human rights.[61] Along these lines, the conciliation of environmental protection with economic growth and progressive industrialisation remains the main focus of environmental diplomacy in the Kyoto and post-Kyoto negotiations on climate change.[62]

Despite this stand, several soft law instruments acknowledge the right to a sustainable environment. Indeed, the 1989 Hague Declaration on the Environment recognises a fundamental right to a healthy environment as strictly interconnected with the right to life.[63] Similarly, UN GA Resolution 45/94 of 14 December 1990 on the Need to Ensure a Healthy Environment for the Well-being of Individuals acknowledges that 'all individuals are entitled to live in an environment adequate for their health and well-being'.[64] Consequently, the Resolution calls upon UN Member States and relevant governmental and non-governmental organisations, to 'enhance their efforts towards ensuring a better and healthier environment'.[65]

In 1994, the now defunct UN Sub-Commission on the Prevention of Discrimination and Protection of Minorities, which had been mandated by UN GA Resolution 45/94 to work on the relationship between the environment and human rights, elaborated some Draft Principles on Human Rights and the Environment.[66] The Principles recognise the intergenerational 'right to a secure, healthy and ecologically sound environment', interdependent and indivisible with respect to all other human rights.[67] Of particular interest to climate change is the fact that the Draft Principles further spell out a general 'right to freedom from pollution, environmental degradation and activities that adversely affect the environment . . . across or outside national boundaries' and 'to protection and preservation of the air, soil, water, sea, ice, flora and fauna . . . [and] biological diversity of ecosystems.'[68] Along the lines of the ICE Draft Statute, the right to an ecologically sound environment entails corresponding universal duties,[69] and

61 UN Conference on Sustainable Development, Report, Rio de Janeiro, Brazil, UN Doc A/Conf/216/16 (2012); World Summit on Sustainable Development, Report, Johannesburg, South Africa, UN Doc A/Conf.199/20 (2002).

62 See *UNFCCC-Kyoto Protocol*, <http://unfccc.int/kyoto_protocol/items/2830.php>.

63 *Hague Declaration on the Environment* (11 March 1989).

64 Principle 1.

65 Principle 2.

66 Ksentini (1994) 22 ff, Annex 1, Draft Principles on Human Rights and the Environment.

67 Principles 2 and 4.

68 Principles 5 and 6.

69 Principle 21:

> All persons, individually and in association with others, have the duty to protect and preserve the environment.

Principle 22:

> All States shall respect and ensure the right to a secure, healthy and ecologically sound environment. Accordingly, they shall adopt the administrative, legislative and other measures necessary to effectively implement the rights in this Declaration.

procedural rights complement substantive provisions.[70] Although this project received little support from the UN Human Rights Commission, in 1999 the UNESCO Bizkaia Declaration on the Right to the Environment recognised that 'everyone has the right, individually or in association with others, to enjoy a healthy and ecologically balanced environment.'[71] Such a right is matched by a correlative duty of everyone 'to protect the environment and to foster environmental protection at both national and international levels'.[72]

Concerning the claim to 'environmental protection' of indigenous peoples, the relevant Draft American Declaration of 1997 states that these have 'the right to a safe and healthy environment, which is an essential condition for the enjoyment of the right to life and collective well-being'.[73] The Declaration also includes a complete set of procedural rights.[74] This approach culminated in the 2007 Declaration on the Rights of Indigenous Peoples, adopted by means of UN GA Resolution 61/295 of 13 September 2007, which proclaims the 'right to the conservation and protection of the environment and the productive capacity of [indigenous peoples'] lands or territories or resources'.[75]

Also different Resolutions adopted by the UN ECOSOC Commission on Human Rights recognise the right to a sustainable environment. In particular, a Draft Resolution on the Promotion of a Democratic and Equitable International Order provides that a democratic order requires 'the right of every person and all peoples to a healthy environment.'[76] Another Resolution on the Effects of

70 Principles 15–20.

71 Article 1.

72 Article 2.

73 *Proposed American Declaration on the Rights of Indigenous Peoples* (1997) Article XIII(1). Para 6 of this provision also asserts:

> The States shall prohibit and punish, and shall impede jointly with the indigenous peoples, the introduction, abandonment, or deposit of radioactive materials or residues, toxic substances and garbage in contravention of legal provisions; as well as the production, introduction, transportation, possession or use of chemical, biological and nuclear weapons in indigenous areas.

74 Article XIII(2)(4)(7):

> 2. Indigenous peoples have the right to be informed of measures which will affect their environment, including information that ensures their effective participation in actions and policies that might affect it . . . 4. Indigenous peoples have the right to participate fully in formulating, planning, managing and applying governmental programmes of conservation of their lands, territories and resources . . . 7. When a State declares an indigenous territory as protected area [any lands, territories and resources under potential or actual claim by indigenous peoples], conservation areas shall not be subject to any natural resource development without the informed consent and participation of the peoples concerned.

75 Article 29.

76 ECOSOC, Commission on Human Rights, Draft Resolution, *Promotion of a Democratic and Equitable International Order*, UN Doc E/CN.4/2005/L.73 (15 April 2005) 5, [4]; OHCHR, *Promotion of a Democratic and Equitable International Order*, Res 2001/65 (25 April 2001) [3.k].

Structural Adjustment Policies and Foreign Debt on the Full Enjoyment of All Human Rights acknowledges that:

> the exercise of the basic rights of the people of debtor countries to food, housing, clothing, employment, education, health services and a *healthy environment* cannot be subordinated to the implementation of structural adjustment policies, growth programmes and economic reforms arising from the [foreign] debt.[77]

Finally, the 2007 Malé Declaration on the Human Dimension of Climate Change, significantly adopted by Small Island Developing States, refers to a 'fundamental right to an environment capable of supporting human society and the full enjoyment of human rights'. This instrument is interesting, because it creates a direct link between climate change, the fundamental right to a sustainable environment, and all other human rights. The Declaration is also particularly important, since it triggered a debate on the relationship between climate change and human rights at the UN HRC.[78]

2.3. Hard law

At the regional level, a few normative instruments on human rights explicitly recognise the claim to a sustainable environment. Article 24 ACHPR provides that 'all peoples shall have the right to a general satisfactory environment favourable to their development'. This right is not formulated as an individual claim, but rather as a collective one. Besides, the right to a 'general satisfactory environment' remains indeterminate and also qualified by the phrase 'favourable to their [peoples'] development'. This obviously leaves a wide margin of appreciation for the States parties to determine the extent to which environmental integrity has to yield to development goals. Similarly, the 2003 AU Protocol to the ACHPR on the Rights of Women in Africa provides that women 'shall have the right to live in a healthy and sustainable environment', as well as 'the right to fully enjoy their right to sustainable development'.[79]

In *Ogoniland*, the AComHPR construed the general language of Article 24 ACHPR in strict environmental terms and held that States must take reasonable measures to 'prevent pollution and ecological degradation, to promote conservation, and to secure an ecologically sustainable development and use of natural resources'.[80] This decision is remarkable, not only for considering human

77 ECOSOC, Commission on Human Rights, *Effects of Structural Adjustment Policies and Foreign Debt on the Full Enjoyment of All Human Rights, Particularly Economic, Social and Cultural Rights*, Res 2003/21 (22 April 2003), in *Report to the Economic and Social Council on the Fifty-Ninth Session of the Commission*, UN Doc E/CN.4/2003/L.11/Add.3 (23 April 2003) 25, [11], emphasis added.

78 See Aktypis, Decaux and Leroy in this volume.

79 Articles 18–19.

80 *Social and Economic Rights Action Center* (2001) [52].

rights guarantees in broad collective terms as legitimate claims of the community to the quality of the environment against unsustainable exploitation of natural resources, but also for indicating remedial measures, preparation of impact assessments, and provision of information and public participation.[81] The focus is clearly on the collective right to a sustainable environment of local populations, which often draw little material benefit from the harmful exploitation of environmental resources.[82]

Within the Inter-American context, Article 11 of the San Salvador Protocol Additional to the ACHR embeds the 'right of everyone' to 'live in a healthy environment and to have access to basic public services', which must be promoted, protected, preserved and improved by States.[83] This provision therefore takes an individualistic approach to environmental protection. The wording of this rule also establishes a connection between environmental safeguards and health, which is regulated in other human rights provisions, in particular, Article 10 of the same Protocol and Article 12(2) ICESCR.

As to general international law, protection of the natural environment is particularly perceived as a *sine qua non* condition for the enjoyment of human rights by members of vulnerable groups. Along these lines, Article 24(2)(c) CRC provides for safeguarding against 'the dangers and risks of environmental pollution', within the context of the right of the child to the enjoyment of the highest attainable standard of health. The 2011 Nagoya Protocol on Access to Benefit Sharing is centred on the rights of indigenous and local traditional communities and compels the sustainable use of benefits arising from biodiversity resources, with a particular focus on procedural rights.[84] Outside the human rights context, the ICJ has recognised the *erga omnes* nature of the obligation to respect the environment, holding that environmental concerns relate to an 'essential interest . . . for the whole of mankind' since the 1970s.[85] The Protocol on Environmental Protection to the Antarctic Treaty provides that the 'Antarctic environment and associated ecosystems' are '*fundamental considerations* in the planning and conduct of all activities in the Antarctic Treaty area'.[86]

81 Ibid, [53] and Findings.

82 Ibid, [54]–[69].

83 Article 11 (Right to a Healthy Environment):

> 1. Everyone shall have a right to live in a healthy environment and to have access to basic public services. 2. The States Parties shall promote the protection, preservation, and improvement of the environment.

84 *Nagoya Protocol on Access to Genetic Resources and the Fair and Equitable Sharing of Benefits Arising from the Utilisation of the Convention on Biological Diversity* (29 October 2010); CBD COP-10, Decision 10/1 (20 January 2011) UNEP/CBD/COP/10/27.

85 *Gabčíkovo-Nagymaros Project (Hungary v Slovakia)* [1997] ICJ, 41, [53]. See also *Pulp Mills on the River Uruguay (Argentina v Uruguay)* [2010] ICJ [76]; *Legality of the Threat or Use of Nuclear Weapons* (1996) [29].

86 Article 3 (emphasis added).

From a different perspective, the right to a sustainable environment can be regarded as part of the right to sustainable development, which is generally acknowledged in the UN Declaration on the Right to Development.[87] In fact, not only is the right to a satisfactory environment incorporated into the right to development under Article 24 of the Banjul Charter, but development must also be economically and environmentally sustainable. In this respect, the 1987 Brundtland Report includes an Annex on legal principles, providing that 'all human beings have the fundamental right to an environment adequate for their health and well-being.'[88] Thus, the right to development of developing countries is matched by the correlative obligation of developed countries to reduce poverty and respect the environment.[89] Moreover, this claim entails an integral duty of developing countries to pursue an ecologically sustainable development.[90] However, the right to development is also a third generation human right, and thus its status is at least as uncertain as that of the right to a sustainable environment.[91]

At the domestic level, currently 177 out of the 193 UN Member States recognise the right to a sustainable environment via their constitutions, statutes, court decisions, or ratification of international agreements.[92] The only holdouts are the US, China, Canada, Japan, Australia, New Zealand, Oman, Afghanistan, Kuwait, Brunei Darussalam, Lebanon, Laos, Myanmar, North Korea, Malaysia and Cambodia. In 110 countries the right to a sustainable environment is constitutionally acknowledged and 120 States have signed international agreements embedding the claim in issue.[93] It is therefore sensible to assume that the right to a sustainable environment is already a general principle of law.[94] Arguably, in light of the fact that only 49 States have ratified the Aarhus Convention and that most

87 UN GA Res 41/128 of 4 December 1986.

88 Annex 1 to the Brundtland Report, *Summary of Proposed Legal Principles for Environmental Protection and Sustainable Development Adopted by the WCED Experts Group on Environmental Law* (1987). The body of the Brundtland Report refers to the notion of 'unhealthy environment' (World Commission on Environment and Development (1987) [9] and [72]).

89 See Knox (2014) VI.B.vii, Obligations to Developing Countries.

90 *Centre for Minority Rights Development* (2010) [298].

91 See further Varayudej in this volume.

92 For instance, Article 14 of the Constitution of Ecuador recognises 'the right of the population to live in a clean and ecologically balanced environment'.

93 UNEP *et al*, Report, *Regional Consultation on the Relationship between Human Rights Obligations and Environmental Protection, with a Focus on Constitutional Environmental Rights*, Johannesburg, South Africa (23–4 January 2014) [11] ff; David R Boyd, *The Environmental Rights Revolution* (UBC Press, 2012) 45 ff.

94 Knox (2012) [12] and [14]: 'were the Universal Declaration to be drafted today, it is easy to imagine that it would include a right [to environment] recognised in so many national constitutions and regional agreements.' See also David R Boyd, *The Constitutional Right to a Healthy Environment* (2013) <www.lawnow.org/right-to-healthy-environment>; Weston and Bollier (2013) 327–8; Johnas Ebbeson, *Participatory and Procedural Rights in Environmental Matters: State of Play*, Unpublished Draft Paper presented at a High Level UNEP and OHCHR Expert Meeting on the New Future of Human Rights and Environment, Nairobi (30 November–1 December 2009).

constitutions refer to a substantive right to a sustainable environment rather than procedural environmental rights, it is probably more sensible to recognise a human right to a sustainable environment, instead of a human right to information, participation and access to justice in environmental matters. It is perhaps the case that the late recognition of the right to a sustainable environment in the 1970s has not provided for sufficient time (*diuturnitas*) for its acknowledgment (*opinio juris*) as a general principle of law.[95]

2.4. A criminally protected human right to a sustainable environment?

The preparatory work for some as yet un-achieved proposals in the field of criminal law are useful to shed light on the nature of the right to a sustainable environment. The question of whether criminal protection is suitable for the right to a sustainable environment is relevant to its categorisation as a human right. Indeed, to a large extent criminal law regulates nothing other than breaches of fundamental rights and consequent sanctions, particularly in the international arena. The issue is currently not given attention in the work of the Special Rapporteur on human rights and the environment and other scholarly analysis. Arguably, this is due to the fact that institutional work on the matter has not led to positive conclusions and effective regulation. However, a criminal law approach to environmental protection is considered by scholars working on the notion of 'ecocide' and its human rights implications.[96]

Three main instruments must be taken into account to get a grasp of the question: (1) the Draft Code of Crimes against the Peace and Security of Mankind; (2) the Statute of the ICC; and (3) the ILC's Draft Articles on State Responsibility. It goes without saying that the first two initiatives concern (mainly) individual responsibility, whereas the latter deals exclusively with State responsibility.

With regard to the individual criminalisation of environmental protection, the first three Drafts of the Code of Crimes against the Peace and Security of Mankind included environmental crimes, in particular, conduct consisting of wilful and severe damage to the environment. Under Article 26 of the 1991 Draft, the Code comprised the conduct of 'an individual who wilfully causes or orders another individual to cause widespread, long-term and severe damage to the natural environment'.[97] Article 26 was withdrawn from the last version of the Code adopted in 1996 for three main reasons. First, although the environment

95 For a customary interpretation of the general principles of law see Georges Scelle, *Précis de droit des gens – Principes et systématique* (Dalloz, 11th edn, 2008).

96 Eradicating Ecocide Global Initiative, *Ecocide Act*, <http://eradicatingecocide.com/overview/ecocide-act>; Anja Gauger, Mai Pouye Rabatel-Fernel, Louise Kulbicki, Damien Short and Polly Higgins, *Ecocide Is the Missing 5th Crime against Peace* (2012) <www.sas.ac.uk>.

97 ILC, *Report of the ILC on the Work of Its Forty-Third Session* (29 April to 19 July 1991) Draft Code of Crimes against the Peace and Security of Mankind, Article 26. The Code also criminalises environmental damage in time of war under Article 22.

was considered essential to human life,[98] it was assumed that the Code could not take an 'ecosystem' approach to environmental protection, but should stick to an anthropocentric interpretation.[99] Second, as to *mens rea*, it was maintained that the requisite wilful conduct does not fit with the notion of widespread 'environmental pollution'.[100] Third, with particular regard to the burning of fossil fuels, it was concluded that the Code is concerned with specific damage, but not with cumulative environmental harm.[101] Finally, domestic jurisdictions were deemed to be the most appropriate fora to deal with environmental pollution from a criminal perspective.[102]

The debate continued within the framework of the subsequent work that led to drafting the Statute of the International Criminal Court (ICC). Indeed, along the lines of several proposals by the International Criminal Law Association,[103] the Preparatory Committee for the Establishment of the ICC discussed the opportunity of including 'serious threats to the environment' in the jurisdiction of the Court,[104] but the question did not have a follow-up.

With regard to States, in 1976 Roberto Ago presented his Fifth Report on State Responsibility to the ILC, including draft Article 19.[105] This norm provided for

98 Christian Tomuschat, *Document on Crimes against the Environment*, ILC(XLVIII)/DC/CRD.3, *Draft Code of Crimes against the Peace and Security of Mankind (Part II) – Including the Draft Statute for an International Criminal Court*, Extract from the Yearbook of the International Law Commission (1996) 22, [18].

99 Ibid, [29]:

> although there is a modern tendency to distance oneself from an anthropocentric approach, attempting to preserve existing ecosystems as such, the draft Code remains committed to maintaining peace and security among human beings . . . protection of the environment is not tantamount to protection of human life in a direct sense. Protecting the environment, even through penal law as a defence of last resort, means securing the survival of humankind from a long-term perspective. The human being is the ultimate beneficiary of the efforts undertaken, but the disruptive effect of damage to the environment does not necessarily need to be measured in terms of injury to human life and physical integrity.

100 Ibid, 25–6, [41].

101 Ibid, 22, [20] and 36, [44]:

> The draft Code cannot possibly deal with harm to the environment by accumulation, where an infinite multitude of separate actions cause damage not individually, but conjunctively in their combination. It would be absurd to postulate that humankind constitutes nothing more than a society of criminals.

102 Ibid, 22, [20].

103 Association internationale de droit pénal, *Résolution adoptée au Colloque international sur les atteintes à l'environnement*, Ottawa, Canada, 2–6 November 1992 (1994) RIDP 1212, Titre IV, [4]; Id, *Résolution adoptée au XV Congrès international de droit pénal*, Rio de Janeiro, Brazil, 4–10 September 1994 (1995) RIDP 26, [28].

104 Preparatory Committee on the Establishment of the ICC, Report, UN GA 51st Session Official Records, Suppl No 22A, A/51/22 (1996) Vol I (Proceedings of the Preparatory Committee during March–April and August 1996) [114].

105 Roberto Ago, *Fifth Report to the ILC*, UN Doc A/CN.4/291 (1976) 24–54.

a distinction between State delicts and crimes, including among the latter 'a serious breach of an international obligation of essential importance for the safeguarding and preservation of the human environment, such as those prohibiting massive pollution of the atmosphere or of the seas'. This could easily encompass GHG emissions. State 'crimes' were envisaged as 'breach[es] by a State of an international obligation . . . essential for the protection of fundamental interests of the international community'.[106] Draft Article 19 remained unchanged until the first draft of the ILC's Articles on State Responsibility, adopted in 1996 with Gaetano Arangio-Ruiz as third Rapporteur.[107]

Although the ILC's Draft Articles are not directly concerned with human rights, there are interesting links between Article 19 and fundamental rights. Indeed, under Article 19 the recognition of the criminal nature of a wrongful act, injuring 'all other States',[108] should have come from the international community 'as a whole'.[109] This disclosed the possibility of envisaging massive atmospheric pollution, including GHG emissions, as a violation of an *erga onmes* obligation, possibly peremptory, matching the nature of the environment and the atmosphere as 'common goods'.[110] Furthermore, environmental pollution was embedded in a list of crimes also involving aggression and serious violations of international obligations of essential importance for safeguarding peace, the right to self-determination and human rights.[111]

The 1996 reading of the Project also regulated the procedural enforcement of State crimes. More specifically, States were obliged to 'cooperate with other States in the application of measures designed to eliminate the consequences of the crime.'[112] Professor Arangio-Ruiz went as far as to envisage a mechanism allowing States to unilaterally bring international crimes to the attention of the UN General Assembly or the Security Council, vested with the power to permit subsequent action before the ICJ.[113] The final version of the Articles on State Responsibility, adopted in 2001 with James Crawford as Fourth Rapporteur, replaced the notion of 'State crimes' with the general notion of 'jus cogens', thus excluding the crime of 'massive environmental pollution of the atmosphere or the sea'.[114]

106 Article 19(2).

107 See the text of the 1996 *Draft Articles on State Responsibility* in ILC, *Report to the General Assembly*, UN Doc A/51/10 (1996) 125–151.

108 Article 40(3).

109 Article 19(2).

110 See Eva M Kornicker Uhlmann, 'State Community Interest, Jus Cogens and Protection of the Global Environment: Developing Criteria for Peremptory Norms' (1998–99) 11 *Geo Int'l Envtl L Rev* 101.

111 Article 19(3).

112 Article 53(d).

113 Gaetano Arangio-Ruiz, *Seventh Report on State Responsibility*, UN Doc A/CN.4/46 and Add.1–2 (1995) *Yearbook of the ILC* 3, 20, [87].

114 ILC (2001) Articles 40–41.

Overall, these efforts demonstrate that a human rights approach to environmental protection *per se* is possible and that the gravest environmental offences might even fall within the category of international crimes. Yet, the compromises reached by the ILC and ICC Preparatory Committees prove that the criminal approach remains technically problematic.

3. Climate change, human right to a sustainable environment and the general interest of the international community

In spite of the undeniable progress marked by the above examined practice of human rights courts and regulatory instruments, the development of a fundamental right to a sustainable environment is still limited. *De lege lata*, environmental protection has its main *raison d'etre* in the interference that environmental disturbances produce within the sphere of specific individual or group rights protected by applicable human rights treaties, not in the harm to the environment as a value in itself for an affected community or for society as a whole. Instead, climate change is by nature a global phenomenon and has an indistinct impact on the life of many people. Therefore, an individualistic approach, based on the existence of specific interference of environmental impacts with the sphere of well-defined rights and clearly identified 'victims' should be a last resort.

This generates the present uncertainties as to how conventional human rights may be used as an effective legal tool to counter the devastating effects of climate change. Uncertainties emerge in the language used in the most important documents addressing the link between climate change and human rights, including the HRC resolutions on climate change and human rights and the various reports of the OHCHR. The tendency is to avoid the precise terminology of 'human rights violations' or 'breaches' arising from States' failure to comply with global warming mitigation and adaptation. The language used is rather that of climate 'adverse effects', 'negative impacts', 'implications', which obviously tend to elude the problem of responsibility for human rights violations related to climate change.[115] Effectively grappling with climate change requires a radical re-conceptualisation of human rights and environmental damage. Such re-conceptualisation is necessary to the extent that climate change is a global problem for humanity as a whole. This approach may be facilitated by certain general principles that have characterised the development of the law of global commons in the past three decades, which are captured by the notion of 'human right to a (climatically) sustainable environment', as proved, in particular, in *Ogoniland*.[116]

Subjectively, a human rights approach to climate change must be conceived of in such a way as to transcend the traditional notion of fundamental rights as individual entitlements and capture the collective dimension of the impact of

115 See above section 1.1.
116 See Knox (2012) [35].

climate change on society and humanity as a whole. The principle of 'common concern of humanity' proclaimed in the 1992 UNFCCC thus emerges as crucial. This tenet implies that climate policy is no longer seen as a matter of free choice, but as a mandatory requirement in view of the protection of the general interest of humanity. The *omnium* nature of the duty to ensure a (climatically) sustainable environment,[117] binding all (State and non-State) members of the international community, facilitates this approach. Furthermore, the *omnium* characterisation of a correlative right would unbundle the relationship between anthropogenic GHG emissions and individual human rights. Within this framework, the duty to ensure a (climatically) sustainable environment simplifies the causal link between GHG emitters, in particular, States, and human right violations.[118] This might permit holding States responsible for breaching the right to a (climatically) sustainable environment when they do not respect binding GHG reduction targets outlined in the UNFCCC-Kyoto Protocol system, triggering international human rights procedures.

Spatially, a complementary concept is that of 'common goods', which is progressively emerging in international law.[119] The different components of the environment, that is, air, the oceans, drinkable water and climate are common goods in the sense that they are not 'excludable' nor subject to privatisation. In the absence of a fully established global governance regime, respect for and protection of common goods such as climate depend on decentralised action by States and their function as agents for the enforcement of the international public interest. From this perspective, adapting the concept of State sovereignty to the present needs of a globalised and interdependent world becomes vital. As public interest is a limit to property in domestic law, common goods, such as climate, should limit sovereignty in international law. In an interdependent and globalised world sovereignty should no longer be considered the old unfettered dominion over a territory, but rather a governmental power functional to the achievement of the general interest of the international community, including the fundamental concern of preventing the aggravation of global warming with its devastating impact on human rights. This is also well captured by the '*erga omnes*' structure of the right to a sustainable environment, which deals with transboundary pollution, and is thus extraterritorial by nature. [120]

From a temporal perspective, another crucial concept is that of 'inter-generational justice'. This principle is more the product of environmental law than human rights law. It was inaugurated by the Brundtland Report in 1987,

117 The notion of '(climatically) sustainable environment' brings the ambiguity of the concept of sustainability within the geometric limits of legality.

118 See the contribution by Quirico, Bröhmer and Szabó in this volume.

119 On this concept, see Federico Lenzerini and Ana Vrdoljak (eds), *International Law for Common Goods* (Hart Publishing, 2014).

120 Ksentini (1994) [29]. For a full theoretical elaboration of the impact of the *erga omnes* concept on public interest, see Giorgio Gaja, 'The Protection of General Interests in the International Community' (2013) 364 *Hague Academy Collected Courses* 9, and the earlier set of essays by Paolo Picone, *Comunità internazionale e obblighi 'erga omnes'* (Jovene, 2006).

subsequently proclaimed in the Stockholm Declaration,[121] and reiterated 20 years later in the Rio Declaration,[122] Nonetheless, this notion acquires increasing importance in the field of human rights because of the way in which it can influence the concept and scope of the obligations that States have with regard to respect for and protection of fundamental rights. It is undeniable that certain developments in technology and the increasing gravity of the environmental crisis caused by climate change tend to project human rights protection into the future. This is happening in the context of the present debate over the patentability of life (embryos, stem cells) and more evidently with regard to climate change and the right of future generations to receive from the current generation an environment that permits their continued existence on the planet.[123] In this case, the impending disaster of global warming is not only a threat to the stability of climate as a global environmental good. It can further signal a devastating rupture of the social contract which presupposes that the State is the custodian of a legal order that extends its binding force to those who have lived in the past – because of already emitted GHGs – and those who will be born tomorrow. Unlike individualised first and second generation human rights, the 'omnium' nature of the right to a (climatically) sustainable environment can also encapsulate this dimension.[124]

121 Principle 1:

> Man . . . bears a solemn responsibility to protect and improve the environment for present and future generations.

122 Principle 3, which, however, obliquely frames the concept of intergenerational equity within the 'right to development', provides:

> The right to development must be fulfilled so as to equitably meet the developmental and environmental needs of present and future generations.

123 On the need to reconceptualise the human rights obligations of the State beyond the limited time span of the persons living under its current jurisdiction, see Francesco Francioni, 'Epilogue', in Lenzerini and Vrdoljak (2014) 443. See also UN Conference on Sustainable Development (2012).

124 This potential is embedded in Principle 1 of the Stockholm Declaration (see Ksentini (1994) [235]–[237]; Louis B Sohn, 'The Stockholm Declaration on the Human Environment' (1973) 14 *Harvard Int'l LJ* 451).

Article 3 (Right to the Environment and Future Generations) of the UNESCO Bizkaia Declaration provides:

> 1. Future generations have the right to inherit a healthy and ecologically balanced environment. 2. States have an obligation to monitor the quality and diversity of the environment and, specifically, to assess in advance long-term environmental consequences of the implementation of major projects.

See also Article 1 of the Aarhus Convention and Principle 2 of the Legal Principles annexed to the Brundtland Report.

In *Gabčíkovo-Nagymaros Project* (1997) 78, [140], the ICJ held:

> Throughout the ages, mankind has, for economic and other reasons, constantly interfered with nature. In the past, this was often done without consideration of the effects upon the environment. Owing to new scientific insights and to a growing awareness of the risks for mankind – for present and future generations – of pursuit of such interventions at an unconsidered and unabated pace, new norms and standards have been developed, set forth in a great number of instruments during the last two decades.

Thus, shall climate change ultimately lead to a re-conceptualisation of the relationship between environment and fundamental rights in favour of a human right to a (climatically) sustainable environment?[125] The experience of domestic jurisdictions demonstrates that the recognition of such a right effectively leads to reducing ecological footprints.[126] However, there is no sufficient support for this re-conceptualisation, especially in light of the hasty dismissal of the Inuit petition by the IAComHR,[127] and the uncertainty surrounding the Athabaskan petition on black carbon.[128] In any case, the human right to a (climatically) sustainable environment can only be helpful to the extent that the substantive limits and procedural implications of the notion are precisely determined. Starting with the definition of a right to a sustainable environment, this is not the case for the time being. Substantively, within the context of the *omnium* and *erga omnes* structure of this right, the content of the claim in issue should be clarified. This would also help to shed light on the relationship between the right to a (climatically) sustainable environment and greening implications of existing human rights,[129] along the lines of the indivisibility and interdependence of fundamental rights.[130] Procedurally, as to decision-making the right to a (climatically) sustainable environment might prompt an international shift from the current governance system, which is based on State sovereignty, and thus specific treaty regimes,[131] to a global multilevel system of delegated sovereign powers for common goods.[132] The discussed transformation of UNEP from a Programme into a Specialised Agency of the UN, along the lines of the proposals of the ICE Coalition and the Francioni–Dupuy Report are a step in this direction.[133]

As to adjudication and enforcement, new *ad hoc* procedures should possibly be outlined. This would require a fundamental choice between supranational enforcement and domestically coordinated implementation. With regard to already existing procedures, the interaction between the right to a (climatically) sustainable environment and existing human rights bodies should be clarified. The HRC

125 See Ron Dudai, 'Climate Change and Human Rights Practice: Observations on and around the Report of the OHCHR on the Relationship between Climate Change and Human Rights' (2009) 1 *Journal of Human Rights Practice* 294, 303–5.

126 Boyd (2013); Id (2012) 117 ff.

127 IAComHR, *Letter to Sheila Watt Cloutier et al*, Petition P-1413–05 (16 November 2006).

128 Arctic Athabaskan Council on Behalf of All Arctic Athabaskan Peoples of the Arctic Regions of Canada and the United States, *Petition to the Inter-American Commission on Human Rights Seeking Relief from Violations of the Rights of Arctic Athabaskan Peoples Resulting from Rapid Arctic Warming and Melting Caused by Emissions of Black Carbon by Canada* (23 April 2013).

129 Knox (2012) [51]–[52].

130 See, in particular, HRC, Resolution 10/4 of 25 March 2009, Preamble.

131 See, for instance, the Antarctic Treaty.

132 Along these lines, see International Centre for Comparative Environmental Law, *Vers un nouveau droit de l'environnement vers Rio+10*, Déclaration de Limoges II, Recommandations de la Réunion mondiale des juristes et des Associations de droit de l'environnement (9–10 November 2001).

133 *Should UNEP Become a Specialised Agency?* (2012) <lullabyofthecommons.wordpress.com>.

should, of course, play a crucial role. In light of the fact that current human rights adjudication and enforcement procedures are largely ineffective,[134] and in view of the increasing impact of climate change on global security, the role of the Security Council could also be re-assessed. Arguably, after the active role it assumed in dealing with the global threat posed by terrorism after 11 September 2001, the UN Security Council could be considered able to face global climate change as a threat to international security. From this perspective, the responsibility of unblocking the deadlock experienced in the Kyoto and post-Kyoto negotiations falls to a large extent into the hands of the permanent members of the Council.[135]

These issues are currently considered by Professor John Knox, who was appointed by the HRC as an Independent Expert on Human Rights and the Environment in 2012. Knox was mandated to identify the obligations related to the protection and promotion of a safe and healthy environment worldwide and corresponding best practices for the fulfillment of such duties.[136] A crucial part of Knox's work focuses on the relationship between climate change and human rights.[137]

Conclusion

There is common consensus that a human right to a sustainable environment is developing globally, at the domestic, regional and international level.[138] However, such a right is not yet satisfactorily defined nor fully recognised with respect to its substantive content and methods for adjudication and enforcement. It is also not completely clear whether this right is recognised procedurally *de lege lata.*[139]

A systemic acknowledgement of the right to a sustainable environment would facilitate a correct approach to climate change from a human rights perspective. Indeed, it would permit transcending the traditional notion of human rights as an individual entitlement and capturing the intrinsic collective and global dimension of climate change and its inter-generational nature.

134 See Olivier De Schutter, *International Human Rights Law* (CUP, 2014) 807 ff.

135 For a critical discussion of a reform of the global environmental governance system see Francesco Francioni and Christine Bakker, 'The Evolution of the Global Environmental System: Trends and Prospects in the EU and the US', in Bakker and Francioni, *The EU, the US and Global Climate Governance* (Ashgate, 2014) 9, 32–5.

136 By means of HRC Resolution 19/10 on Human Rights and the Environment, John H Knox (US) was appointed as Independent Expert on the issue of 'human rights obligations relating to the enjoyment of a safe, clean, healthy and sustainable environment'.

137 So far, the Independent Expert (since March 2015 Special Rapporteur) submitted his first report to the HRC on 24 December 2012 (Knox (2012)), a second report on 30 December 2013 (Knox (2013)), a third report specifically focusing on human rights and climate change in June 2014 (Knox (2014)), and a fourth report in February 2015 (UN DOC A/HRC/28/61). See also John Knox *et al*, *The Effects of Climate Change on the Full Enjoyment of Human Rights*, Report to the Climate Vulnerable Forum (30 April 2015).

138 Boyd (2012) 105–6.

139 See Knox (2012) [41], who does not speak of a 'human right' to information, participation and access to justice in environmental matters.

Anthropogenic GHG emissions should therefore prompt an intensification of our efforts to assess the impact of climate change on traditional ideas of human rights, focusing more on the new collective dimension of a human right to a climatically sustainable environment. How and if, however, this right encompasses the global and inter-temporal dimensions of climate change depends on the specific way in which the claim is framed both substantively and procedurally. The work of the UN HRC Independent Expert on Human Rights and the Environment is a step towards untying such a Gordian Knot.

Part III

Specific regimes

9 A double-edged sword

Climate change, biodiversity and human rights

Federico Lenzerini and Erika Piergentili

Introduction

The Preamble to the Biodiversity Convention emphasises the significance of biodiversity for humanity. In addition to its 'intrinsic value', the 'ecological, genetic, social, economic, scientific, educational, cultural, recreational and aesthetic values of biological diversity and its components' are underscored.[1] The Preamble also stresses 'the importance of biological diversity for evolution and for maintaining life sustaining systems of the biosphere',[2] to the point of considering biodiversity a 'common concern of humankind'.[3]

At the same time, the Biodiversity Convention also expresses the concern that 'biological diversity is being significantly reduced by certain human activities',[4] noting that it is consequently 'vital' that the international community 'anticipate[s], prevent[s] and attack[s] the causes of significant reduction or loss of biological diversity at source'.[5] Among the causes leading biodiversity to be significantly reduced, climate change – to be imputed to human activities – undoubtedly plays a particularly important role, and is therefore to be attacked 'at source'. Indeed, there is no reasonable doubt that climate change represents today the main cause of biodiversity loss. It is estimated that approximately 20–30 per cent of known plant and animal species are likely to be at increased risk of extinction if the global average temperature rises more than 1.5–2.5°C.[6] This figure appears particularly upsetting if one considers that, according to some estimates, warming of up to 5.5°C could be possible by 2100.[7] By 2050, climate change alone is expected to

1 *Biodiversity Convention*, Preamble, first recital.
2 Ibid, second recital.
3 Ibid, third recital.
4 Ibid, sixth recital.
5 Ibid, eighth recital.
6 IPCC, 'Summary for Policymakers', in Martin L Parry *et al* (eds), *Climate Change 2007: Impacts, Adaptation and Vulnerability. Contribution of Working Group II to the Fourth Assessment Report of the Intergovernmental Panel on Climate Change* (2007) 7, 11; *AR5, Synthesis* (2014) 72.
7 See *Two Degrees, One Chance. The Urgent Need to Curb Global Warming*, 6, <www.tearfund.org/webdocs/Website/Campaigning/Policy%20and%20research/Two_degrees_One_chance_final.pdf>.

threaten with extinction approximately one quarter or more of all species on land, becoming the most serious threat to life on the planet. Of course, plants and animals living in the oceans and in fresh water are also at risk of extinction as a result of global warming, especially those inhabiting ecosystems particularly sensitive to rising temperatures, like coral reefs.[8] Increasing ocean acidification – caused by high levels of CO_2 emissions arising from the burning of fossil fuels, about 50 per cent of which have been absorbed by oceans – is likely to have devastating effects over the fragile marine ecosystems, including drastic changes in species' habitats, extinction of the weakest species and, consequently, disentanglement of food webs.[9]

The effect of climate change on the conservation of biodiversity is so serious that, in articulating its main objective, the UNFCCC emphasises the need to achieve 'stabilisation of greenhouse gas concentrations in the atmosphere at a level that would prevent dangerous anthropogenic interference with the climate system . . . [to] be achieved within a time frame sufficient *to allow ecosystems to adapt naturally to climate change* [and] to ensure that food production is not threatened [. . .]'.[10] This sentence clarifies that biodiversity-related adverse effects arising from climate change are not confined to the environmental sector *per se*; on the contrary, a solid argument may be advanced to assert that they are likely to produce – and are already producing at present – huge adverse implications for the actual enjoyment of *human rights* by individuals and communities. In fact, depletion of ecosystems and destruction of the food chain inescapably translate into living conditions resulting in the violation of a notable number of internationally recognised human rights standards.

1. Human rights implications of biodiversity loss determined by climate change

Principle 1 of the 1972 Stockholm Declaration on the Human Environment[11] affirms that '[m]an has the fundamental right to freedom, equality and adequate conditions of life, in an environment of a quality that permits a life of dignity and well-being'. This provision offers a clearly anthropocentric vision of the environment, the good quality of which is essential for allowing a decent life for human beings. As a consequence, human life and the environment are

8 Harvard School of Public Health, *Climate Change and Biodiversity Loss* (2012) <http://chge.med.harvard.edu/topic/climate-change-and-biodiversity-loss>. See also Center for Biological Diversity, *Petition to List Acropora Palmata (Elkhorn Coral), Acropora Cervicornis (Staghorn Coral), and Acropora Prolifera (Fused-Staghorn Coral) as Endangered Species under the EPA* (4 March 2004); *Endangered and Threatened Species; Proposed Threatened Status for Elkhorn Coral and Stagkhorn Coral* (9 May 2005) 70 Fed Reg 26852.

9 Svitlana Kravchenko, 'Climate Change, Biodiversity and Human Rights: Can Synergy Help?', in Frank Maes *et al* (eds), *Biodiversity and Climate Change. Linkages at International, National and Local Levels* (Edward Elgar, 2013) 209, 213–15.

10 See Article 2 (emphasis added).

11 *Declaration of the United Nations Conference on the Human Environment* (16 June 1972).

interdependent, the proper preservation of the environment being crucial for ensuring the realisation of internationally recognised human rights standards. This was confirmed in 2008 by the HRC, which emphasised that 'climate change poses an immediate and far-reaching threat to people and communities around the world and has implications for the full enjoyment of human rights'.[12] In elaborating this point, the Council also stated:

> climate change-related impacts have a range of implications, both direct and indirect, for the effective enjoyment of human rights, including, *inter alia*, the right to life, the right to adequate food, the right to the highest attainable standard of health, the right to adequate housing, the right to self-determination and the right to safe drinking water and sanitation . . . in no case may a people be deprived of its own means of subsistence.[13]

Today empirical evidence shows that a number of internationally recognised human rights are affected by biodiversity-related detrimental effects of climate change. For instance, since loss of biodiversity resulting from climate change determines the extinction or deterioration of animal and plant species, it leads to disruption of the food chain and, *a fortiori*, to shortage of food for human beings. Climate change is leading crop productivity to decreasing at low latitudes, threatening 600 million people with malnutrition, especially in developing countries, where their survival is in danger on account of their dependency on food resources at risk of extinction due to increasing temperatures.[14] This results in the violation of fundamental human rights such as the right to life, the right to adequate standards of living, the right to health and the right to food. These human rights standards include not only the right to physical survival, but also all prerogatives which are essential for a dignified existence, including – as emphasised by the IACtHR – the rights to access water of good quality, food, adequate healthcare services as well as satisfactory educational opportunities.[15]

Further, the right to private and family life, the right to property, the right to adequate housing and the right to a shelter may be prejudiced by loss of biodiversity caused by climate change. This is crystal clear if one considers, for example, that – as specified by the CESCR – the right to adequate housing is 'the right to live

12 Res 7/23 of 28 March 2008, Preamble, first recital.

13 Res 18/22 of 17 October 2011, Preamble, thirteenth recital. See also Res 10/4 of 25 March 2009, Preamble, seventh recital.

14 *Report of the Office of the United Nations High Commissioner for Human Rights on the Relationship between Climate Change and Human Rights*, UN Doc. A/HRC/10/61 (2009) 26.

15 *Case of the Xákmok Kásek Indigenous Community v Paraguay* (Series C No 214, 24 August 2010) 195–217. For a comprehensive assessment of the interpretation of the right to life in the jurisprudence of the IACtHR see Alexandra R Harrington, *Life as We Know It: the Expansion of the Right to Life under the Jurisprudence of the Inter-American Court of Human Rights* (Albany Law School Working Papers Series No 6, 2012–2013) <http://papers.ssrn.com/sol3/papers.cfm?abstract_id=2142796>.

somewhere in security, peace and dignity',[16] presupposing, among other things, that '[a]ll beneficiaries of the right to adequate housing should have sustainable access to natural and common resources, safe drinking water, energy for cooking, heating and lighting, sanitation and washing facilities, means of food storage, refuse disposal, site drainage and emergency services'.[17] Similar considerations may be developed with respect to the other rights just listed. Loss of biodiversity may lead to scarce (or no) availability of food, and, *a fortiori*, to hunger, and it is not possible to assert that one who is hungry lives 'in security, peace and dignity'. In some cases, lack of food may force entire communities to relocate. Also, erosion of living resources is increasing rural to urban migration, bringing many people to live in settlements, like slums or hazardous areas, in conditions which are far from being respectful of the minimum requirements of human dignity.[18]

Other rights extensively affected by climate change-related depletion of biodiversity are the right to a safe environment, which can be currently considered an emerging human rights standard,[19] the right to development and the right to natural resources. The loss of natural resources, while an environmental harm in and of itself, may also have a deleterious impact on the means available to local populations. This, in turn, prevents harmonious development of individuals and communities.

2. Indigenous peoples and cultural rights

As noted by the HRC, the detrimental human rights-related implications determined by loss of biodiversity induced by climate change do not affect all human societies to the same extent. On the contrary, such implications – although 'affect[ing] individuals and communities around the world' – 'will be felt most acutely by those segments of the population that are already in vulnerable situations owing to factors such as geography, poverty, gender, age, indigenous or minority status and disability'.[20] For instance, the effects of loss of food resources – determining threats to the enjoyment of, inter alia, the right to adequate food, the right to health as well as the rights to life and to adequate standards of living – are exacerbated in areas that are geographically disadvantaged, where climate change-related loss of biodiversity has a particularly huge impact. Indeed, when animal or plant species are threatened with extinction in areas characterised by

16 See General Comment No 4, *The Right to Adequate Housing (Article 11(1) of the Covenant)* (1991) 7.

17 Ibid, 8(b).

18 OHCHR, Report, UN Doc A/HRC/10/61 (2009) 37.

19 For example, the ECtHR has affirmed that, even when it does not seriously endanger their health, 'environmental pollution may affect individuals' well-being and prevent them from enjoying their homes in such a way as to affect their private and family life adversely'. See *Case of López Ostra v Spain*, Appl No 16798/90 (ECtHR, 9 December 1994) [51]; *Guerra and Others v Italy*, Appl No 14967/89 (ECtHR, 19 February 1998) [60]. See also the contribution by Francesco Francioni and Ottavio Quirico in this volume.

20 See Res 18/22 of 17 October 2011, Preamble, fourteenth recital. See also Res 10/4 of 25 March 2009, Preamble, eighth recital.

a huge abundance of food, people living there may obtain what is necessary for nutrition from other species. But when the existence of certain human communities living in depressed areas depends on one or a few particular species, their disappearance makes it impossible for those communities to survive.

Indigenous peoples – especially those living in environmentally disadvantaged areas – are probably the communities most prejudiced by depletion of biodiversity caused by climate change. This is due to the special – often spiritually based – relationship they have with (animal and plant) natural resources, by reason of which loss of biodiversity is likely to disrupt fundamental elements of their very cultural identity, in addition to various other human rights. This reality is epitomised by the well-known Inuit petition of 2005, submitted to the IAComHR by an Inuk woman, Chair of the Inuit Circumpolar Conference, on behalf of herself, 62 other named individuals, and all Inuit of the Arctic regions of the US and Canada, claiming redress for assumed violations of human rights committed by the US resulting from GHG emissions.[21] The IAComHR found it impossible to adjudicate upon the petition, because 'the information provided [did] not enable [the Commission] to determine whether the alleged facts would tend to characterise a violation of rights protected by the American Declaration [of the Rights and Duties of Man]'.[22] However, the petition clearly explains how '[t]he potential impacts of climate change *on harvesting wildlife resources* are of fundamental concern for the social and economic well-being, the health, and the cultural survival of indigenous peoples throughout the Arctic'.[23] In more detail,

> [t]he quality of the available country food is also deteriorating. For example, game animals are undernourished because of changes in habitat caused by climate change ... Several other health problems have increased among caribou ... problems generally affecting increasing numbers of game animals ... Fish meat has become soft, unpalatable, and difficult to process. The meat from unhealthy animals is discarded, wasting scarce resources and decreasing the supply of subsistence protein.[24]

21 *Petition to the Inter-American Commission on Human Rights Seeking Relief from Violations Resulting from Global Warming Caused by Acts and Omissions of the United States* (Inuit Petition) (2005). On the Inuit Petition see, inter alia, Sarah Nuffer, 'Human Rights Violations and Climate Change: the Last Days of the Inuit People?' (2010) 37 *Rutgers Law Record* 182; Jessica Gordon, 'Inter-American Commission on Human Rights to Hold Hearing after Rejecting Inuit Climate Change Petition' (2007) 2(2) *Sustainable Development Law and Policy* 55; Hari M Osofsky, 'The Inuit Petition as a Bridge? Beyond Dialectics of Climate Change and Indigenous Peoples' Rights' (2006–2007) 31 *American Indian Law Review* 675.

22 *Letter from Ariel E Dulitzky, Assistant Executive Secretary, Inter-American Commission on Human Rights, to Paul Crowley, Legal Representative Inuit Petition* (16 November 2006).

23 *Inuit Petition* (2005) 39 (emphasis added).

24 Ibid, 63. Similar effects are experienced by other animal species which are fundamental for the physical and cultural survival of the Arctic people, including reindeer, fish and birds (see Ministry of the Environment of Finland and Secretariat of the Convention on Biological Diversity, *Indigenous Peoples and Traditional Knowledge Related to Biological Diversity and Responses to Climate Change in the Arctic Region* (2009) 3 <www.cbd.int/doc/publications/tk-cc-arctic-en.pdf>).

According to the petition, the rights of the Inuit people affected by global warming and climate change are the rights to life, health, physical integrity and security, the right to residence and movement, the right to property and to use and enjoy the lands they have traditionally used and occupied, the right to inviolability of the home, as well as the rights to enjoy the benefits of culture, to health and to means of subsistence.

In sum, the Inuit petition claimed that, in addition to rights related to the harmonious 'physical' existence of human beings, loss of biodiversity caused by climate change also leads to the infringement of rights which have a predominant *cultural* connotation. This holds particularly true with respect to indigenous peoples, as their vision of life is grounded in their cultural identity and specificity, and cultural rights are therefore of special importance to them, to the point of making their cultural existence, in most cases, inseparable from their physical survival.[25]

To an equivalent extent, as far as indigenous communities are concerned, it is not possible to draw a clear distinction between cultural rights and other categories of human rights, for the reason that *all* indigenous peoples' rights attain a marked cultural connotation. To have a clear idea of this reality, it is sufficient to consider the right to self-determination of peoples. This right is recognised in favour of all peoples by Article 1 common to the UN Covenants on Human Rights of 1966; by virtue of it peoples 'freely determine their political status and freely pursue their economic, social and cultural development'. With respect to indigenous peoples in particular, the right to self-determination is proclaimed by Article 3 of the United Nations Declaration on the Rights of Indigenous Peoples (UNDRIP),[26] which also specifies that, through exercising the right in question, the peoples concerned 'freely determine their political status and freely pursue their economic, social and cultural development'. Indeed, for indigenous communities the right to self-determination has an evident cultural and spiritual characterisation; at the same time, its very existence is based on the presence of a 'material' element, which is a territory where the communities concerned may concretely realise their own cultural identity and transmit it to future generations.

More generally, the proper enjoyment of cultural rights inescapably depends on the maintenance of the relationship indigenous communities have with their traditional lands. Land rights of indigenous peoples – which encompass their right to the natural resources located in their traditional territories – are of decisive significance for the very existence of such peoples. They are explicitly or implicitly safeguarded by a number of treaty or soft law provisions, which are interpreted

25 See also Arctic Athabaskan Council on Behalf of All Arctic Athabaskan Peoples of the Arctic Regions of Canada and the United States, *Petition to the Inter-American Commission on Human Rights Seeking Relief from Violations of the Rights of Arctic Athabaskan Peoples Resulting from Rapid Arctic Warming and Melting Caused by Emissions of Black Carbon by Canada* (2013); Centre for Biological Diversity, *Petition to List the Polar Bear (Ursus Maritimus) as a Threatened Species under the Endangered Species Act* (16 February 2005); *Determination of Threatened Status for the Polar Bear (Ursus maritimus) throughout Its Range ('Listing Rule')* (15 May 2008) 73 Fed Reg 28211.

26 GA Res 61/295 of 13 September 2007.

by human rights monitoring bodies taking into account the cultural significance of the land for the communities concerned.[27] Such an evolutionary interpretation is based on the awareness that loss of ancestral territories by indigenous peoples (including climate change and related biodiversity loss) represents a particularly severe violation of their internationally recognised human rights, which are mainly grounded in a cultural rationale. For indigenous communities, the significance of land is not limited to its value as commodity and/or as the place providing shelter, but is qualified by its special spiritual significance and by the fact of representing the element around which all other rights internationally recognised to them revolve. As explained by the IACtHR,

> the close ties of indigenous peoples with the land must be recognised and understood as the fundamental basis for their cultures, their spiritual life, their integrity, and their economic survival. For indigenous communities [their relationship with] the land is not merely a matter of possession and production but a material and spiritual element, which they must fully enjoy . . . to preserve their cultural legacy and transmit it to future generations.[28]

Similarly, the HRCte, in interpreting Article 27 of the 1966 ICCPR,[29] has held that 'one or other aspect of the rights of individuals protected under that article – for example, to enjoy a particular culture – may consist in a way of life which is closely associated with territory and use of its resources. This may particularly be true of members of indigenous communities'.[30] The Committee has further emphasised that 'culture manifests itself in many forms, including a particular way of life associated with the use of land resources, especially in the case of indigenous peoples. That right may include such traditional activities as fishing or hunting'.[31] With respect to the latter example, it is crystal clear that, once the cultural identity of a community depends, for instance, on hunting or fishing, and the animal species which are usually hunted or fished disappear (owing to climate change), such identity is ultimately affected. More generally, denying indigenous peoples the possibility of living on their traditional lands – which may actually happen as a

27 On indigenous peoples' land rights see International Law Association, Committee on the Rights of Indigenous Peoples, The Hague Conference – Report (2010), 20 ff, and Sofia Conference – Report (2012), 23 ff (both reports are available at <www.ila-hq.org/en/committees/index.cfm/cid/1024>). On the culturally driven interpretation of the relevant provisions see Federico Lenzerini, *The Culturalisation of Human Rights Law* (OUP, 2014) 131 ff.

28 *Case of the Mayagna (Sumo) Awas Tingni Community v Nicaragua*, Series C No 79 (31 August 2001) 149.

29 999 UNTS 171. Article 27 states that, '[i]n those States in which ethnic, religious or linguistic minorities exist, persons belonging to these minorities shall not be denied the right, in community with the other members of their group, to enjoy their own culture, to profess and practice their own religion, or to use their own language.'

30 General Comment No 23, *Article 27*, UN Doc CCPR/C/21/Rev.1/Add.5 (1994) [3.2] (footnotes omitted).

31 Ibid, 7.

result of climate change-related loss of biodiversity – is tantamount to seriously threatening their cultural integrity and identity, as well as, in some cases, even their very physical survival.

In the same way, the CESCR has found that the right of everyone to take part in cultural life – proclaimed by Article 15(1)(a) of the 1966 ICESCR[32] – encompasses the right to 'follow a way of life associated with the use of cultural goods and resources such as land, water, [or] biodiversity'.[33] In particular, as regards indigenous peoples, the 'strong communal dimension' of their cultural life – which is 'indispensable to their existence, well-being and full development' – includes, inter alia, 'the right to the lands, territories and resources which they have traditionally owned, occupied or otherwise used or acquired'.[34] As a consequence, '[i]ndigenous peoples' cultural values and rights associated with their ancestral lands and their relationship with nature should be regarded with respect and protected, in order to prevent the degradation of their particular way of life, including their means of subsistence, *the loss of their natural resources* and, ultimately, their cultural identity'.[35] The concrete cases of cultural loss determined by depletion of biodiversity are innumerable. In addition to the example of hunting and fishing there are so many other traditions which are linked to the use of plant or animal species, and loss of the latter translates into violations of cultural rights of the community concerned. A number of provisions included in the UNDRIP refer to the right of indigenous peoples to the natural resources existing in their traditional lands. In particular, Article 29 affirms the right of indigenous peoples to 'the conservation and protection of the environment and the productive capacity of their lands or territories and resources'. In this respect, Article 8(j) of the Biodiversity Convention establishes that, in the context of *in-situ* conservation of biodiversity, each State party shall, 'as far as possible and as appropriate',

> [s]ubject to its national legislation, respect, preserve and maintain knowledge, innovations and practices of indigenous and local communities embodying traditional lifestyles relevant for the conservation and sustainable use of biological diversity and promote their wider application with the approval and involvement of the holders of such knowledge, innovations and practices and encourage the equitable sharing of the benefits arising from the utilization of such knowledge, innovations and practices.

The matter is regulated by the Nagoya Protocol to the Biodiversity Convention,[36] which recognises the special nature and importance of genetic resources 'for

32 993 UNTS 3.

33 General Comment No 21, *Right of Everyone to Take Part in Cultural Life (Article 15, [1(a)] of the International Covenant on Economic, Social and Cultural Rights)*, UN Doc E/C.12/GC/21 (2009) [15(b)].

34 Ibid, 36.

35 Ibid (emphasis added). See also paras 49(d) and 50(c).

36 *Nagoya Protocol on Access to Genetic Resources and the Fair and Equitable Sharing of Benefits Arising from Their Utilization to the Convention on Biological Diversity* (2010) <www.cbd.int/abs/doc/protocol/nagoya-protocol-en.pdf>.

sustainable development of agriculture in the context of poverty alleviation and climate change',[37] as well as certain rights of indigenous peoples over genetic resources and traditional knowledge related to it. In particular, the Protocol affirms, inter alia, the right of indigenous and local communities to the fair and equitable sharing of the benefits arising from the utilisation of genetic resources held by them,[38] as well as from the utilisation of related traditional knowledge.[39] In addition, an obligation is established for States parties to ensure that 'the prior informed consent or approval and involvement of indigenous and local communities is obtained for access to genetic resources where they have the established right to grant access to such resources',[40] as well as for access to 'traditional knowledge associated with genetic resources that is held by indigenous and local communities'.[41]

Among the many examples of indigenous traditions related to the use of biodiversity, the traditional medicinal use of plants or animals or their utilisation as dressing ornaments are included. In particular, with respect to the former, loss of biodiversity may lead to disruption of practices of traditional medicine – particularly diffused within indigenous peoples – consequently worsening the quality of healthcare services within the community. The fundamental importance of conservation of biodiversity for traditional medicine, and, *a fortiori*, for access to healthcare by the members of the communities which base their health systems on such kind of medicine, is emphasised by Article 24 UNDRIP, which affirms the right of indigenous peoples 'to their traditional medicines and to maintain their health practices, including the *conservation of their vital medicinal plants, animals and minerals*. Indigenous individuals have also the right to access, without any discrimination, to all social and health services'.[42]

To summarise, the proper exercise and enjoyment of the right of indigenous peoples to the natural resources located in their traditional lands actually depends on the correct conservation of such resources. As a consequence, loss of biodiversity, which is in most cases determined by climate change, is likely to make this right totally ineffective.

It is important to emphasise that, while for the purposes of the present investigation the situation of indigenous peoples – as a result of their special culturally driven relationship with natural resources – is of particular significance, biodiversity-related detrimental effects of climate change may also result in the violation of cultural rights of non-indigenous individuals and/or communities, to the extent that they also attribute a cultural connotation to those resources. Even the international community as a whole may be affected. This happens,

37 Preamble, sixteenth recital.
38 Article 5(2).
39 Article 5(5).
40 Article 6(2).
41 Article 7.
42 Emphasis added.

in particular, with respect to World Heritage properties threatened by climate change.[43] Since those properties are protected in the interest of humanity, it is actually mankind – as a whole – that will be prejudiced by their degradation. Several ad hoc studies have shown the adverse impact of climate change on World Heritage properties,[44] which includes depletion of biodiversity.[45] In this respect, humanity is not only affected by the loss (or impairment) of irreplaceable natural sites of unique worth, but also by the cultural values associated with them.

3. Human rights-related effects of climate change response measures

In a paradoxical fashion, it is not only climate change per se that may lead to disruption of biodiversity and consequent infringement of human rights. The same outcome may also occur as a consequence of response measures adopted to mitigate the effects of climate change itself.

For instance, among the measures taken to that end, biofuel production is today increased to reduce the concentration of greenhouse gases in the atmosphere.[46] However, this practice is resulting in adverse consequences for human rights, especially the right to food. As stressed by the UN Special Rapporteur on the Right to Food, '[r]ushing to turn food crops – maize, wheat, sugar, palm oil – into fuel for cars, without first examining the impact on global hunger is a recipe for disaster. It is estimated that to fill one car tank with biofuel (about 50 litres) would require about 200 kg of maize – enough to feed one person for one year'.[47] Owing to an increasing demand for biofuels, global grain prices multiply, and they can no longer be afforded by a huge number of people whose nutrition depended on them. In addition, '[t]he production of biofuels will require substantial amounts of water, diverting water away from the production of food crops'.[48] Last but not least, biofuel production requires an increasing amount of land; this might lead to

43 World Heritage properties are the cultural and/or natural properties of outstanding universal value inscribed on the World Heritage List pursuant to the 1972 Convention Concerning the Protection of the World Cultural and Natural Heritage (1037 UNTS 151).

44 See, inter alia, World Heritage Committee, *Issues related to the State of Conservation of World Heritage Properties: the Impacts of Climate Change on World Heritage Properties*, UNESCO Doc WHC-06/30.COM/7.1 (2006); World Heritage Centre, *Case Studies on Climate Change and World Heritage* (2007) <http://whc.unesco.org/uploads/activities/documents/activity-473-1.pdf>; Id, *Policy Document on the Impacts of Climate Change on World Heritage Properties* (2008) <http://whc.unesco.org/document/10046>.

45 See, for instance, Australia's *Great Barrier Reef*, at <http://whc.unesco.org/en/list/154>, discussed in *Case Studies on Climate Change and World Heritage* (2007) 30, or Nepal's *Sagarmatha National Park* (<http://whc.unesco.org/en/list/120>) ibid, 18. See also Fang Jing and Brigitte Leduc, *Potential Threats from Climate Change to Human Wellbeing in the Eastern Himalayan Region* (MacArthur Foundation, 2010).

46 OHCHR (2009) 65.

47 *Report of the Special Rapporteur on the Right to Food*, UN Doc A/62/289 (2007) 21.

48 Ibid, 41.

deforestation and, to the extent that such an operation concerns lands traditionally belonging to indigenous peoples, translates into violations of the rights of those peoples to their lands, natural resources as well as, *a fortiori*, self-determination and cultural identity.[49]

A related problem caused by response measures is represented by the fact that reliance on agricultural products which are resistant to climate change may determine an increase in monoculture crops and plantations, with a consequent deterioration of biodiversity and worsening of food security.[50]

In this respect, Article 2 UNFCCC specifies that stabilisation of GHGs must be carried out in a way that will 'ensure that food production is not threatened and to enable economic development to proceed in a sustainable manner'. Article 4(8) requires States parties to minimise the 'adverse effects of climate change and/or the impact of the implementation of response measures'. Similar provisions are included in the Kyoto Protocol of 1997.[51] Furthermore, in 2008 the CESCR recommended that States develop '[i]mplementing strategies to combat global climate change that do not negatively affect the right to adequate food and freedom from hunger, but rather promote sustainable agriculture'.[52]

4. Looking for strategies to mitigate climate-induced depletion of biodiversity

Climate change is a global phenomenon which is barely reversible, unless a huge part of humanity drastically changes its current behaviour. However, a number of measures may be adopted in order to mitigate its effects. In addition to measures of general character – including reduction of GHG emissions or carbon trading – some specific initiatives have proven effective in reducing loss of biodiversity determined by climate change, and, *a fortiori*, consequent human rights-related harmful effects.

First of all, improvement of legal action in the field – at all levels (that is, 'universal', regional and national) – would be necessary, accompanied by proper implementation of existing pertinent legal instruments, including those adopted at the international level (for instance, UNFCCC, Kyoto Protocol, Biodiversity Convention, Nagoya Protocol, UNDRIP). In developing such legal action, a human rights-based approach should be adopted,[53] integrating human rights considerations in the context of implementation of biodiversity and climate change-

49 OHCHR (2009) 67–8.

50 UN Permanent Forum on Indigenous Issues, *Climate Change and Indigenous Peoples. Backgrounder* (2008) 2 <www.un.org/en/events/indigenousday/pdf/Backgrounder_ClimateChange_FINAL.pdf>.

51 *Kyoto Protocol to the United Nations Framework Convention on Climate Change* (1997) 37 ILM 22 (1998). See Articles 2(3), and 3(14).

52 *The World Food Crisis – Statement*, UN Doc E/C.12/2008/1 (2008) 13.

53 Elisa Morgera, 'No Need to Reinvent the Wheel for a Human Rights-Based Approach to Tackling Climate Change: the Contribution of International Biodiversity Law', in Erkki J Hollo *et al* (eds), *Climate Change and the Law* (Springer, 2013) 359, 383 ff.

related policies. This course of action has been recommended, inter alia, by the ACom HPR, which has urged the Assembly of Heads of State and Government of the African Union 'to ensure that human rights standards safeguards, such as the principle of free, prior and informed consent, be included into any adopted legal text on climate change as preventive measures against forced relocation, unfair dispossession of properties, loss of livelihoods and similar human rights violations',[54] as well as to guarantee that 'special measure of protection for vulnerable groups such as children, women, the elderly, indigenous communities and victims of natural disasters and conflicts are included in any international agreement or instruments on climate change'.[55]

In terms of operational measures, adaptation is particularly advocated. With specific regard to climate change-related biodiversity loss, this might consist, for instance, in replacing certain crops originally used for human nutrition, or for breeding, with others that are more resistant to the effects of climate change. However, this would be appropriate to neutralise the effects of climate change with respect to certain human rights (for instance, the right to food or to adequate nutrition), but not with regard to others, especially cultural rights (when harvesting of specific crops has a cultural connotation). Also, adaptation to new crops may require huge costs that may not be affordable by human communities living in poverty, or re-organisation of cultivating cycles of the land that may take several years, leaving the community without food for a long time. The example just described shows that, whatever initiative is intended to be adopted in the field, it is to be taken only after a careful and competent assessment of the particular needs of a specific area and of the people living there, depending on the relevant geographical, environmental and cultural characteristics. This implies, inter alia, that local communities are granted full access to all relevant information, are properly consulted and are ensured full participation in decision-making concerning the areas where they live. Important decisions relating to specific areas should not be taken without the free, prior and informed consent of the communities concerned. Management plans should integrate local traditional knowledge and community concerns.[56] Cultural considerations have a crucial significance in this respect; environmental impact assessments should therefore be accompanied by cultural impact assessments.[57]

Provided that the conditions just described are respected, in a number of cases adaptation has already proven effective in efficiently facing biodiversity loss caused by climate change. For instance, in a number of Central and Latin American territories, local indigenous communities are shifting their agricultural activities

54 *Resolution No 153 on Climate Change and Human Rights and the Need to Study Its Impact in Africa* (2009) 1.

55 Ibid, 2. See also the analysis on the incorporation of procedural human rights in the UNFCCC context developed by Kravchenko (2013) 232 ff.

56 Morgera (2013) 382.

57 See Morgera (2013) 373, asserting that 'biodiversity-inclusive and socio-cultural assessments . . . have great importance from an adaptation and mitigation perspective' (footnotes omitted).

and relocating to new areas less exposed to the effects of climate change.[58] In the Arctic, fishing sites have been relocated closer to shores in response to reductions in thickness of lake ice, and winter grazing patterns for livestock have been shifted in response to changes in ice and snow conditions.[59] Similarly, indigenous peoples of Alaska have started changing their hunting habits and food-storage methods.[60] With respect to these practices, however, the problem remains that the adjustments imposed by climate change may not be consistent with the cultural needs of the communities concerned, therefore resulting in breaches of their cultural rights.

Another concrete strategy which has proven effective to face climate change-related impairment of biodiversity is ecosystem restoration (particularly in the forms of reforestation, revegetation and reintroduction of endangered animal species), especially when adopted at the local level. For example, since 2000, in the State of Orissa (India), women's groups from local villages have developed very successful initiatives to restore forest and coastal biodiversity, leading to over 60 per cent growth in local vegetation adversely affected by climate change-related effects.[61] Similar activities have been developed in other States of India.[62]

Initiatives of this kind should be boosted, promoted, supported and protected by law. At the same time, as previously stressed, when such programmes are developed at the governmental level, they should be carried out with the involvement of local communities (who know the needs of the land better than anybody else) and must be culturally sustainable.

Conclusion

> Biodiversity is not an abstract concept. It is life itself: populations, nature and our planet. Biodiversity is made up of people, wild and cultivated plants, domesticated and wild animals. Biodiversity is made up of natural climates and environments, languages and cultures and food. Its custodians are herders, farmers, fishermen, and artisans . . . 300,000 plant varieties have become extinct in the last century and one more is lost every six hours.[63]

These few sentences perfectly epitomise the value of biodiversity for humanity, and make it clear how severely human rights (especially indigenous peoples'

58 See UN Permanent Forum (2008) 2.

59 Ministry of the Environment of Finland and Secretariat of the Convention on Biological Diversity (2009) 4.

60 Ibid, 5.

61 Sweta Mishra, 'Conserving Local Biodiversity through Community Conservation Initiatives in Orissa, India', in Shamsul Arefin *et al*, *Community Action in a Climate Change Regime. Managing Biodiversity* (Regional Centre for Development Cooperation, Bhubanesvar, 2011) 36, 39 ff.

62 Kanna K Siripurapu, 'Climate Change: Loss of Biodiversity and Rural Livelihoods . . . ! What are we doing?', in Shamsul Arefin *et al* (2011) 46, 59 ff.

63 Slow Food Beirut, *Biodiversity in Words* (2009) <http://slowfoodbeirutb.blogspot.it/2009/09/please-take-time-to-read-this-very.html>.

rights) are infringed when biodiversity is depleted, as is happening on a huge scale as a result of climate change. Therefore, protecting biodiversity against the effects of climate change is not only a matter of preserving environmental commodities, but is essential for saving the very life and dignity of human beings and communities through the protection of different human rights standards, ranging from the right to life, health and food to cultural rights of indigenous peoples.

A complicated balance between competing human rights standards is to be achieved when adopting measures taken in response to the detrimental effects of climate change, including adaptation and ecosystem restoration. Human rights considerations should play a primary role in the context of the implementation of biodiversity and climate change-related policies adopted by the international community. Any initiative taken with the purpose of mitigating or adapting to the adverse effects of climate change should integrate local traditional knowledge and community concerns; in particular, local communities should be granted full access to all relevant information, should be properly consulted and should be ensured full participation in decision-making concerning the territories where they live.

Today there is no more time for thinking about the possible measures to be taken. Now is the time to act and efficiently respond to the ongoing destruction of biodiversity caused by climate change. Our very survival and inherent dignity are at stake.

10 Climate change, migration and human rights

Towards group-specific protection?

Benoît Mayer and Christel Cournil[1]

Introduction

Much has recently been written on the question of the protection of environmental migrants, that is, people displaced in connection with environmental phenomena, and, more specifically, 'climate' migrants, where such phenomena can be regarded as effects of climate change. The idea of reforming international refugee law through an amendment to the 1951 Convention relating to the Status of Refugees was swiftly rejected, in part through fear of undermining the fragile achievement of the protection of refugees, but also partly because substantial differences between environmental migrants and refugees do not seem to allow such an analogy.[2] Unlike persecution, environmental factors mostly generate internal migration (that is to say, migration within national borders) and the emerging standards of protection and assistance to internally displaced persons (IDPs) are of great relevance. By focusing on international migrants, however, some authors have preferred to discuss possible mechanisms of complementary or subsidiary protection, previously developed for people who could not return to their country of origin because of armed conflicts or widespread violence, which, they argue, could extend to environmental migrants or, at least, to some of them.[3]

In addition to the lack of support by critical States, these proposals face technical obstacles. It is particularly difficult to identify environmental migrants, and *a fortiori* 'climate' migrants, because of the indirect and complex causal links between

1 Christel Cournil received financial support from ANR CIRCULEX (ANR-12-GLOB-0001–03 CIRCULEX).

2 Benoît Mayer, 'Pour en finir avec la notion de "réfugiés environnementaux": critique d'une approche individualiste et universaliste des déplacements causés par des changements environnementaux' (2011) 7(1) *McGill International Journal for Sustainable Development Law and Policy* 33.

3 Frank Biermann and Ingrid Boas, 'Preparing for a Warmer World: towards a Global Governance System to Protect Climate Refugees' (2010) 10(1) *Global Environmental Politics* 60; Bonnie Docherty and Tyler Giannini, 'Confronting a Rising Tide: a Proposal for a Convention on Climate Change Refugees' (2009) 33 *Harvard Environmental Law Review* 349; CRIDEAU, 'Projet de Convention relative au statut international des déplacés environnementaux' (2008) 39 *Revue de droit de l'Université de Sherbrooke* 451.

environmental phenomena and migration.[4] Few people spontaneously identify themselves as environmental or 'climate' migrants. Since climate change is a change in the probability of specific weather events, it is quite problematic to attribute a specific environmental phenomenon to it. More generally, the concepts of climate or environmental migration often seem too broad to be operational, as they include internal migrants and international migrants, as well as voluntary migrants and forced migrants, regardless of the duration or distance of the displacement. This combination of political and more technical obstacles makes the adoption of a convention relating to environmental or 'climate' migrants quite unrealistic, and thus the ongoing Nansen Initiative of Switzerland and Norway seems to be moving toward the promotion of a non-binding, guiding document on the protection of people displaced across borders in the context of disasters and other environmental phenomena possibly related to climate change.[5]

This chapter consists of an assessment of the need for specific protection of environmental or 'climate' migrants. It is based on an apparent paradox between the existence of universal human rights instruments (which, being universal, apply equally to environmental migrants) and the development of group-based protection mechanisms. Each State has an obligation, under relevant international human rights law instruments, to respect and protect the rights of any individuals within its jurisdiction, including environmental migrants. In the past, group-specific mechanisms have, however, been established to protect the rights of women and children, racial, national or religious minorities, refugees and Stateless persons, migrant workers and members of their families, and people with disabilities – and debates are now extending to the rights of indigenous peoples and the elderly. Should environmental or 'climate' migrants also be the object of specific protection?

The discussion is organised in three sections. The first section explores the applicability and limits of general international human rights law in responding to the need for protection highlighted by the debate on environmental migration. The second section explores how existing group-specific protection applies to environmental or 'climate' migrants. Finally, the third section assesses the risks and opportunities of moving toward a category-specific protection of environmental or 'climate' migrants.

1. The applicability of general human rights instruments to environmental or 'climate' migrants

International human rights law recognises certain fundamental individual rights that are considered essential to human dignity. For the purpose of this chapter, general human rights protection is defined by reference to non-group-specific instruments such as the UDHR, ICCPR and ICESCR, and regional instruments

4 FORESIGHT, *Migration and Global Environmental Change: Final Project Report* (2011).

5 See Walter Kälin, 'From the Nansen Principles to the Nansen Initiative' (2012) 41 *Forced Migration* 48.

such as the ECHR. General human rights protection also includes rights-specific instruments, such as the 1986 Declaration on the Right to Development. The rights that these instruments recognise are applicable to any persons within the jurisdiction of a State bound by them: every State has the duty to protect all persons within its jurisdiction, irrespective of nationality. As long as States have ratified relevant treaties, environmental or 'climate' migrants should therefore benefit from general human rights protection.

It is now well understood that climate change can have diverse consequences on multiple human rights, particularly on economic, social and cultural rights.[6] Environmental phenomena that generate migration may affect the enjoyment of a multitude of rights, such as the rights to life, health and food, along with partially recognised rights, such as the right to a healthy environment and to development. Within this context, migration may also challenge the rights to housing, property, security and non-discrimination, among others.[7]

The right to substantive equality is of particular importance in the context of our analysis. *Formal* equality implies the right not to be discriminated against, that is, that persons in similar situations should not be treated differently without justification. However, sometimes the application of similar treatment in very different situations can result in an unfair difference of treatment. Thus, *substantive* equality goes further: it requires different treatment for significantly different situations. Accordingly, particularly vulnerable persons must be given special attention,[8] and thus the right to substantive equality can be an argument for specific protection of the rights of vulnerable populations under general human rights law.

It is, however, difficult to consider environmental or 'climate' migrants as a distinct category of vulnerable individuals. Environmental phenomena can create situations of vulnerability that may cause migration, but populations that do not migrate are also vulnerable – sometimes even more so if they are unable to migrate owing to a lack of resources.[9] Migration itself can be a source of vulnerability, but the vulnerability of migrants does not depend on the cause of migration (vulnerability and precariousness are rather a product of the circumstances where migration takes place). Some experts have preferred concepts that highlight the vulnerability of migrants, such as 'survival migration'[10] or 'crisis migration',[11] which are discussed in the last section of this chapter.

6 See Human Rights Committee, *The Position of Aliens under the Covenant*, General Comment 15 (1986).

7 See Sébastien Jodoin and Katherine Lofts (eds), *Economic, Social, and Cultural Rights and Climate Change: a Legal Reference Guide* (Centre for International Sustainable Development Law, Academics Stand against Poverty Network, and the Governance, Environment and Markets Initiative at Yale University, 2014); Christel Cournil and Anne-Sophie Tabau (eds), *Politiques climatiques de l'Union européenne et droits de l'homme* (Larcier, 2013) Annex 4, 316–17.

8 François Crépeau, *Report of the Special Rapporteur on the Human Rights of Migrants*, UN Doc A/67/299 (2012), featuring a thematic section on climate change and migration.

9 FORESIGHT (2011).

10 Alexander Betts, *Survival Migration: Failed Governance and the Crisis of Displacement* (Cornell University Press, 2013).

11 Susan F Martin, Sanjula Weerasinghe and Abbie Taylor, *Humanitarian Crises and Migration: Causes, Consequences and Responses* (Routledge, 2014).

Some general human rights, recognised by non-specific instruments, only become relevant in the context of migration. Thus, under Article 13(1) UDHR, '[e]veryone has the right to freedom of movement and residence within the borders of each State.' Similarly, Article 12(1) ICCPR provides that '[e]veryone lawfully within the territory of a State shall, within that territory, have the right to liberty of movement and freedom to choose his residence.' This right is also recognised in the framework of the Council of Europe, under Article 2(1) of the Fourth Protocol (1963) to the ECHR, where it is also limited to '[e]veryone lawfully within the territory of a State.'

Restrictions to Article 12(1) ICCPR are only allowed if they 'are provided by law, are necessary to protect national security, public order, public health or morals or the rights and freedoms of others, and are consistent with the other rights recognised in the present Covenant'.[12] Thus, any person within the jurisdiction of a State that is party to the ICCPR has the right to migrate within the State where he or she is present, unless the conditions for a restriction are met. However, the concepts of national security, public order or the protection of the rights and freedoms of others are notoriously flexible.

Persons wanting to migrate also benefit from Article 13(2) UDHR, which states that '[e]veryone has the right to leave any country, including his own.' This is reaffirmed in similar terms in Article 12(2) ICCPR and in Article 2(2) of the Fourth Protocol to the ECHR. Thus, people from a State that is particularly affected by environmental phenomena which may be related to climate change cannot be prevented from leaving their country. Permissible restrictions on the right to freedom of movement and residence are also applicable to the right to leave one's country (ICCPR, Art 12(3)). In practice, however, it is most often the absence of the right to enter a third country which limits the right to leave one's country.

Furthermore, Article 14 UDHR states that '[e]veryone has the right to seek and to enjoy in other countries asylum from persecution.' Along these lines, Article 3(1) of the 1967 Declaration on Territorial Asylum not only prohibits expulsion (refoulement), but also 'rejection at the frontier' of 'persons entitled to invoke Article 14 of the Universal Declaration of Human Rights.' Unlike the aforementioned rights, this one has not been subsequently confirmed by the ICCPR nor by the ECHR and its protocols. The 1951 Convention relating to the Status of Refugees provides that refugees have the right not to be expelled or forcibly returned 'to the frontiers of territories where [their] life or freedom would be threatened',[13] but it stops short of an explicit recognition of the right of an asylum-seeker to enter the territory of a third State. More importantly, the Refugee Convention only protects individuals in cases of 'persecution'.[14] Persecution necessarily implies the will to harm: those displaced as a result of environmental

12 Article 12(3).
13 Article 33.
14 See section 2.1 below.

phenomena are not generally 'persecuted' and cannot convincingly invoke the Refugee Convention. While diverse regional instruments have extended the protection of refugees, including the implicit right to asylum, it has generally been in situations where displacement is directly caused by human agency, such as armed conflicts or generalised violence.

The possibility of amending the law to include a right to asylum in circumstances other than persecution has been considered. An Australian political party has thus proposed the recognition of a 'right to environmental asylum' for populations of small island States whose atolls are directly threatened by rising sea levels.[15] However, given the reluctance of States to respect the right to asylum in conventional situations of persecution, it is questionable whether the right to asylum on environmental grounds would be generally recognised. The acknowledgement of a right of entry on a case-by-case basis for small island populations seems much more realistic.

Finally, Article 13(2) UDHR recognises to everyone 'the right to return to his country'. Similarly, the ICCPR states that '[n]o one shall be arbitrarily deprived of the right to enter his own country'.[16] More restrictively, Article 3(2) of the Fourth Protocol to the ECHR states that '[n]o one shall be deprived of the right to enter the territory of the State of which he is a national'. Of course, this right applies to any categories of environmental or 'climate' migrants as much as it does to any other conceivable category of persons or, indeed, to anyone.

Overall, general human rights instruments providing non-group protection afford some foundations for the protection of any migrants – including environmental or 'climate' migrants. General human rights instruments define relatively abstract principles that States commit to respect in all circumstances, unless restrictions, limitations or derogations apply. Owing to their general nature, these principles do not always exactly match the specific protection needs of all. More generally, law, by its abstract nature, often strives to deal with the infinite range of unique individual circumstances.

2. The suitability of group-specific human rights instruments

In the field of the international protection of human rights, specific standards have been adopted in several ways. There has been a trend towards the clarification of generally applicable rights, or the definition of new rights (especially third generation rights). Another trend, which is of greater interest here, is towards the

15 Anthony Albanese and Bob Sercombe, *Labor Calls for International Coalition to Accept Climate Change Refugees*, Press Release (9 October 2006) <http://anthonyalbanese.com.au/labor-calls-for-international-coalition-to-accept-climate-change-refugees>.

16 'His country' or 'his own country' may extend not only to the country of the person's nationality, but also to a host country, especially in the case of refugees and Stateless persons.

definition of the rights of specific categories of individuals, from women to migrant workers and from indigenous peoples to children.[17]

Group-specific protection instruments were developed well before the adoption of general instruments such as the UDHR, aimed in particular at the protection of national minorities and refugees since the 19th century. Yet, the adoption of general protection instruments has not put an end to specific protective tools. Major pre-existing specific mechanisms have remained relevant, such as the protection of refugees defined by the 1951 Convention relating to the Status of Refugees. Moreover, new mechanisms for group-specific protection have been adopted following the implementation of general protective measures, for example, the 1989 Convention on the Rights of the Child, the 1990 International Convention on the Protection of the Rights of All Migrant Workers and Members of Their Families, and the 2006 Convention on the Rights of Persons with Disabilities. Other soft law instruments or ongoing proposals relate to the protection of internally displaced persons, indigenous peoples, the elderly and prisoners. Proposals for the protection of environmental or 'climate' migrants would also be a form of group-specific protection.

How far should group-specific protection extend? Too many laws may diminish the authority of laws and reduce their intelligibility. While a number of specific rules may contribute to a fairer law, equally, too many specific rules run the risk of making the law arbitrary and unintelligible.[18] A balance must therefore be found between rules that are too general and rules that are too numerous and obscure. Above all, it is important to consider the relevance of group-specific protection not only in terms of the individuals it includes, but also of those it excludes – in relation to its direct benefits as well as its unintended consequences.

Before considering the relevance of group-specific protection for the benefit of environmental or 'climate' migrants, it is useful to explore experiences from the protection of other vulnerable groups and question the applicability of these regimes to environmental migration. Four group-specific human rights treaties are particularly relevant to the situation of environmental or 'climate' migrants, concerning the protection of refugees, Stateless persons, migrant workers and IDPs.

17 What Frédéric Mégret described as the 'pluralisation of human rights' not only involves an analysis of existing rights or a learning process that enables the implementation of an international advocacy effort, but also the adaptation of certain rights to the particular circumstances of certain categories of people with more specific needs. The Convention on the Rights of Persons with Disabilities, for example, includes an affirmation of existing rights, some reformulations and also some extensions and legal innovations. See Frédéric Mégret, 'The Disabilities Convention: Human Rights of Persons with Disabilities or Disability Rights?' (2008) 30(2) *Human Rights Quarterly* 494.

18 To control the proliferation of rights, Philip Alston, for example, has proposed a mechanism for the monitoring of new rights, a form of 'appellation contrôlée' (certified designation of origin) that would at least guarantee compliance with a certain design process before the approval of new rights by international institutions, particularly facilitating broad and transparent consultation. See Philip Alston, 'Conjuring up New Human Rights: a Proposal for Quality Control' (1984) 78(3) *American Journal of International Law* 607.

First, it needs to be determined whether, and to what extent, such group-specific treaties can apply to environmental or 'climate' migrants. Second, these historical examples illustrate crucial issues relating to group-specific protection.

2.1. Refugees?

Building on a system of bilateral treaties for the protection of specific groups of refugees developed during the interwar period, the Convention relating to the Status of Refugees was adopted in 1951. Although the UDHR and the ECHR had just been adopted, the Refugee Convention constitutes the first human rights treaty with universal scope (even if States initially had the option to limit their obligation to refugees from Europe).

The Refugee Convention defines a set of rights for refugees, most of which nonetheless do not go beyond what the ICESCR and the ICCPR now provide for everyone, refugee or not. The Refugee Convention does however provide a specific right which is not guaranteed in other treaties, as mentioned above: the right for refugees not to be expelled or returned.[19] For the purpose of the Convention, a refugee is a person who, 'owing to well-founded fear of being *persecuted* for reasons of race, religion, nationality, membership of a particular social group or political opinion, is outside the country of his nationality and is unable or, owing to such fear, is unwilling to avail himself of the protection of that country'.[20]

Some States have extended the personal scope for the protection of refugees, either unilaterally or through regional cooperation.[21] For example, the 1969 AU Convention Governing the Specific Aspects of Refugee Problems in Africa offers additional protection to people that have been displaced 'owing to external aggression, occupation, foreign domination or event seriously disturbing public order.'[22] The 1984 Cartagena Declaration on Refugees adopted by a number of Latin American States promotes a similar type of additional protection for those who 'have fled their country because their lives, safety or freedom have been threatened by generalised violence, foreign aggression, internal conflicts, massive violation of human rights or other circumstances that have seriously disturbed public order.'[23] Since 1997, the EU has also extended the protection of refugees, particularly in cases of 'serious and individual threat to a civilian's life or person by reason of indiscriminate violence in situations of international or internal

19 Article 33.

20 Article 1(A)(2), emphasis added. Although the condition of 'persecution' is mentioned in the UDHR, the negotiators of the Refugee Convention found it necessary to add a condition relating to the ground of persecution. The narrow definition of a refugee reflects the great caution of the drafters of the Convention.

21 See Jane McAdam, *Complementary Protection in International Refugee Law* (Oxford University Press, 2007).

22 Article 1(2).

23 III(3).

armed conflict.'[24] However, rather than offering tangible protection, some of these provisions seem more a façade. The AU Convention, for example, has not been implemented systematically.[25] In the framework of European legislation, the condition of *indiscriminate* violence posing an *individual* threat is vague and somewhat paradoxical, which means that there is a risk of arbitrary interpretations.

More fundamentally, environmental or 'climate' migrants cannot be considered 'persecuted' and thus do not fulfil the conditions to qualify as refugees under Article 33 of the Refugee Convention. These migrants should therefore not qualify as refugees for the purpose of the Refugee Convention. Nor can they rely, generally speaking, on additional protection mechanisms, except perhaps when they are displaced by violence generated by environmental phenomena. Thus, unsurprisingly, courts have denied refugee protection when it was invoked on the ground of environmental factors. For instance, the High Court of New Zealand considered inadmissible a request for refugee protection of individuals who highlighted environmental circumstances in their country of origin, Kiribati. The Court noted growing scientific evidence on the impacts of climate change on this small island developing State, but rejected the application as:

> [a] person who becomes a refugee because of an earthquake or growing aridity of agricultural land cannot possibly argue, for that reason alone, that he or she is being persecuted for reasons of religion, nationality, political opinion, or membership of a particular social group.[26]

2.2. Stateless persons?

In order to assess whether group-specific human rights protection provided to Stateless persons is relevant in the context of environmental migration, a fundamental distinction must be drawn between *de facto* and *de jure* Statelessness.

On the one hand, *de facto* Statelessness relates to a situation where a person is not effectively protected by his or her State of nationality.[27] Refugees are *de facto*

24 *Parliament and Council Directive 2011/95/UE of 13 December 2011 on Standards of Qualification of Third Country Nationals or Stateless Persons as Beneficiaries of International Protection*, OJ L 337/9 (2011) ('directive qualification').

25 See, for instance, Betts (2013) 14.

26 *Ioane Teitiota v the Chief Executive of the Ministry of Business, Innovation and Employment*, CIV-2013-404-3528 [2013] NZHC 3125 (High Court of New Zealand, 26 November 2013) [11]. For further details on this case, see the contribution by Ben Boer in this volume.

27 United Nations Department of Social Affairs, *Study of Statelessness* (1949, re-edited by the UNHCR in 1995) 9:

> *Stateless persons de facto* are persons who, having left the country of which they were nationals, no longer enjoy the protection and assistance of their national authorities, either because these authorities refuse to grant them assistance and protection, or because they themselves renounce the assistance and protection of the countries of which they are nationals.

See also Hugh Massey, *UNHCR and De Facto Statelessness*, LPPR/2010/01 (2010).

Stateless, but other persons who are *de facto* Stateless do not fall within the narrow definition of a refugee. *De facto* Stateless persons are not the object of any specific protection in international law. When a State is unable to protect its citizens because of environmental phenomena, for instance in the occurrence of severe adverse impacts of climate change, it is possible to consider these nationals as *de facto* Stateless, but, in the absence of any specific protection of *de facto* Stateless persons, this categorisation has no legal effect.[28]

On the other hand, *de jure* Statelessness relates to the situation of persons who are not considered nationals by any State.[29] The protection of *de jure* Stateless persons, which was first debated in conjunction with the protection of refugees, eventually became the subject of a specific treaty, the 1954 Convention relating to the Status of Stateless Persons. This instrument has a low level of ratification and includes few rights beyond the prohibition of expulsion not justified by reasons of 'national security or public order.'[30]

To be considered *de jure* Stateless, and to fall into the scope of the 1954 Convention relating to the Status of Stateless Persons, environmental or 'climate' migrants would need to cease to be recognised as nationals by their State of origin. If a State ceases to exist, as has been envisaged in the case of some small island States (possibly because territory becomes uninhabitable, even before the whole State is submerged by sea-level rise), its citizens would arguably become Stateless. It is, however, difficult to determine how a State could cease to exist (outside of the classical scenarios defining State succession), even when its habitable territory disappears. Such an event is unprecedented and is not necessarily governed by rules that are applicable to the creation of a State.[31]

In any case, formal protection as *de jure* Stateless persons would be of little material effect for environmental or 'climate' migrants. The protection of *de jure* Stateless persons provides no right of entry or residence in the territory of a third State, and rights under the 1954 Convention hardly go beyond those guaranteed by the general protection of human rights. General international human rights instruments are also more largely ratified and more systematically implemented.

Yet, from a political perspective, the existence of a regime for the protection of Stateless persons may make a difference. The categorisation of environmental or 'climate' migrants as possibly Stateless persons may come in support of political arguments for the recognition of the need for international protection. Along these

28 See UNHCR, *Climate Change and Statelessness: an Overview* (15 May 2009) 2.

29 Article 1 of the 1954 Convention relating to the Status of Stateless Persons: 'For the purpose of this Convention, the term "Stateless person" means a person who is not considered as a national by any State under the operation of its law.'

30 Article 31.

31 See UNHCR (2009) 2. See also Jane McAdam, *Climate Change, Forced Migration, and International Law* (Oxford University Press, 2012) 119 ff; Id, '"Disappearing States", Statelesness and the Boundaries of International Law', in Id, *Climate Change and Displacement: Multidisciplinary Perspectives* (Hart, 2010) 105 ff; Walter Kälin, 'Conceptualising Climate Change-Induced Displacement', ibid, 81 ff. Further see the contribution by Cameron Moore in this volume.

lines, the specific protection of Stateless persons has come to have mostly an expressive function, helping to set protection for certain categories on a political agenda, rather than a strictly legal function based on positive norms.

2.3. Migrant workers?

Every State has the duty to protect all persons within its jurisdiction, irrespective of nationality. Apart from the right to return to one's country and electoral rights, foreigners and Stateless persons have the same rights as nationals of the State in which they are situated.[32] However, the adoption of instruments for the general protection of human rights has not prevented a growing demand for group-specific protection of migrant workers, which – paradoxically – has been met with hostility from many States. The International Convention on the Protection of the Rights of All Migrant Workers and Members of Their Families, adopted on 18 December 1990, did not come into force until 2003, and only includes around 50 State parties.

For the most, the Migrant Workers Convention does not define new rights (although it clarifies a few), but it reaffirms the rights of migrants embedded in general human rights instruments. It includes in particular a list of fundamental rights that must be guaranteed for migrant workers lacking documentation or who are in an undocumented situation. While the universal and irrevocable nature of rights leaves no doubt as to their application to undocumented migrants, such provisions are extremely controversial, and explain the reluctance of many States to ratify the Migrant Workers Convention. However, despite its lack of ratification, the Migrant Workers Convention – just like the Convention relating to the Status of Stateless Persons – provides a significant basis for domestic rights advocacy in favour of migrant workers. It participates in a pedagogical endeavour of advancing the effective implementation of general international human rights law with regard to a specific category of individuals that States tend not to protect effectively.

Environmental factors may incentivise displacement either through affecting living conditions at the place of origin ('push factor', for instance in the case of a drought) or through providing new opportunities at the place of destination ('pull factor', for instance when warming allows agricultural development). Some environmental or 'climate' migrants may not qualify as migrant workers or members of their families, most obviously isolated individuals who do not work (such as kids travelling alone). Yet, the group-specific protection of migrant workers is likely to be numerically much more relevant to these migrants than the protection of refugees and Stateless persons.

2.4. Internally displaced persons?

The general protection of human rights applies to persons who move within their State. A trend has developed, particularly since the end of the Cold War, to

32 See Human Rights Committee, *The Position of Aliens under the Covenant*, General Comment 15 (1986).

promote the group-specific protection of IDPs. It has emerged that many States have been unwilling or unable to provide sufficient protection to IDPs, in particular with regard to the specific protection needs of these populations.

Unlike refugees, IDPs remain on the territory of States affected by the cause of departure, whose resources may already be strained by the cause of displacement (such as a conflict or a natural disaster). Large-scale internal displacement frequently results in situations of vulnerability, in particular because of a lack of capacity of the State to provide protection.

Advocacy in the 1990s called for specific efforts to effectively protect the rights of IDPs. Like any group-specific protection, however, the risk was that attention and resources could be arbitrarily diverted from other issues – here, most evidently, from the protection of non-displaced individuals. As James Hathaway argued in 1996, 'we ought not [to] privilege those who are displaced, effectively doing a disservice to those who are trapped in their own homes, and we ought simply to get about the business of enforcing international human rights law internally.'[33] The enhancement of the protection of displaced persons may be part of a humanitarian dynamic in response to perceived suffering: the media in particular find it more straightforward to portray the suffering of displaced persons than that of non-displaced people, although their suffering may be equally or more severe. This illustrates an important caveat for group-specific protection, which may be based on arbitrary distinctions and, by diverting attention from those most in need, may result in an inadequate allocation of limited protection resources.

The protection of IDPs developed in particular from the 1990s in two parallel ways: an effort at reaffirming the primary obligation of each State to protect the population within its jurisdiction on the one hand, and an attempt at upgrading assistance from third States on the other hand. Consultations were undertaken by Francis Deng, on behalf of the Secretary-General of the UN and with significant material support from the Brookings Institute, leading to the adoption of the Guiding Principles on Internal Displacement in 1998. Many countries have incorporated these principles into national law or have drawn considerable inspiration from them, within a dynamic that contributed to an increased domestic and international focus on the vulnerability of IDPs. The 2009 Convention on the Protection and Assistance of Internally Displaced Persons in Africa, which came into force in 2012, is also inspired by the Guiding Principles. On the other hand, efforts were made to promote international solidarity through the creation of the UN Office for the Coordination of Humanitarian Affairs in 1991 and a better definition of the responsibilities of humanitarian organisations (cluster approach) in 2005, as well as the progressive development of the 'responsibility to protect' doctrine.

The objective of a group-specific protection of IDPs is not to recognise new rights, but to reaffirm existing rights and to create a momentum for their effective

33 James Hathaway, 'Discussion', in 'Are International Institutions Doing their Job?' (1996) 90 *Proceedings of the Annual Meeting of the American Society of International Law* 558, 562.

protection. The Guiding Principles do not constitute a formally binding document. They do not create new rights, but solely 'codify' the rights already recognised to IDPs – in particular in international human rights law, as well as international humanitarian law and, by analogy, international refugee law. The IDPs – like anyone else – have a right to life,[34] to physical, mental and moral integrity,[35] to liberty and security of the person,[36] and to an adequate standard of living,[37] among others. The definition of an IDP in the Guiding Principles, which include 'natural and human-made disasters'[38] among the causes of departure, should not be approached as a restrictive definition, but rather as an illustrative one. The persons displaced in relation to slow-onset environmental changes, if they are in a similar situation of vulnerability, should, under international human rights law, be protected in the same way as IDPs. Yet, beyond promoting international assistance, the Guiding Principles have done little to address the issue of a lack of capacity of the State concerned, on the territory of which large movements of populations occur, to protect displaced persons. When a State is severely affected by the adverse impacts of climate change, in particular by slow-onset environmental changes, its protection capacities are likely to decrease (for instance, diminished fiscal revenues) while the protection needs of its population increase.

3. Opportunities and risks of a new group-specific protection

In light of the current international protection of human rights and the experience of the four group-specific protection mechanisms discussed above, what are the risks and opportunities associated with group-specific protection of environmental or 'climate' migrants?

3.1. Approaching migration through green lenses

The rationale for group-specific protection to the benefit of environmental or 'climate' migrants may relate to sheer political expediency. The four examples of group-specific protection presented in the previous section show that such protective measures are only implemented in very specific political circumstances, in particular when an international crisis paves the way for structural changes in global governance. Thus, the first steps for the international protection of refugees followed the Great War, and the systematic protection provided by the Refugee Convention appeared necessary and possible in the post-Second World War reconstruction era. The protection of Stateless persons broadly followed the same

34 Principle 10.
35 Principle 11.
36 Principle 12.
37 Principle 18.
38 Introduction, [2].

dynamic. Likewise, the specific protection of migrant workers and IDPs was developed in the context of a growing influence of the third world and of the new beliefs in global governance that accompanied the dissolution of the USSR. The needs for governance were not new, but new circumstances created windows for policy change. In turn, climate change creates a political momentum for innovation – an opportunity for important and much-needed reforms that address pre-existing needs. Linking migration to climate change might constitute an opportunity to bring the protection of migrants to the forefront of international political debates.

The appearance of the migration issue within climate change negotiations provides a taste of the political expediency for the protection of environmental or 'climate' migrants. The Cancun Agreements adopted by the 16th COP of the parties to the UNFCCC have thus called for 'measures to enhance understanding, coordination and cooperation with regard to climate change-induced displacement, migration and planned relocation, where appropriate, at the national, regional and international levels.'[39] Two years later, COP 18 cited 'human mobility' in the definition of a research programme on the approaches to address loss and damage associated with climate change impacts in developing countries that are particularly vulnerable to the adverse effects of climate change.[40] These negotiations did not set specific obligations for the protection of the human rights of migrants, but they have certainly helped to draw attention to the plight of migrants in the context of climate change or, more broadly, to environmental changes.

Yet, even though a linkage between migrants' protection and negotiations on climate change may appear politically expedient, this approach is not without risk. A particular problem is that of a non-optimal allocation of resources for protection – a problem similar to the one highlighted by James Hathaway in the broader context of IDP protection. While the adverse impacts of climate change and other environmental phenomena may exacerbate the vulnerabilities of certain populations, in particular migrants, it remains that environmental or 'climate' migrants are certainly not the sole or the most vulnerable population that needs protection. As a mechanism for coping with environmental phenomena, in principle, migration enables migrants to reduce their risks. The persons who are unable to migrate are by definition more vulnerable than those who are able to cope through migration. Politically, stationary populations are less visible and more easily forgotten than large populations of 'refugees' gathered in makeshift camps, and also have less access to organisations that provide protection and assistance. The protection of stationary populations is often a greater challenge both logistically and financially.

The resources of any State are limited and environmental phenomena that put vast populations at risk have the concomitant effect of reducing the protection capacity of the affected States (in particular by means of a reduction of fiscal

39 UNFCCC, *Cancun Agreements*, Decision 1/CP.16 (2010) [14(f)].
40 UNFCCC Decision 3/CP.18 (2012) [7].

revenues). A State severely affected by environmental phenomena such as adverse impacts of climate change cannot commit to providing a better level of protection to its migrants without running the risk of diverting critical resources away from the protection of non-displaced populations.

3.2. Environmental or 'climate' migrants and their others

Any legal category draws a circle in the sand: some people are within the circle, others are outside. Group-specific protection of environmental or 'climate' migrants would exclude both non-environmental migrants (or environmental migrants who are not 'climate' migrants) and persons affected by environmental phenomena who do not move, possibly for lack of financial or social resources. Yet, these other populations may be more vulnerable than environmental or 'climate' migrants.

In principle, a group-specific protection mechanism should address specific protection needs. Yet, there seem to be no specific protection needs attached to the vague category of environmental migrants, and, *a fortiori*, 'climate' migrants, given in particular the many different ways in which environmental phenomena may cause migration. At most, it can be argued that certain *subgroups* of environmental or 'climate' migrants have specific needs for protection. In particular, in the eventuality where an island State ceases to exist and needs to be evacuated, its population would be in a unique situation of Statelessness: it would then be important to define new obligations for the protection of this population and to enable the naturalisation of such a population, or to establish a mechanism for the maintenance of a specific collective identity that would guarantee respect for individual rights. Similarly, but in a less clear-cut manner, it would probably be appropriate to allow the international migration of persons living in a State that has gone beyond its carrying capacity and become subject to major environmental changes, even if that State does not cease to exist. However, in both cases, given the specificity of such situations and the small number of people concerned (the populations of small island States, in particular, only amount to tens of thousands, rarely hundreds of thousands), one may question the value of any abstract legal protection regime, rather than *ad hoc* political resettlement agreements.

More fundamentally, instead of looking at climate change as the cause of migration, a more coherent endeavour would consist of protecting migrants who share the same type of vulnerability. Alexander Betts has formulated a proposal to this effect, involving the meeting of protection needs relating to 'survival migration', which is a broad concept that may also extend, for instance, to those displaced because of extreme economic deprivation or violence, on the sole condition that migration appears necessary for the survival of individuals concerned.[41] A similar approach, but extended to internal migration, has been

41 Betts (2013).

developed through the concept of 'crisis migration'.[42] Broader perspectives would nonetheless protect the human rights of all migrants, without attempting to draw a line between migrants 'deserving' or 'undeserving' specific protection.

Yet, politically, these proposals face significant resistance. While protecting *all* 'survival' or 'crisis migrants' would require a paradigm change in international migration governance, incremental changes in the existing paradigm could allow a group-specific protection of environmental or 'climate' migrants. In other words, from a medium-term perspective, protecting environmental or 'climate' migrants is a more realistic project. One should, however, wonder whether, in the long term, this incremental change would facilitate a structural change – a protection of all migrants, or of all individuals affected by environmental phenomena – or whether it is more likely to delay much-needed structural changes.

In a pessimistic hypothesis, the debate on environmental migration may contribute to a hardening of the positions on migration and create an additional hurdle to the protection of other migrants. Accordingly, alarmist talk focusing on the security consequences of mass migration in the context of environmental change could exacerbate concerns about migration and general xenophobia in many countries, thus preventing the successful integration of migrants. Because it is difficult to attribute an individual decision to migrate to an environmental factor in isolation from social, political, economic, cultural and demographic dynamics, and States would probably adopt a very narrow definition of environmental or, *a fortiori*, 'climate' migrants, protection of these migrants is likely to be the exception that proves the rule – a legal development that would perpetuate the dichotomy between those migrants who 'deserve' international protection, and the rest. In other words, there is a significant risk that the specific protection of certain narrow categories of migrants may eclipse the application of the protection of the human rights of migrants in general.

Alternatively, an optimistic scenario is that the protection of environmental or 'climate' migrants would constitute a first step towards a more systemic protection of all migrants. At best, it would trigger a progressive extension of mechanisms for the *ad hoc* protection of certain categories of forced migrants (refugees as well as environmental or 'climate' migrants) to all migrants, regardless of the cause of migration. At least, recognising environmental or 'climate' migrants could contribute to a renewal of currently entrenched political debates on the role of migration and migrants in our societies, where migration is still too often seen as a social anomaly rather than a normal phenomenon occurring in any human society.

There is no clear ground to think that either scenario is more likely to materialise. However, in terms of research and advocacy, a better understanding not only of the opportunities but also of the risks associated with a group-specific protection of environmental or 'climate' migrants is required. While arguments

42 Martin, Weerasinghe and Taylor (2014).

on the protection of these migrants may play a useful role in raising awareness of larger causes, they must be advanced carefully. Arguments based on the fears of an invasion of 'climate refugees', in particular, are unlikely to contribute to emancipatory causes.

Conclusion

Environmental or 'climate' migrants are naturally human beings protected by general instruments such as the UDHR, the ICESCR and the ICCPR. Some of them may also, to a certain extent, be covered by existing group-specific protection mechanisms, in particular as migrant workers and IDPs. Proposals have recently emerged for a group-specific protection. Yet, when approaching such proposals, one needs to ask how group-specific protection would impact the protection of other migrants and of those who, being affected by adverse environmental phenomena, are unable to adapt or cope through migration.

Every migrant and every person affected by environmental phenomena – and, in fact, every human being – deserves his or her rights to be protected. Human rights concerns raised in relation to migration intensify public scrutiny of existing tensions and contradictions between the international management of migration and the international human rights movement, shedding new light on questions that have been evaded for too long. This debate is an opportunity to raise awareness and to demand the advancement of the protection of the human rights of migrants more generally, via a reconsideration of the role of migration in all human societies. In this context, group-specific protection of environmental or 'climate' migrants runs the risk of being only a palliative measure delaying much-needed responses to structural deficiencies of the international governance of migration.

11 Balancing human rights, climate change and foreign investment protection

Valentina Vadi

Introduction

Originally perceived as an environmental issue only, climate change – meant as any change in climate over time, whether due to natural factors or as a result of human activity – is now deemed to affect a number of human rights.[1] In fact, climate change can affect diverse determinants of human well-being such as access to water, energy supplies and public health, and can cause social disruptions such as migration due to drought or rising sea levels and loss of traditional livestock and habitat.[2] This has prompted the adoption of climate-friendly measures within the key regulatory area of international investment protection. However, these measures can (and have) affect(ed) a range of investors' rights.[3]

Against such a background, this chapter examines the specific question as to whether regulatory measures relating to climate change can infringe investors' rights, including property rights and the fair and equitable treatment standard. In the end, the chapter argues that a balance needs to be struck between different human rights: while the right to property is a classical first generation fundamental right, environmental claims are considered third generation human rights. This area is largely unexplored and in need of systematisation. In particular, no study has yet focused on the burgeoning arbitrations in which foreign investors have alleged breaches of relevant investment treaty provisions by States adopting or repealing climate change-related regulation.

The chapter will proceed as follows. First, it explores critical legal issues raised by the complex interplay between climate change, foreign direct investments and human rights in light of international investment law. Second, some relevant investor–State arbitrations will be examined in which climate change-related measures clash with investors' rights. These arbitrations raise a number of critical

1 OHCHR, *Report on the Relationship between Climate Change and Human Rights*, UN Doc A/HRC/10/61 (15 January 2009).

2 See Sumudu Atapattu, 'Climate Change, Differentiated Responsibilities and State Responsibility: Devising Novel Legal Strategies Caused by Climate Change', in Benjamin J Richardson, Yves Le Bouthillier, Heather McLeod-Kilmurray and Stepan Wood (eds), *Climate Law and Developing Countries* (Edward Elgar, 2009) 37.

3 See generally Bradley Condon and Tapen Sinha, *The Role of Climate Change in Global Economic Governance* (Oxford University Press, 2013).

issues because foreign investors have articulated their claims, inter alia, in terms of human rights. This study aims to feed into and inform the ongoing debate on the role of foreign direct investments in the climate change discourse. The chapter concludes that a comprehensive regulatory change is necessary in order to foster beneficial economic growth, prevent climate change and ensure respect for human rights.

1. Climate change and human rights in investment governance

The interplay between foreign direct investment and climate change on the one hand, and the consequences of this interplay on human rights on the other has not been fully investigated yet, despite the evident connection between these phenomena. On the one hand, corporations are the main GHG emitters,[4] and GHG emissions contribute to climate change, which, in turn, can affect a range of human rights. On the other hand, private actors and especially multinational corporations can play an important role in 'greening' the economy. Climate-friendly investments benefit fundamental rights at large, because they can reduce GHGs. Supporting investment in low-carbon technology in capital-intensive sectors such as energy and infrastructure can contribute to fighting climate change. Reducing emissions through investing in energy efficiency improvement and clean energy is supposed to limit global warming, bring significant economic benefits, and contribute to respecting, protecting and fulfilling human rights.[5]

Scholars suggest that the investment regulatory framework is not prepared to cope with climate change.[6] However, in principle, human rights obligations should compel States to include climate change issues in investment treaties, at least within the limits outlined by the UNFCCC regime and the Kyoto Protocol.[7] In this respect, a human rights interpretation of the concept of sustainable development would be particularly useful. Indeed, although most investment treaties do not include references to environmental concerns, since they tend to be short texts, in recent years international investment law has gone through a period of re-balancing aimed at aligning investment protection with other policy objectives.[8]

From the perspective of international climate change law, the UNFCCC[9] aims at 'control[ling] greenhouse gas emission,'[10] enabling 'economic development

4 John von Doussa, Allison Corkery and Renée Chartres, 'Human Rights and Climate Change' (2007) 14 *Austl Int'l LJ* 161, 170.

5 Elisa Morgera and Annalisa Savaresi, 'A Conceptual and Legal Perspective on the Green Economy' (2013) 22 *RECIEL* 14.

6 Philipp Aerni and Others, 'Climate Change and International Law: Exploring the Linkages between Human Rights, Environment, Trade and Investment' (2010) 53 *Ger Yb Int'l L* 139, 183–6.

7 See, more extensively, the considerations developed in the first part of this volume.

8 José E Alvarez, 'The Return of the State' (2011) 20 *Minn J Int'l L* 223.

9 *United Nations Framework Convention on Climate Change (UNFCCC)*, adopted in New York on 9 May 1992, opened for signature on 20 June 1992, and entered into force on 21 March 1994, 31 ILM 848.

10 UNFCCC, Preamble.

... in a sustainable manner'.[11] While the UNFCCC acknowledges that 'human activities have been substantially increasing the atmospheric concentration of GHGs, thereby enhancing the natural greenhouse effect',[12] it does not address individuals; rather, like other international treaties, it encourages States to stabilise GHG reduction. This does not mean that corporations do not have any role to play. According to the UN, 'nearly 90 per cent of the funds needed to address global warming will derive from the private sector'.[13]

The Kyoto Protocol[14] legally binds States to a reduction of their GHG emissions and calls for the involvement of private entities, such as foreign investors, to pursue its targets. Private actors can play an important role in helping States limit their emissions under two of the Kyoto Flexibility Mechanisms.[15] In particular, under the joint implementation mechanism, industrialised countries can authorise legal entities, such as private investors, to participate in green investments in other industrialised countries to meet the home country's Kyoto targets. Under the clean development mechanism (CDM), industrialised countries and their investors can participate in projects in developing countries, thus contributing to the home country's Kyoto targets. Industrialised countries obtain certified emission reductions (CERs) credits that can be used to meet their targets or sold on the carbon market, and developing countries receive low carbon technology and financial flows that contribute to their sustainable growth. Foreign investment can thus represent an effective tool to address climate change.[16]

As the international economic order has become increasingly intertwined with concerns for climate change,[17] the international investment regime has begun to play a role in governing clean energy strategies, fostering climate change mitigation and sustainable development.[18] As is well known, there is no comprehensive multilateral framework governing foreign direct investments. Rather, more than

11 UNFCCC, Article 2.

12 UNFCCC, Preamble.

13 *Private Investment Key in Tackling Climate Change, Says Assembly President*, <www.un.org/apps/news/printnewsAr.asp?nid=26954>.

14 *Kyoto Protocol to the UNFCCC*, adopted in Kyoto on 11 December 1997, opened for signature on 16 March 1998, and entered into force on 16 February 2005, 37 ILM 22.

15 For a detailed analysis of the mechanisms, see Freya Baetens, *Foreign Investment Law and Climate Change* (International Development Law Organisation, Sustainable Development Law on Climate Change, Legal Working Paper No 1, 2010) 6–9.

16 Abbe Brown (ed), *Environmental Technologies, Intellectual Property and Climate Change: Accessing, Obtaining and Protecting* (Edward Elgar, 2013) 128–51.

17 On the linkage between international trade law and international climate law, through, inter alia, the interpretation of Article XX GATT, see, for instance, Patrick Messerlin, 'Climate and Trade Policies: from Mutual Destruction to Mutual Support' (2012) 11 *World Trade Review* 53, 77; Cinnamon Carlarne, 'The Kyoto Protocol and the WTO: Reconciling Tensions Between Free Trade and Environmental Objectives' (2006) 17 *Colorado J International Environmental Law & Policy* 45. See also De Schutter in this volume.

18 Edna Sussman, 'The Energy Charter Treaty's Investor Protection Provisions: Potential to Foster Solutions to Global Warming and Promote Sustainable Development' (2008) 14(2) *ILSA Journal of International & Comparative Law* 391.

3,000 bilateral investment treaties (BITs) regulate this vital area of international law. This does not necessarily entail that the system is fragmented; instead, scholars have suggested that a *de facto* multilateral evolution of the system has taken place.[19] In fact, many BITs share common and/or similar provisions, and arbitral tribunals do tend to refer to earlier awards, despite the absence of *stare decisis* in international investment law. The Energy Charter Treaty (ECT)[20] is also an example of a multilateral agreement governing energy investment and trade.

Both BITs and the ECT provide a number of substantive standards of protection, including the fair and equitable treatment standard and the prohibition of unlawful expropriation. At the procedural level, investors are allowed to file arbitration claims directly against host States for violations of these standards. Investor–State arbitrations can be *ad hoc* or institutionalised. In the latter case, litigants can refer disputes to a variety of institutional settings, including the International Centre for the Settlement of Investment Disputes (ICSID), the Permanent Court of Arbitration (PCA), and the Arbitration Institute of the Stockholm Chamber of Commerce (SCC).

Recent investment treaties have expressly included references to climate change in their preambles,[21] or have included environmental measures in carve-outs. Embedding common concerns such as climate change and/or referring to multilateral environmental agreements in the preambles of international investment agreements is a welcome new approach that can foster cross-pollination of ideas and an increased coherence between different branches of international law. The same is true of carve-outs. General carve-outs clarify that *bona fide* regulation applied to environmental protection does not amount to indirect expropriation. Specific provisions exclude environmental measures from the scope of conventional dispute settlement mechanisms.[22] The ECT adopts a middle way, as it refers to climate change in its preamble and other provisions.[23]

However, given that renegotiating investment treaties is a lengthy process, some scholars and practitioners have proposed the adoption of multilateral declarations to help enhance coherence between international investment law and the climate change regime.[24] According to these authors, multilateral declarations would

19 Stephan Schill, *The Multilateralisation of International Investment Law* (Cambridge University Press, 2009).

20 *Energy Charter Treaty*, signed on 17 December 1994, in force on 16 April 1998, 34 ILM 360, 385 (1995).

21 See, for instance, the preamble to the Japan–Switzerland FTA, where the parties reaffirm their determination 'in implementing this Agreement, to seek to preserve and protect the environment, to promote the optimal use of natural resources in accordance with the objective of sustainable development and to adequately address the challenges of climate change'.

22 Recent Belgian BITs exemplify this trend. See, for instance, the *Belgium–Barbados BIT*, signed on 29 May 2009, Article 11(3).

23 ECT, Preamble and Article 19(3)(b).

24 See Wolfgang Alschner and Elisabeth Tuerk, 'The Role of International Investment Agreements in Fostering Sustainable Development', in Freya Baetens (ed), *Investment Law within International Law – Integrationist Perspectives* (Cambridge University Press, 2013) 217.

clarify that international investment treaties do not constrain climate change measures enacted in good faith. As is well known, the World Trade Organisation adopted a similar declaration – the Doha Declaration on the TRIPS Agreement and Public Health – to clarify the interplay between the protection of intellectual property, including pharmaceutical patents and public health.[25] Yet this has not prevented States from starting complaints on different aspects of that interplay. Therefore, it may be more appropriate for States to issue binding interpretations, along the lines of what has been done in the context of the NAFTA with regard to the reading of the fair and equitable treatment clause.[26]

Do investment treaties chill and/or impede endeavours to mitigate the effects of climate change?[27] Some scholars have pointed out that international investment treaties can foster investments in renewable energy, highlighting that these treaties 'should not lead to a chill on environmental regulation nor obstruct measures that are introduced in an attempt to mitigate climate change'.[28] Yet, other scholars have expressed concerns that foreign companies can file investor–State arbitrations, contending that climate change-related regulatory measures breach relevant investment treaty provisions.[29] In fact, the adoption (or non-adoption) of climate change-related regulatory measures can affect the economic interests of private actors, by requiring (or not requiring) technological upgrades and specific economic behaviour.[30] Where the affected economic activities are owned by foreign investors, such measures may give rise to investor–State arbitrations under the relevant treaties.[31] Foreign investors can argue that these measures violate investment provisions, in particular, the prohibition on unlawful expropriation, and the fair and equitable treatment standard.[32]

Therefore, on the one hand, foreign investment can be an effective tool against climate change. Indeed, foreign investors can, and do, transfer technology and invest in renewable energy projects. On the other, despite the potential for synergy between foreign investment and environmental policies addressing climate change, foreign investment in carbon-intensive sectors remains common. Carbon-free investments, for instance, in nuclear energy, hydropower dams and shale gas are controversial for various reasons.[33] Therefore, repealing regulation governing

25 *Declaration on the TRIPs Agreement and Public Health*, WT/MN(01)/DEC2, adopted on 14 November 2001 at the Doha Ministerial Conference of the WTO.

26 See Anthea Roberts, 'Power and Persuasion in Investment Treaty Interpretation: the Dual Role of States' (2010) 104 *Am J Int'l L* 179.

27 See, for instance, Stephan Schill, 'Do Investment Treaties Chill Unilateral State Regulation to Mitigate Climate Change?' (2007) 24(5) *Journal Int'l Arb* 469.

28 Ibid, 477.

29 Kate Miles, *The Origins of International Investment Law* (Cambridge University Press, 2013) 187.

30 See, for instance, at the EU level, *Arcelor SA v European Parliament and Council*, Case T-16/04 (ECJ, 2 March 2010).

31 Schill (2007) 470.

32 Ibid, 471–6.

33 See, for instance, Valentina Vadi, 'Foreign Investment in the Energy Sector and Public Health', in Eric De Brabandere and Tarcisio Gazzini (eds), *Foreign Investment in the Energy Sector: Balancing Private and Public Interests* (Brill, 2014) 238.

investment in carbon-neutral technology may impinge upon the rights of those who have invested in these technologies.

2. Climate change and human rights in investment arbitration

Until recently, foreign investors had not challenged climate change-related measures before arbitral tribunals.[34] This was in stark contrast with a number of national, regional and even international cases concerning the interplay between the regulatory autonomy of the State with regard to environmental matters and other international law norms.

Currently, a number of arbitrations are emerging concerning regulatory measures relating to climate change and human rights. This is not to say that investment treaty tribunals are the best venues for this type of disputes, let alone the only venue. Rather, investment treaty arbitration has become the last frontier for these controversies. As these disputes have been filed only recently, there is no relevant available literature on these matters yet, and consequently, this chapter necessarily has an explorative character. Moreover, most disputes are pending at the time of writing, and thus it is not possible to foresee whether these disputes will be settled by the parties and, if so, on what terms, or whether and, if so, how they will be adjudicated by the relevant arbitral tribunals. Nonetheless, a brief overview of the relevant issues that have arisen in these emerging disputes is both timely and appropriate.

Investors have claimed that by adopting or repealing climate change-related regulatory measures the host State has affected their investments, inter alia in breach of the fair and equitable treatment standard. On the one hand, investors in carbon-polluting activities have threatened to file investor–State arbitrations, deeming Kyoto-related measures a violation of the host State conventional investment obligations. On the other hand, investors in clean technology have challenged climate-related regulatory changes allegedly affecting their investments. Therefore, arbitrators will have to balance human rights considerations, in particular environmental protection, and investors' rights in assessing the validity of climate change-related regulatory measures in the field of foreign investment.[35]

2.1. Clean investment regulation

A first type of investment dispute is brought by investors in polluting activities. Foreign investors have argued that Kyoto-related measures violate host State

34 Rather, investors used to challenge other types of environmental measures. See, for instance, Rahim Moloo and Justin Jacinto, 'Environmental and Health Regulation: Assessing Liability under Investment Treaties' (2011) 29 *Berkeley J Int'l Inv L* 1.

35 Anatole Boute, 'The Potential Contribution of International Investment Protection Law to Combat Climate Change' (2009) 27 *J En & Nat Res L* 333, 342.

obligations under international investment treaties, including basic standards such as non-discrimination and the prohibition on unlawful expropriation. Reportedly, investors have threatened to file investor–State arbitration claims against climate change regulation not including compensation mechanisms for alleged losses in reducing carbon emissions.[36] In addition, foreign companies could contend that moratoria on polluting activities amount to a form of indirect expropriation and require compensation. For instance, an oil and gas firm has filed an investment arbitration claim against Canada over a moratorium on controversial drilling techniques ('fracking') in Quebec.[37] Although the prohibition of fracking is not a climate change-related measure, similar cases may arise with other moratoria related to climate change. Furthermore, foreign investors outside the Kyoto system could bring discrimination claims against host States' regulatory measures promoting green investment under the Kyoto system.[38] Even if this seems a remote hypothesis in light of the almost universal ratification of the UNFCCC and the Kyoto Protocol, it cannot be excluded that incentives offered by a State to renewable energy projects are perceived as discriminatory against carbon-intensive business.

It has been persuasively argued that the police power doctrine should apply to these claims.[39] According to this doctrine, general regulation adopted *bona fide* and in a non-discriminatory manner to protect public health or safety or to prevent a public nuisance, does not amount to expropriation and cannot be compensated.[40] In fact, States do have the right – and, some would argue, the duty – to restrict private property to prevent a public nuisance.[41] Few would deny that climate change is a severe type of nuisance and that States have the duty to prevent and/or mitigate its effects. The legitimate purpose of climate change-related measures can also be inferred from the fact that they are based on scientific evidence as recognised by several international organisations.[42] Express references to environmental protection and climate change in investment treaties would further support this argument.

In this respect, a critical issue will be distinguishing the climate-friendly and carbon-intensive projects in light of the host State's Kyoto commitments. Arguably,

36 Luke Eric Peterson, 'Coal-Fired Power Plant Investors Reportedly Threaten Australia with BIT Claims', *Investment Arbitration Reporter* (28 December 2009) <www.iareporter.com/articles/20091228_3>.

37 *Lone Pine Resources Inc v The Government of Canada*, Notice of Intent to Submit a Claim to Arbitration under Chapter 11 NAFTA (UNCITRAL, 8 November 2012).

38 Baetens (2010) 10.

39 Schill (2007) 472.

40 See, for instance, Article 18 ECT.

41 The ECHR system also allows for legitimate restrictions to property rights. See *Jahn and Others v Germany*, Applications Nos 46720/99, 72203/01 and 72552/01 (Grand Chamber, Judgment, 30 June 2005).

42 On the role of science in international investment law and arbitration, see Valentina Vadi, *Public Health in International Investment Law and Arbitration* (Routledge, 2012) 153–5.

the Kyoto requirements constitute a legitimate ground for distinguishing different economic activities.[43] Carbon-intensive investments and climate-friendly economic activities are not 'like investments' because they have different impacts on climate change. Therefore, the host State will be able to defend its regulatory measures on the ground that no discrimination is at stake, but rather a legitimate distinction between economic activities which have different impacts on climate change.

Other claims could be raised against climate-friendly investment measures in relation to stabilisation clauses embedded in contracts between host States and foreign investors. Stabilisation clauses aim to isolate investment from adverse regulatory changes. While stabilisation clauses take different forms, they aim at immunising investment from political risks, freezing the applicable investment law at the time when the parties entered into a contract.[44]

Finally, foreign investors could contend that changes in the regulatory framework of the host State amount to a violation of the fair and equitable treatment standard. Since a number of arbitral tribunals have interpreted the fair and equitable treatment standard extensively so as to include the obligation on the part of the State to protect an investor's legitimate expectations and to provide a stable legal environment, it is possible that investors claim that climate-friendly regulatory measures violate the fair and equitable treatment standard. For instance, in an ECT arbitration, a company won a case against the host State for a change of government policy that altered an incentive system for green investment.[45]

However, although the parameters of the fair and equitable treatment standard are particularly vague, this standard certainly does not protect foreign investors against any regulatory changes. Authors have cautioned that public welfare regulation may be accommodated through appropriate drafting.[46] Protected legitimate expectations stem from specific statements by the relevant State authorities, or can arise from the host State regulatory framework in the event that the State in question has induced a given investor's confidence on the stability of the legal framework. It is difficult to argue that climate change mitigation measures are not foreseeable, since climate change has made headlines and has come to the forefront of the legal debate. Furthermore, as Schill points out, 'the protection of legitimate expectations does not make the domestic legal framework unchangeable or subject any change to a compensation requirement'.[47] In this regard, the *Saluka* Tribunal held that ascertaining whether fair and equitable treatment is breached requires 'a weighting of the Claimant's legitimate and

43 For analogous considerations in the light of the World Heritage Convention, see *Parkerings Compagniet AS v Republic of Lithuania*, Award, Case No ARB/05/8 (ICSID, 11 September 2007).

44 Lorenzo Cotula, 'Reconciling Regulatory Stability and Evolution of Environmental Standards in Investment Contracts: towards a Rethink of Stabilization Clauses' (2008) 1 *J W En L Bus* 158.

45 *Nykomb Synergistics Tech Holding AB v The Republic of Latvia*, Case No 118/2001 (Arb Inst of the SCC, 16 December 2003).

46 Cotula (2008) 158.

47 Anthony Crockett, 'Stabilisation Clauses and Sustainable Development: Drafting for the Future', in Chester Brown and Kate Miles (eds), *Evolution in Investment Treaty Law and Arbitration* (Cambridge University Press, 2011) 516; Schill (2007) 476.

reasonable expectations on the one hand, and the Respondent's legitimate regulatory interests on the other'.[48]

2.2. Repealing climate-related investment regulation

In order to increase the production of clean energy, a number of countries, including Spain, the Czech Republic, Italy and Bulgaria, have adopted incentives to attract investments in the renewable energy sector. Among these incentives was a 'feed-in tariff' (FIT), that is, a guaranteed electricity purchase price set higher than market rates. Reportedly, many investments were made relying on the duration of these incentives. After the advent of the global financial crisis, however, host countries began to repeal these incentives, including by means of the reduction of FITs.[49] In a number of arbitrations, foreign investors are contending that these regulatory changes amount to a violation of relevant investment treaty obligations, in particular, protection against unlawful expropriation, the principle of fair and equitable treatment and stabilisation clauses.

For instance, Bulgaria recently imposed a 20 per cent tax on the income of solar energy producers, many of which are foreign-owned, and considered changing its renewable energy policies by cutting by 50 per cent the preferential rates the country is obliged to pay for electricity generated by wind and solar power plants. As a consequence, an industry association declared that multinational companies would be 'compelled to protect their rights before an international arbitral tribunal'.[50] In parallel, reportedly, more than 50 solar companies have lodged a complaint with Bulgaria at the ECtHR, alleging violations of the right to property in breach of Article 1 Protocol 1 to the ECHR.[51]

Spain is facing a steadily lengthening number of investment treaty claims in relation to its reductions to incentives previously offered to investors in renewable energy production.[52] Reportedly, these changes were introduced to cope with 'an electricity tariff deficit in the country's energy market that the International Monetary Fund and the European Commission have considered a significant drag on the Spanish economy'.[53] For instance, in *InfraRed Environmental Infrastructure GP Limited and Others v Spain*[54] the claimant – a UK-based investment fund which had acquired equity participation in solar projects in Spain – alleges that reforms

48 *Saluka Investments BV v Czech Republic*, Partial Award (UNCITRAL, 17 March 2006) [306].

49 Joseph Tirado and Jerry R Bloom, 'Renewable Energy Reforms in Europe: Growing Threats to International Investors' (2014) *Lexology*, <www.lexology.com/library/detail.aspx?g=ab40069a-7f6a-4c84–9291–3bda41429af4> 1.

50 Jarrod Hepburn, 'Bulgaria May Face BIT and Human Rights Claims over Renewable Energy Measures' (4 June 2014) *IA Reporter* <www.iareporter.com/articles/20140604_2>.

51 Ibid.

52 Tirado and Bloom (2014).

53 Kyriaki Karadelis, 'Spain Faces More Claims from Renewable Investors' (29 November 2013) *Global Arbitration Review* <http://globalarbitrationreview.com>.

54 *InfraRed Environmental Infrastructure GP Limited and Others v Spain*, Case No ABR/14/12 (ICSID, registered on 3 June 2014).

affecting the renewable energy sector are in breach of several legal provisions, including the fair and equitable treatment standard under Article 10 ECT.[55] A number of companies have brought analogous cases against Spain before the ICSID,[56] the Arbitration Institute of the Stockholm Chamber of Commerce,[57] or *ad hoc* arbitral tribunals pursuant to the UNCITRAL rules.[58]

Other Member States of the European Union are facing similar investment treaty claims. On 21 February 2014, ICSID registered the first known claim filed against Italy for alleged violations of the ECT.[59] The claimants are investors in a photovoltaic energy generation project. The Czech Republic is also facing several claims in relation to repealing its favourable treatment of solar-generated energy.[60]

Analogous investment disputes are arising under Chapter 11 of NAFTA.[61] In *Mesa Power v Canada*,[62] a Texas-based energy company brought an arbitral claim against Canada in relation to the province of Ontario's 2009 *Green Energy Act*.[63]

55 Luke Eric Peterson, 'Another Week, Another Arbitration Claim against Spain: Infrared Environmental Infrastructure Sues at ICSID' (4 June 2014) *IA Reporter* <www.iareporter.com/articles/20140604_1>.

56 *NextEra Energy Global Holdings BV and NextEra Energy Spain Holdings BV v Kingdom of Spain*, Case No ARB/14/11 (ICSID, registered on 23 May 2014); *Masdar Solar & Wind Cooperatief UA v Kingdom of Spain*, Case No ARB/14/1 (ICSID, registered on 11 February 2014); *Eiser Infrastructure Limited and Energía Solar Luxembourg Sà Rl v Kingdom of Spain*, Case No ARB/13/36 (ICSID, registered on 23 December 2013); *Antin Infrastructure Services Luxembourg Sàrl and Antin Energia Termosolar BV v Kingdom of Spain*, Case No ARB/13/31 (ICSID, registered on 22 November 2013); *RREEF Infrastructure (GP) Limited and RREEF Pan-European Infrastructure Two Lux Sà rl v Kingdom of Spain*, Case No ARB/13/30 (ICSID, registered on 22 November 2013).

57 *Isolux Infrastructure Netherlands BV v Spain* (registered at the Arbitration Institute of the Stockholm Chamber of Commerce (SCC) in 2013); *Charanne (the Netherlands) and Construction Investments (Luxembourg) v Spain* (registered at the Arbitration Institute of the SCC in 2013); *CSP Equity Investment Sàrl v Spain* (registered at the Arbitration Institute of the SCC in 2013).

58 *The PV Investors v Spain* (ad hoc, UNCITRAL Arbitration Rules, registered in November 2011). See Luke Eric Peterson, 'Tribunals Finalized in UNCITRAL and SCC Claims Arising out of Solar-Power Controversies' (16 April 2014) *IA Reporter* <www.iareporter.com/articles/20140416>.

59 *Blusun SA, Jean-Pierre Lecorcier and Michael Stein v Italian Republic*, Case No ARB/14/3 (ICSID, registered on 21 February 2014).

60 *Antaris Solar and Dr Michael Göde v Czech Republic* (UNCITRAL administered by the PCA, registered in 2013); *Natland Investment Group NV, Natland Group Limited, GIHG Limited, and Radiance Energy Holding SARL v Czech Republic* (UNCITRAL *ad hoc*, registered on 8 May 2013); *Voltaic Network GmbH v Czech Republic* (UNCITRAL *ad hoc*, registered on 8 May 2013); *ICW Europe Investments Limited v Czech Republic* (UNCITRAL *ad hoc*, registered on 8 May 2013); *Photovoltaik Knopf Betriebs-GmbH v Czech Republic* (UNCITRAL *ad hoc*, registered on 8 May 2013); *WA Investments-Europa Nova Limited v Czech Republic* (UNCITRAL *ad hoc*, registered on 8 May 2013); *JSW Solar (zwei) v Czech Republic* (UNCITRAL *ad hoc*, registered in June 2013).

61 North American Free Trade Agreement, 32 ILM 289 (1993) (table of contents, preamble, parts I–III) and 32 ILM 605 (1993) (parts IV–VII, annexes).

62 *Mesa Power Group LLC v Government of Canada*, Notice of Intent to Submit a Claim to Arbitration (6 July 2011).

63 The Ontario FIT Programme was also the subject of a WTO dispute initiated by Japan against Canada: *Canada – Certain Measures Affecting the Renewable Energy Generation Sector, Canada – Measures Relating to the Feed-In Tariff Program*, Reports of the Appellate Body WT/DS412/AB/R; WT/DS426/AB/R (6 May 2013).

The company, which owns four wind farms in Ontario, contends that the province changed the rules by which renewable energy producers can obtain power purchase agreements, favouring other investors and thus breaching the fair and equitable treatment standard.

3. Challenges and prospects

From a procedural perspective, the general lack of transparency in investment treaty arbitration is of particular concern. Owing to the specific features of investment arbitration and the fact that none of the above-mentioned disputes have been settled yet, very little information is available. While the number of investment disputes relating to renewable energy is growing steadily as reported by the news, very little is known about the claims and arguments of the parties. A list of the relevant investor–State disputes related to energy – and thus also renewable energy – is compiled and updated regularly by the Energy Charter Secretariat.[64] However, 'there is no requirement that such disputes be notified to the Secretariat', nor is the Secretariat involved in the administration of investment disputes. Therefore, the information available on the Energy Charter's website 'relies on various public sources, and includes links to publicly available documents' but 'completeness cannot be guaranteed'.[65] The ICSID website lists all of the cases that are registered with the Centre,[66] yet the list provides very little information, generally mentioning the sector of the investor activity, the date of registration, and details about the constitution of the arbitral tribunals. Moreover, the ICSID website does not generally publish the notice of claim, let alone the statement of defence and subsequent documents submitted by the parties. If the parties so agree, ICSID publishes the award on jurisdiction and the final award. The parties, however, may also opt for confidentiality or request the redaction of only specific parts of the awards to protect personal data, business information and the like. The NAFTA is more transparent where this information is concerned.

More substantively, investment disputes show that climate policies may have a varied impact on different actors and it is worth exploring these implications. Generally, climate policies benefit the public at large because they reduce GHGs, which not only worsen people's quality of life but can also determine drought, famine and rising sea levels. At the same time they can affect the economic interests of corporations which need to invest in technological upgrades or even convert a given business into a more eco-friendly economic activity. However, in specific cases there may be mutual supportiveness between climate change-related measures, the economic interests of businesses and the human rights of local communities. This scenario is enhanced by the recent establishment of State incentives

64 Energy Charter, Investor–State Dispute Settlement Cases, available at <www.encharter.org/index.php?id=213&L=0#Isolux>.

65 Ibid.

66 ICSID's website, available at <https://icsid.worldbank.org/ICSID/FrontServlet>.

for the promotion of renewable energy. These policies entail investment opportunities but also significant economic risks. In fact, the profitability of clean energy projects often depends on subsidies and feed-in tariffs and there is a risk that 'once investments are made, public authorities will be tempted to reconsider their commitments'.[67] In this context, arbitrators will have to take into account the various circumstances – financial crisis, state of necessity and even human rights considerations – which may eventually justify a change in the relevant regulatory measures.

Conclusion

Our climate is a global public good that defies traditional notions of territorial sovereignty. It is a common and shared resource that is both within and beyond the jurisdiction of every State. The interplay between foreign direct investment, climate change and human rights has not yet been explored exhaustively. This chapter has contributed to mapping some aspects of the interplay, assessing the recent boom of investment treaty arbitrations in the field. In particular, cuts in subsidies for renewable energy have been criticised by both investors and environmentalists. They may prevent other investors from investing in the renewable energy sector. Yet they may be indispensable to prevent a financial crisis, with a foreseeable impact on a range of human rights. Therefore, the adoption of a human rights lens compels consideration of the range of policy objectives that the host States may be pursuing in a given case.

67 Anatole Boute, 'The Potential Contribution of International Investment Protection Law to Combat Climate Change' (2009) 27 *Journal of Energy & Natural Resources Law* 333, 342.

12 Linking trade and climate change

What room for human rights?

Olivier De Schutter[1]

Introduction

This chapter explores the role of human rights in shaping the relationship between trade and climate change. It asks whether the linkages between these regimes should be strengthened, in order to prevent the massive threat to human rights that will result from accelerated global warming and associated climate disruptions.

Trade liberalisation has made significant progress since the 1947 General Agreement on Tariffs and Trade (GATT),[2] particularly after an agreement was reached on the establishment of the World Trade Organization (WTO) in 1994.[3] In parallel, the 1992 UNFCCC[4] and the Kyoto Protocol adopted in 1997 at the Third Conference of Parties (COP-3) to the UNFCCC[5] impose on governments a number of duties to mitigate climate change and adapt to it, including quantified targets for emissions reduction agreed to by most industrialised countries. It is argued here that these agendas are incompatible, and that only by linking trade to climate change more explicitly could both be reconciled with human rights objectives. In this sense, a human rights approach to climate change requires that we overcome the current state of fragmentation of international law, with its associated inconsistencies.[6]

1 This chapter develops an articulated human rights stand on considerations I formulated in an article published in vol 5 (2014) of the *Journal of Human Rights and the Environment*.

2 *General Agreement on Tariffs and Trade* (GATT 1947), Geneva, 30 October 1947, entered into force on 1 January 1948 (55 UNTS 187).

3 *Agreement establishing the World Trade Organization*, Marrakesh, 15 April 1994, entered into force on 1 January 1995 (33 ILM 1125 (1994)).

4 The UNFCCC was opened for signature on 9 May 1992, after an Intergovernmental Negotiating Committee produced the text of the Framework Convention as a report following its meeting in New York from 30 April to 9 May 1992. It was signed by 154 countries on 12 June 1992 and entered into force on 21 March 1994 (1771 UNTS 107; 31 ILM 851 (1992)). Currently, the UNFCCC has 196 States parties.

5 *Kyoto Protocol to the United Nations Framework Convention on Climate Change*, Kyoto, 11 December 1997, entered into force on 16 February 2005 (UN Doc FCCC/CP/L.7/Add.1, 10 December 1997, (1998) 37 ILM 22).

6 On fragmentation generally, see the Report of the Study Group of the International Law Commission, *Fragmentation of International Law: Difficulties Arising from the Diversification and Expansion of International Law*, UN Doc A/CN.4/L.702 (18 July 2006) [8]; Bruno Simma, 'Self-contained Regimes' (1985) 16 *Netherlands Yearbook of International Law* 111.

The chapter proceeds in three steps. Section 1 explains how the trade and climate change agendas interfere with one another, within the context of the relationship between developed and developing countries. Section 2 examines the potential of a sanctions-based mechanism to protect human rights, consisting of a ban or the imposition of higher tariffs on products or services that do not comply with certain requirements to use clean technologies in the production process. It offers a general presentation of the substantive legal framework under the main WTO Agreement, the GATT, applicable to trade in goods, which has been replicated in a similar form in the General Agreement on Trade in Services (GATS). Section 3 discusses the role of fundamental procedural rights, in particular, eco-labelling schemes.

1. Trade and climate change within the context of the right to sustainable development

The relationship between trade and climate change may be considered from two perspectives. We may ask, first, to what extent international competition is a disincentive for the adoption of regulatory standards that, albeit aiming to mitigate climate change, may raise costs for enterprises operating from within the State seeking to adopt such standards. It is at least arguable that the more economies are characterised by a high degree of openness to trade – depending more on exports to be able to import more in order to satisfy their needs – the more difficult it will be to adopt measures imposing constraints on companies, in the form of stronger environmental requirements, which are perceived as reducing competitiveness on global markets.[7] The debates that followed the establishment of the carbon emissions trading system within the EU illustrate this, as they included an important discussion on how not to penalise EU-based companies in international competition, in the sectors that are relatively exposed to such competition and where compliance with the requirement to reduce emissions would be relatively costly.[8]

We may also ask how the expansion of trade as such affects the growth of GHG emissions. On the one hand, in many cases trade favours the diffusion of cleaner technologies which, once taken up, can lead to less carbon-intensive types of growth in the importing country. This is the 'technology effect' of international trade, that the WTO Members have pledged to accelerate through prioritising the

7 See *Trade and Climate Change. A Report by the United Nations Environment Programme and the World Trade Organization* (World Trade Organization, 2009) 99.

8 See, for instance, Commission Decision of 24 December 2009 (2010/2/EU) *Determining, Pursuant to Directive 2003/87/EC of the European Parliament and the Council, a List of Sectors and Subsectors which are Deemed to Be Exposed to a Significant Risk of Carbon Leakage* (OJ L 1 of 5.1.2010) 10 (identifying the lines of production within the EU that are at a significant risk of losing markets as a result of having to pay for carbon emission quotas under the EU's emissions trading scheme ('carbon leakage'), as organised by Directive 2003/87/EC of 13 October 2003 *Establishing a Scheme for GHG Allowance Trading within the Community* (OJ L 275 of 25.10.2003) 32)).

liberalisation of trade in environmentally friendly goods and services.[9] On the other hand, however, international trade favours increased economic growth and higher levels of consumption, as resources are freed-up from their less productive uses to be reinvested or spent elsewhere. This is the 'scale effect' of trade; it is built into the very idea of trade having to improve allocative efficiency, thus leading to increased levels of outputs and reduced prices for the end consumer.

Studies are now converging to show that the 'scale effects' of international trade outweigh 'technology effects'.[10] In other words, the increased consumption favoured by trade expansion raises the levels of GHG emissions more than the technological spillover effects of trade. It follows that we cannot pretend to be simultaneously pursuing a free trade agenda leading to the expansion of North–South trade flows while also combating climate change. Indeed, with the expansion of trade, consumers in industrialised countries continue to have access to cheap manufactured products imported from developing countries, in breach of the economic implications of (the right to) development. These products are cheap, both because they are relatively labour-intensive and produced in low-wage countries, and because the negative externalities that result from polluting technologies are not taken into account in their retail prices. But this postpones the need for a change in affluent lifestyles, despite the urgency of such changes; it may in fact favour the runaway consumption that increases our *per capita* ecological footprint, in breach of the environmental implications of the right to development.[11]

Perversely, because of how GHG emissions are computed in the global climate change regime, it is highly tempting for industrialised countries to ignore these impacts of trade. This is the problem of 'carbon leakage', or 'virtual emissions'. As trade volumes grow, a larger proportion of the GHGs emitted in the production processes of commodities that are consumed will be emitted outside the national territory, and thus 'externalised' – or outsourced to trading partners – in the importing country. Thus, it has been calculated that in 2001 the 'virtual emissions' of the EU amounted to 992 megatonnes (Mt) CO_2, representing the emissions from products imported by the EU, whereas the emissions from EU exports represented 446 Mt CO_2.[12] In effect therefore, the EU displaced over 500 Mt of CO_2 emissions overseas that year: by the simple magic of trading more, it could

9 See Doha Ministerial Declaration, WTO Doc WT/MIN(01)/DEC/1, [31(iii)].

10 See Matthew Cole and Robert Elliott, 'Determining the Trade-Environment Composition Effect: the Role of Capital, Labor and Environmental Regulations' (2003) 46–3 *Journal of Environmental Economics and Management* 363; Mark Heil and Thomas Selden, 'International Trade Intensity and Carbon Emissions: a Cross-Country Econometric Analysis' (2010) 10–1 *Journal of Environment and Development* 35–49. For an overview, see *Climate and Trade. Why Climate Change Calls for Fundamental Reforms in World Trade Policies*, Report authored by Tilman Santarius for the German NGO Forum on Environment and Development and Heinrich Böll Foundation (2009).

11 On the economic and environmental implications of the right to development see the contribution by Same Varayudej in this volume.

12 Santarius (2009) 9.

make progress towards meeting its reduced GHG emissions targets with almost no impact on people's consumption levels and habits. This trend still continues. Researchers from the Carnegie Institute estimated in 2010 that 23 per cent of the GHG emissions linked to the goods consumed in developed countries – for a total of 6.4 billion tonnes of CO_2 – had in fact been emitted elsewhere, and that 22.5 per cent of the GHG emissions from China were for the production of export goods – to satisfy the tastes of consumers in the North.[13]

2. Trade, substantive human rights provisions and climate change

In order to facilitate free trade, both under the 1994 GATT and GATS, WTO Members may not discriminate either between 'like' goods from different countries (most-favoured nation clause – Articles I:1 GATT and II:1 GATS) or between foreign goods and domestic goods (national treatment – Articles III:4 GATT and XVII:1 GATS); nor may they impose quantitative restrictions, for example by limiting market access through a system of quotas (Articles XI GATT and XVI GATS). On their face, measures banning certain goods that are produced with highly polluting technologies or imposing higher import tariffs on such goods thus appear highly suspect under these rules.

The non-discrimination requirement included in the most-favoured nation and national treatment clauses prohibits both *de jure* and *de facto* discrimination:[14] the prohibition extends to apparently neutral measures that have the *effect* of disadvantaging goods from a particular country, or that have the effect of disadvantaging foreign goods vis-à-vis domestic goods. The decisive question is whether, considering the nature of the industry in question, consumers' preferences, and historic trade patterns, the introduction of the measure in issue changes competitive conditions in such a way as to represent a disadvantage for the goods originating in one country, or for foreign goods.[15]

As opposed to *product-based* regulations, which address the physical characteristics of the goods concerned, regulations concerning *production or process methods* (PPMs) applied to imported goods are treated with particular suspicion. Such regulations make it more difficult for foreign goods to penetrate domestic markets. They are thus seen as potentially protectionist either in intent, or in effect, because they relate to preferences that are 'outward-looking', as they concern *how* goods are produced

13 Steven J David and Ken Caldeira, 'Consumption-based Accounting of CO_2 Emissions' (2010) 107(12) *Proceedings of the National Academy of Sciences* 5687.

14 Appellate Body Report, *United States – Measures Affecting the Production and Sale of Clove Cigarettes (US – Clove Cigarettes)*, WTO Doc WT/DS406/AB/R (24 April 2012) [175]; Appellate Body Report, *United States – Measures Concerning the Importation, Marketing and Sale of Tuna and Tuna Products (US – Tuna II (Mexico))*, WTO Doc WT/DS381/AB/R (13 June 2012) [225].

15 Appellate Body Report, *Korea – Measures Affecting Imports of Fresh, Chilled and Frozen Beef*, WTO Doc WT/DS161/AB/R, WT/DS169/AB/R (10 January 2001) DSR 2001:I, 5, [139].

outside the importing country. Though this position has not gone unchallenged,[16] panels have generally treated the imposition of such regulations on foreign goods as discriminatory. Indeed, there has been a tendency in the jurisprudence of the Dispute-Settlement Bodies to treat regulatory requirements that concern PPMs, when extended to foreign goods, as a quantitative measure, comparable to a quota or an import ban, to be examined under Article XI GATT.[17]

This is a particularly penalising approach, since Article XI:1 GATT forbids quantitative restrictions outside the narrow exceptions of Article XI:2: where a quantitative restriction is imposed, then, it can only be allowed by the successful invocation of the General Exception Clause of Article XX GATT (or its equivalent provision under GATS, Article XIV), which contains a closed set of admissible justifications for otherwise prohibited measures. Article XX GATT reads, in its relevant part:

> Subject to the requirement that such measures are not applied in a manner which would constitute a means of arbitrary or unjustifiable discrimination between countries where the same conditions prevail, or a disguised restriction on international trade, nothing in this Agreement shall be construed to prevent the adoption or enforcement by any contracting party of measures:
>
> (a) necessary to protect public morals;
> (b) necessary to protect human, animal or plant life or health;
> [. . .]
> (e) relating to the products of prison labour;
> [. . .]
> (g) relating to the conservation of exhaustible natural resources if such measures are made effective in conjunction with restrictions on domestic production or consumption [. . .]

The equivalent provision in GATS, Article XIV, provides:

> Subject to the requirement that such measures are not applied in a manner which would constitute a means of arbitrary or unjustifiable discrimination between countries where like conditions prevail, or a disguised restriction on trade in services, nothing in this Agreement shall be construed to prevent the adoption or enforcement by any Member of measures:
>
> (a) necessary to protect public morals or to maintain public order;

16 Robert Howse and Donald Regan, 'The Product/Process Distinction – An Illusory Basis for Disciplining "Unilateralism" in Trade Policy' (2000) 11–2 *European Journal of International Law* 249. See also the comments by Joel Trachtman on the first Tuna/Dolphin dispute (1992) 86–1 *American Journal of International Law* 142.

17 *United States – Restrictions on Imports of Tuna ('Tuna I')*, WTO Doc DS21/R, Panel Report (3 September 1991) (not adopted) [5.14]; *United States – Restrictions on Imports of Tuna ('Tuna II')*, WTO Doc DS29/R, Panel Report (16 June 1994) (not adopted) [5.10]; *United States – Import Prohibition of Certain Shrimp and Shrimp Products*, WTO Doc WT/DS58/R, Report of the Panel (15 May 1998) [7.16].

(b) necessary to protect human, animal or plant life or health;
(c) necessary to secure compliance with laws or regulations which are not inconsistent with the provisions of this Agreement including those relating to:
   (i) the prevention of deceptive and fraudulent practices or to deal with the effects of a default on services contracts;
   (ii) the protection of the privacy of individuals in relation to the processing and dissemination of personal data and the protection of confidentiality of individual records and accounts;
   (iii) safety [. . .]

Although the formulations of General Exceptions used are not identical in Article XX GATT and in Article XIV GATS respectively, the wordings are sufficiently close to be treated together.

2.1 Environmental protection

Whereas both Article XX GATT and Article XIV GATS refer to measures 'necessary to protect human, animal or plant life or health', Article XX(g) GATT includes an additional reference to measures that 'relat[e] to the conservation of exhaustible natural resources'. The possibility of justifying certain climate measures is probably most closely related to this latter reference. Indeed, just like 'clean air' in the jurisprudence of the WTO's Dispute Settlement bodies,[18] the atmosphere may be seen as an 'exhaustible natural resource', insofar as the modification of its chemical composition is irreversible. For the exception to apply, it is not necessary that the resource be exhausted or depleted; it is sufficient that it can be exhausted or depleted.[19]

The test under Article XX(g) is whether the contested measure '*relat[es] to* the conservation of exhaustible natural resources'. This is distinct from a requirement of strict necessity. It is a test best described as one of appropriateness: a measure that serves the objective of conserving exhaustible natural resources will pass the test, even if other measures might have had the same effect and if, therefore, the measure concerned was not, strictly speaking, 'necessary' to achieve the aim pursued.[20]

18 *United States – Standards for Reformulated and Conventional Gasoline ('US – Reformulated Gasoline')*, WT/DS2/R (29 January 1996) (for the Panel Report); and WT/DS2/AB/R (29 April 1996) (for the Appellate Body Report).

19 See also *United States – Restrictions on the Import of Tuna* (1994) [5.13].

20 *Canada – Measures Affecting Exports of Unprocessed Herring and Salmon*, BISD 35S/98 (adopted on 22 March 1988) [4.6] (the emphasis is added by the Panel itself). The same reading was applied in two non-adopted Panel reports: *United States – Restrictions on Imports of Tuna* (1994); *United States – Taxes on Automobiles*, DS31/R (11 October 1994).

Alternatively, measures that would otherwise create obstacles to trade may be justified if they are 'necessary to protect human, animal or plant life or health'. Measures to reduce climate change by seeking to discourage trade of goods that result in large avoidable volumes of GHG emissions would appear in principle to correspond to this objective: the human consequences of climate change are well documented and by now have achieved a high degree of scientific consensus. The test of 'necessity', which appears in both Article XX(b) GATT 1994 and Article XIV(b) GATS, seems on its face to impose a stricter requirement than the reference in Article XX(g) to a measure 'relating to the conservation of exhaustible natural resources'. However, even the stricter test of 'necessity' does not require a demonstration that no other measure could have succeeded in fulfilling the objective. In its interpretation of Article XX(d) GATT, which also uses the term 'necessary', the Appellate Body explicitly excluded a reading of 'necessity' that would equate it with 'indispensability'. It explained:

> As used in Article XX(d), the term 'necessary' refers, in our view, to a range of degrees of necessity. At one end of this continuum lies 'necessary' understood as 'indispensable'; at the other end, is 'necessary' taken to mean as 'making a contribution to'. We consider that a 'necessary' measure is, in this continuum, located significantly closer to the pole of 'indispensable' than to the opposite pole of simply 'making a contribution to' . . . [Assessing whether a measure is 'necessary' for the fulfilment of a legitimate aim is] a process of weighing and balancing a series of factors that prominently include the contribution made by the compliance measure to the enforcement of the law or regulation at issue, the importance of the common interests or values protected by that law or regulation, and the accompanying impact of the law or regulation on imports or exports.[21]

The test of necessity does not require that the risk to be prevented or the impact from the challenged measure be quantified; nor is it required that the country imposing a restriction to trade, such as an import ban, demonstrates that the ban has been effective in addressing the problem the measure leading to such a restriction was meant to address. Not only would such proof be difficult to provide, as any improvement that would be found to occur may always be traced back to a number of interacting causes, but in addition, trade-restrictive measures may have impacts only after a period of time: this is particularly relevant to the assessment of measures aimed at curbing GHG emissions, which is the very example referred to by the Appellate Body.[22] The key question under this test is

21 Appellate Body Report, *Korea* (2001) [161] and [164].

22 Appellate Body Report, *Brazil – Measures Affecting Imports of Retreaded Tyres*, WT/DS332/AB/R (3 December 2007) [151]:

> the results obtained from certain actions – for instance, measures adopted in order to attenuate global warming and climate change, or certain preventive actions to reduce the incidence of diseases that may manifest themselves only after a certain period of time – can only be evaluated with the benefit of time.

whether the challenged measure has been shown to be 'apt to produce a material contribution to the achievement of the objective': it must neither be indispensable, nor must it have been proven successful; but its potential impacts should not be merely hypothetical.[23]

2.2. The relevance of human rights

Articles XX(a) GATT and XIV GATS mention measures 'necessary to protect public morals'.[24] The notion of 'public morals' has been defined as denoting 'standards of right and wrong conduct maintained by or on behalf of a community or nation'.[25] Most commentators take the view that this includes universally recognised human rights.[26] This can encompass the right to a healthy environment, as such a right is increasingly recognised at both domestic and international levels.[27]

Indeed, it may be argued that an interpretation of 'public morals' that is grounded on internationally recognised human rights should be less controversial than one that simply refers to the deeply held values in one society: by definition, universal human rights are not imposed unilaterally by one State restricting access to its markets; they are a reference to norms that all States, in principle, have recognised to be legitimate and obligatory. In addition, WTO agreements should be interpreted, in so far as possible, consistently with international human rights law, as required under the Vienna Convention on the Law of Treaties.[28] It is relevant in this regard that human rights are considered as imposing on States

23 Ibid, [152].

24 In Article XIV(a) GATS, where the same expression appears, reference is also made to 'public order', and a footnote appended to the Agreement notes that this latter notion can be invoked 'where a genuine and sufficiently serious threat is posed to one of the fundamental interests of society'.

25 *United States – Measures Affecting the Cross-Border Supply of Gambling and Betting Services*, WTO Docs WT/DS285/R (10 November 2004) (Panel Report) [6.465] and WT/DS285/AB/R (7 April 2005) (Appellate Body Report) [296].

26 Sarah Joseph, *Blame It on the WTO? A Human Rights Critique* (Oxford University Press, 2011) 109; Adam McBeth, *International Economic Actors and Human Rights* (Routledge, 2010); Robert Howse, 'Back to Court After Shrimp? Almost but Not Quite Yet: India's Short-lived Challenge to Labour and Environmental Exceptions in the European Union's Generalized System of Preferences' (2003) 18 *Amer Univ International Law Review* 1333, 1338; Steve Charnovitz, 'The Moral Exception in Trade Policy', in Steve Charnovitz (ed), *Trade Law and Global Governance* (Cameron May, 2002) 361; Robert Howse, 'The World Trade Organization and the Protection of Workers' Rights' (1999) 3 *Journal of Small and Emerging Small Business Law* 131, 143.

27 See John H Knox, *Report of the Independent Expert on the Issue of Human Rights Obligations relating to the Enjoyment of a Safe, Clean, Healthy and Sustainable Environment*, presented to the 25th session of the Human Rights Council (UN Doc A/HRC/25/53) (30 December 2013), and the contribution by Francesco Francioni and Ottavio Quirico in this volume.

28 Appellate Body Report, *United States – Standards for Reformulated and Conventional Gasoline Treatment of Imported Gasoline and Like Products of National Origin*, WT/DS2/AB/R (20 May 1996). Article 3.2 of the Dispute Settlement Understanding confirms that WTO norms may be 'clarified . . . in accordance with customary rules of interpretation of international law', which the Vienna Convention codifies. Article 31(3)(c) of the Vienna Convention on the Law of Treaties stipulates

obligations that are *erga omnes*. Such obligations are owed to the international community as a whole, rather than to those States alone whose interests have been prejudiced by the alleged violation. All States therefore may in principle assert a legal interest in taking measures that seek to respond to such a violation: by invoking the need to protect 'public morals' and introducing trade restrictions in order to discourage production processes that may be described as human rights violations due to their human impacts, a WTO Member would simply be acting in accordance with this understanding of the interest that all States have in not leaving human rights unpunished wherever breaches may occur.[29]

Measures aimed at protecting 'public morals' in the context of Article XX(a) GATT and Article XIV(a) GATS must be 'necessary' to the fulfilment of that objective. Again, this test appears stricter than where the challenged measures aim at the conservation of exhaustible natural resources under Article XX(g) GATT. As we have seen, however, the Appellate Body has moved away from a 'strict necessity' test, mentioning instead that the test should be 'whether a WTO-consistent alternative measure which the Member concerned could "reasonably be expected to employ" is available, or whether a less WTO-inconsistent measure is "reasonably available"'.[30] In engaging in what the Appellate Body calls 'a process of weighing and balancing a series of factors' to assess whether a particular trade restriction introduced in the name of the protection of 'public morals' is indeed 'necessary' for that purpose,[31] the importance of the aims pursued by the measure concerned and the importance of the restriction to trade are both assessed: whether the 'necessity' condition is fulfilled will depend on whether, considering the respective weight of the interests at stake, there were alternative measures, less restrictive of trade, that the State could have reasonably been expected to resort to.[32]

This test of necessity, of course, produces paradoxical results when applied to measures that a State adopts in order to put pressure on another State, or on producers of goods or services providers within that State, to improve their shift to less polluting technologies. First, the more important the restriction to trade, the more effective the pressure will be on the exporting side. Therefore, we may expect any less restrictive measure to diminish effectiveness as regards the aim pursued: this would seem to plead in favour of the recognition to the Member State of a large margin of appreciation. Second, the more important the market of the Member imposing the measure, the more plausible it is to expect that its trading partner will bow to the pressure and make the changes necessary to ensure that it will not lose access to markets. This of course explains in part the sensitivity of the topic: within the WTO, the most powerful Members – those on whose

that the interpretation of treaties must take into account 'any relevant rules of international law applicable in the relations between the parties'.

29 See, within the context of the interpretation of WTO agreements, Sarah Cleveland, 'Human Rights Sanctions and International Trade: a Theory of Compatibility' (2002) 5–1 *Journal of International Economic Law* 133, 152–3.

30 Appellate Body Report, *Korea* (2001) [166].

31 *Korea – Various Measures on Beef*, WT/DS169/AB/R (adopted 10 January 2001) [164].

32 See also *United States – Gambling*, WTO Doc WT/DS285/AB/R (10 November 2004) [305]–[307].

markets the other Members depend the most – have the most important bargaining power, to the extent that the unilateral measures they may adopt are not in violation of WTO disciplines. Indeed, it is for this very reason that the requirements concerning the universality of the aims pursued and of non-discrimination are particularly important, insofar as such requirements can prevent the most powerful Members of the WTO from abusing their dominant position.

2.3. Scope of application of human rights protection under the GATT and WTO

The reliance on the General Exception Clauses to justify environmental conditionalities raises another question. Whether they are construed as aiming to 'protect human, animal or plant life or health' under Article XX(g) GATT or Article XIV(b) GATS, to preserve an 'exhaustible natural resource' under Article XX(g) GATT, or to safeguard 'public morals', such conditionalities (restricting access to markets on the basis of environmental considerations) aim to protect interests *outside the territory of the importing country adopting such measures.*

Importantly however, these conditionalities do not seek to protect the environment *only in the exporting country*, as would be the case, for instance, where products originating from that country would have higher tariffs imposed on them if produced by techniques leading to pollution of the soils in that country alone, without the impacts of the pollution being felt in the importing country adopting the measure concerned. Rather, measures that make market access conditional on GHG emission reductions aim to protect a *global public good.* We may reason here by analogy with attempts to protect migratory species: in the *Shrimp-Turtle* dispute, the fact that the sea turtle species the measures adopted by the US sought to protect were known to migrate through the waters of several countries, including waters over which the US exercises jurisdiction, was considered sufficient to justify the possibility of the US invoking Article XX(g).[33] The same reasoning would be valid for measures aiming at mitigating climate change by incentivising a reduction in GHG emissions: 'clean air', we have seen above, is considered an 'exhaustible natural resource', and it is clear that a failure by one State to take measures against the growth of emissions has impacts on the whole community of States. Doubts remain, perhaps, as to the acceptability of measures that are *purely outward-looking*, in other words, all the consequences of which would be located on the territory

33 *United States – Import Prohibition of Certain Shrimp and Shrimp Products ('United States – Shrimp I')*, WTO Doc WT/DS58/AB/R, AB-1998–4 (12 October 1998) (Report of the Appellate Body) [133]. See also the Reports of the Panels in *Canada – Measures Affecting the Exports of Unprocessed Herring and Salmon*, WTO Doc 35S/98 (22 March 1988); *United States – Prohibition of Imports of Tuna and Tuna Products from Canada*, WTO Doc 29S/91 (22 February 1982). In both these disputes, the Panels did not question the reliance on Article XX GATT for the protection of migratory species of fish. They made no distinction between tuna caught within or outside the territorial waters of the Member invoking the exception.

of the exporting Member State.[34] But measures that aim to protect a good that both the importing and the exporting country share as a global common should be seen as covered by the General Exceptions Clause.

Finally, the General Exceptions Clauses also include requirements of even-handedness and non-discrimination, in order to avoid these clauses being used for protectionist purposes. The General Exceptions Clauses of Article XX GATT and Article XIV GATS both include a *chapeau* that makes reliance on these exceptions conditional upon the measures not constituting 'a means of arbitrary or unjustifiable discrimination between countries where the same conditions prevail', or a 'disguised restriction on international trade'. While allowing for the adoption of measures 'relating to the conservation of exhaustible natural resources', Article XX(g) GATT also requires that, in order for an otherwise contestable measure to be saved under this clause, it should be 'made effective in conjunction with restrictions on domestic production or consumption'. No demonstration is required that the measures affecting the production or consumption of domestic products in order to conserve exhaustible natural resources are *successful*, that is, that their impacts are such that they are proven to work. Such a burden would be difficult to meet for the State concerned, both because of the complexity of proving causation (that is, how much in the conservation of the resource is attributable to the challenged measure) and because, as already noted, environmental measures typically necessitate a long time before they can produce any effects. Rather, the requirement is one of 'even-handedness in the imposition of restrictions, in the name of conservation, upon the production or consumption of exhaustible natural resources'.[35]

Insofar as there exists consensus across the international community that climate change is a serious threat that demands action by governments, and insofar as climate change produces a range of human rights impacts, a flexible interpretation of the GATT and GATS agreements is necessary to allow for the adoption of measures protecting the human right to a healthy environment and other relevant human rights – such as the right to health, the right to housing, or the right to food[36] – under Article XX GATT and Article XIV GATS. A flexible interpretation would be justified by the need to read these agreements in accordance with the duty of all States to respect, protect and fulfil human rights. The fact that such trade policy measures seek to influence production processes under the exporting

34 In *Shrimp-Turtle*, the Appellate Body explicitly declined to 'pass upon the question of whether there is an implied jurisdictional limitation in Article XX(g), and if so, the nature or extent of that limitation' (*United States – Shrimp I* (1998) [133]):

> We note only that in the specific circumstances of the case before us, there is a sufficient nexus between the migratory and endangered marine populations involved and the United States for purposes of Article XX(g).

35 *US – Reformulated Gasoline*, Appellate Body Report, WT/DS2/AB/R (29 April 1996) 21.

36 For a systematic review of the human rights impacts of climate change, see *Report of the Office of the United Nations High Commissioner for Human Rights on the Relationship between Climate Change and Human Rights*, A/HRC/10/61 (15 January 2009).

country's jurisdiction should not be seen as an obstacle to the possibility of invoking the General Exceptions Clauses, in particular because the atmosphere is a global public good, that all WTO Members have a legitimate (*erga omnes*) interest in seeking to preserve: this corresponds to the *erga omnes* duty towards human rights, which all States have a legitimate interest in ensuring compliance with. Any such measures would also be easier to justify under the relevant WTO agreements however, if they were accompanied by appropriate burden-sharing towards developing countries. When combined with the transfer of clean technologies to developing countries and the financing of measures that can support reform towards sustainable development in those countries, environmental conditionalities would be less easily denounced as disguised protectionism: they would also be more acceptable politically.

2.4. The special regime of border tax adjustments

Border tax adjustments (BTAs) are subject to a specific regime under the GATT. Article II:2(a) GATT states that the schedule of commitments to reduce tariffs agreed under Article II GATT shall not be interpreted to:

> prevent any contracting party from imposing at any time on the importation of any product . . . a charge equivalent to an internal tax imposed consistently with the provisions of paragraph 2 of Article III in respect of the like domestic product or in respect of an article from which the imported product has been manufactured or produced in whole or in part.

This allows WTO Members to apply a charge on imports equivalent to an internal tax, provided that such a charge complies with the national treatment principle of Article III:2, in other terms, provided the charge is not 'in excess of those applied . . . to like domestic products'. This defines the essential distinction between a BTA and import tariffs.

BTAs aim to compensate for differences between the respective taxation systems of the exporting and importing countries: these taxation systems shall be 'equalised' by ensuring that the taxes that are paid are those applicable in the country of destination of the goods (that is, where the final products are bought by the end consumer), rather than in the country of origin. BTAs are increasingly discussed in the context of climate change mitigation strategies that seek to reduce emissions by carbon taxes or through cap-and-trade schemes: the UNFCCC itself alludes to such measures being adopted, though without prejudging their legality under WTO law.[37] BTAs may thus be a precious tool to protect human rights, particularly the claim to a healthy environment. Indeed, carbon taxes or cap-and-

37 Article 3.5 UNFCCC states that 'measures taken to combat climate change, including unilateral ones, should not constitute a means of arbitrary or unjustifiable discrimination or a disguised restriction on international trade', thus leaving open the question of the conditions under which unilateral measures such as BTAs might be compatible with WTO disciplines.

trade schemes for carbon emissions allowances may expose certain sectors of the economy to competition from foreign producers who will produce in more competitive conditions, because they will not be forced to buy GHG allowances, leading in turn more investment to go to the countries which do not have cap-and-trade schemes, which is the phenomenon described above as 'carbon leakage'. A plausible reaction to such a scenario is for the importing country where a carbon tax or a cap-and-trade scheme is in place to oblige importers of products to join the said scheme, thus compensating for the 'unfair' advantage that the imported products would otherwise benefit from, as a result of not having to buy emissions allowances in the country of origin.[38]

The only system of BTAs that is at once practical to implement and irreproachable from the point of view of its legality, may be one in which foreign products are taxed at a level that assumes that they rely on the cleanest technologies available. However, this is not incentivising foreign producers to switch to better technologies; nor does it encourage the country of origin to improve its climate change mitigation regime, since whatever those producers or that country do, their access to the markets of the importing country will not improve: the most likely outcome is that exporters will simply pay the tax at the border, and the country of origin raises concerns, but takes no action. In contrast, positive changes may be triggered by increased efforts by the importing country adopting border measures to finance climate change mitigation and adaptation strategies in developing countries, with the revenues that would accrue from the introduction of BTAs. Thus understood, BTAs would become more than a tool to equalise conditions of competition in the presence of different taxation systems that may lead to carbon leakage as a form of 'environmental dumping': they would form the basis of a positive policy encouraging reform under the jurisdiction of the trading partners of the WTO Member adopting such measures, accelerating the adoption of cleaner technologies. BTAs would thus be a tool for the fulfilment of the right to a healthy environment and the right to development.

3. Procedural rights: eco-labelling schemes

The former Special Rapporteur on the adverse effects of the illicit movement and dumping of toxic and dangerous products and wastes on the enjoyment of human rights noted that the rights to information and participation are 'both rights in themselves and essential tools for the exercise of other rights, such as the right to life, the right to the highest attainable standard of health, the right to adequate housing and others'.[39] Along the same lines, the Independent Expert on the issue of human rights obligations relating to the enjoyment of a safe, clean, healthy and sustainable environment noted that procedural environmental rights imply 'duties

38 Kateryna Holzer, 'Proposals on Carbon-related Border Adjustments: Prospects for WTO Compliance' (2010) 1 *Carbon and Climate Law Review* 51, 54.

39 Okechukwu Ibeanu, *Promotion and Protection of All Human Rights: Civil, Political, Economic, Social and Cultural Rights, Including the Right to Development – Report*, A/HRC/7/21 (18 February 2008) 2.

(a) to assess environmental impacts and make environmental information public; (b) to facilitate public participation in environmental decision-making, including by protecting the rights of expression and association; and (c) to provide access to remedies for harm'.[40] The UNFCCC dedicates a provision to 'Education, Training and Public Awareness' (Article 6) and the right of the public to have access to information about the environment is addressed in a number of instruments, including in particular the 1998 Aarhus Convention. Insofar as such intiatives provide information to consumers about the environmental impacts of the products they purchase, eco-labelling initiatives may be said to contribute to the right of access to information, and consequently to the rights of access to decision-making and justice.

The proliferation of eco-labelling initiatives in recent years led the Ministerial Declaration adopted on 14 November 2001 at the WTO Doha Ministerial Conference to mention labelling requirements for environmental purposes as one area to which the WTO Committee on Trade and Environment should pay greater attention.[41] Indeed, developing countries have expressed a range of concerns about the eco-labelling trend.[42] By discouraging the import of products from developing countries because of their failure to comply with certain environmental standards and, in particular, because of their carbon footprint, rich countries would be ignoring that they are themselves primarily responsible for the building up of GHG emissions in the atmosphere: denying developing countries growth opportunities by imposing eco-labelling schemes would therefore be in violation of the principle of 'common but differentiated responsibilities and respective capabilities' included in the UNFCCC.[43] It would also be depriving developing countries of what they perceive as their comparative advantage, or misleading consumers where the focus is placed on 'food miles' as a yardstick (albeit a highly misleading one) for the carbon impacts of their purchases.[44]

40 Knox (2013) [29].

41 WTO Doc WT/MIN(01)/DEC/1, [32(iii)].

42 See the summary of the discussions held within the WTO Committee on Trade and Environment, *Report of the Meeting Held on 10 July 2009*, WTO Doc WT/CTE/M/47 (31 August 2009) [3]; Paul Brenton, Gareth Edwards-Jones and Michael Friis Jensen, 'Carbon Labelling and Low-income Country Exports: a Review of the Development Issues' (2009) 27–3 *Development Policy Review* 243.

43 James MacGregor, *Carbon Concerns: How Standards and Labelling Initiatives Must Not Limit Agricultural Trade from Developing Countries*, ICTSD–IPC Platform on Climate Change, Agriculture and Trade, Issue Brief No 3 (International Centre for Trade and Sustainable Development and International Food & Agricultural Trade Policy Council, Washington DC, 2010) 10.

44 Committee on Trade and Environment (2009) [10]. Although food miles are an important concern as regards the transport by air of fresh fruits and vegetables, most of international trade of goods relies on ships as its mode of transportation, the levels of emissions of GHGs of which are much lower per volume transported than transport by road or by air. Furthermore, insofar as carbon labels are concerned, a life-cycle assessment approach (including not only transport but also production, processing, packaging, storage, consumption and recycling) would often work in favour of agricultural products originating from tropical regions, because of the favourable growing conditions they benefit from (see Hans-Jürgen Schmidt, 'Carbon Footprinting, Labelling and Life Cycle Assessment' (2009) 14 *The International Journal of Life Cycle Assessment* 6).

These concerns are compounded by the fact that labels are often costly to acquire, because of the need to collect information and go through a certification process, and because the costs of the monitoring or auditing procedures in place. The introduction of labels may therefore affect the price of the product at the retail level, and it may particularly affect small-scale producers, especially small-scale farmers, who are the least well-equipped to comply with the requirements imposed for the acquisition of labels and cannot easily meet the upfront costs of acquiring such labels.[45] Indeed, the standards that have emerged in recent years have been developed without the participation of stakeholders from developing countries, whether governmental or private, and they generally do not take into account the specific conditions in those countries or the needs of producers within those countries.[46]

Eco-labelling schemes are extremely varied, as they are generally part of retailers' supply chain management and corporate social responsibility schemes designed at company level, without harmonisation at sector level. Some private bodies aim at harmonising environmental standards applicable to private actors: that is the case for the Global Eco-labelling Network (GEN) or for the International Social and Environmental Accreditation and Labelling Alliance (ISEAL Alliance).[47] The International Organization for Standardization (ISO) has also developed standards that should guide eco-labelling efforts, and may be considered to have to be taken into consideration under Article 2.4 of the WTO Agreement on Technical Barriers to Trade (TBT Agreement). However, there still is no scientific consensus on how to measure the carbon footprint of a product using a life-cycle assessment (LCA) approach, and there is no single multilateral body that can act as a standard-setter, which explains in part the great diversity of eco-labelling schemes and the lack of transparency in the implementation of existing schemes.

Against this background, the question of the compatibility of eco-labelling schemes with WTO law takes on an added importance. The provisions of WTO law that are the most relevant to labelling schemes are stipulated in the TBT Agreement.[48] The TBT Agreement applies to all products, whether industrial or agricultural,[49] and covers labels that provide information to consumers about the

45 Olivier De Schutter, *Competing for Standards in Global Food Supply Chains* (Cridho Working Papers Series 2012/2).

46 Jason Potts, *The Legality of PPMs under the GATT: Challenges and Opportunities for Sustainable Trade Policy* (International Institute for Sustainable Development, 2008) 5.

47 See Shane Baddeley, Peter Cheng and Robert Wolfe, 'Trade Policy Implications of Carbon Labels on Food' (2012) 13(1) *The Estey Centre Journal of International Law and Trade Policy* 59, 67–8.

48 For more detailed studies, see Ilona Cheyne, 'Proportionality, Proximity and Environmental Labelling in WTO Law' (2009) 12(4) *Journal of International Economic Law* 927; Manoj Joshi, 'Are Eco-Labels Consistent with World Trade Organization Agreements?' (2004) 38(1) *Journal of World Trade* 69.

49 However, 'purchasing specifications prepared by governmental bodies for production or consumption requirements of governmental bodies are not subject to the provisions of this Agreement but are addressed in the Agreement on Government Procurement' (TBT Agreement, Art 1:4).

social or environmental conditions under which certain goods were produced, insofar as such schemes may result in obstacles ('technical barriers') to trade.

The provisions of the GATT may also be applicable to labelling schemes.[50] Particularly relevant is Article III:4 GATT, which requires that foreign products be granted 'no less favourable treatment' in respect of 'all laws, regulations and requirements affecting their internal sale, offering for sale, purchase, transportation, distribution or use'. This may cover labelling schemes, insofar as such schemes affect the conditions of retail for products. However, this provision will seldom be invoked in this context. First, the TBT Agreement constitutes *lex specialis* with respect to technical regulations and standards, and in the event of a conflict between the two regimes, it should prevail.[51] Second, judicial economy implies that where a finding of violation of the TBT Agreement has been made, the Dispute Settlement Bodies of the WTO shall refrain from examining the same measures under the provisions of the GATT. However, where a particular measure has been considered consistent with the TBT Agreement, it may still be found inconsistent with the GATT, requiring a separate examination under that Agreement.[52]

With regard to the TBT Agreement, the key requirement is that technical regulations do not result in discrimination between domestic and foreign products (Article 2.1 of the TBT Agreement). Less favourable treatment is prohibited even where it does not impose an insuperable obstacle on foreign products' access to markets.[53] What matters is that the regulatory distinction is a valid one – that it is 'even-handed', rather than resulting in the creation of unnecessary obstacles to the entry of foreign products.[54] However, restrictions to trade that would otherwise be prohibited may be justified by legitimate objectives under Article 2.2 of the TBT Agreement, which lists among such objectives 'national security requirements; the prevention of deceptive practices; protection of human health or safety, animal or plant life or health, or the environment'. The adoption of a 'technical regulation' may pass the test of Article 2.2, *even though it only partly or*

50 Indeed, despite important similarities between those provisions and the provisions of the TBT Agreement, it would be incorrect to assume that the two agreements impose identical obligations. See *United States – Tuna II (Mexico)* (Appellate Body Report of 16 May 2012) [405].

51 General Interpretative Note to Annex 1A (Multilateral Agreements on Trade in Goods) to the Marrakech Agreement establishing the World Trade Organisation.

52 Certain labelling schemes can also be addressed under the Agreement on Sanitary and Phytosanitary Measures (SPS), which seeks to ensure that the sanitary and phytosanitary measures that the WTO Members adopt for the protection of human, animal or plant life or health do not arbitrarily or unjustifiably discriminate between Members where identical or similar conditions prevail, and are not applied in a manner which would constitute a disguised restriction on international trade. However, this concerns a narrow range of labelling schemes, and generally would not include eco-labels other than those regarding the use of pesticides or chemical fertilizers. Labels play a relatively marginal role in the disputes that arise under the SPS Agreement (see Baddeley *et al* (2012) 73). Labelling schemes are therefore primarily relevant to the TBT Agreement.

53 *Tuna/Dolphin II*, Appellate Body Report (2012) [221].

54 Ibid [297]–[299].

imperfectly fulfils the objective it is assigned, unless another measure is available that is at least as effective with regard to the said objective, while less restrictive of trade.[55]

Again with a view to limiting the adoption of unilaterally designed measures that might hide protectionist motives or disproportionately affect trade, Article 2.4 of the TBT Agreement supports the adoption of international standards. However, since there is as yet no recognised standard-setting organisation open to all WTO Members, these are free to define their own standards when establishing or supporting eco-labelling schemes that take into account the carbon footprint of products. In this regard, the convergence of carbon footprint labels towards one single methodology adopting a life-cycle assessment approach, and recognised widely enough to stimulate adoption in both public and private labelling schemes, would certainly be a positive step forward. It would be welcomed by many developing countries, which may otherwise be penalised by the use of the 'food miles' concept, excluding a full life-cycle assessment taking into account the different phases of the product's life 'from cradle to grave', or by the attitudes of consumers who, while desiring to reduce the carbon footprint of their purchasing practices, take geographical distance as a yardstick for the volume of GHG emissions involved. Convergence towards a single metric would improve transparency and the position of producers in developing countries who may otherwise have to adapt to a variety of schemes. It could also imply that, in accordance with Article 2.4 of the TBT Agreement, WTO Members wishing to introduce a carbon footprint labelling scheme would have to do so by reference to this metric, at least if the said metric is the result of a standard-setting process open to all WTO Members. The future development of eco-labels should go hand in hand with increased efforts to develop a multilateral approach to this question, therefore with the active participation of WTO Members from developing countries and organisations of producers from the South. This would strengthen the position of countries wishing to introduce such labels, and gradually lead to the identification of the best way to reconcile trade with environmental concerns.

Conclusion

Realising the right to (sustainable) development requires overcoming fragmentation in global governance. Trade and climate change policies in particular should be better coordinated: they are currently at odds with one another and should be made mutually supportive. Grounding such coordination in human rights may facilitate integration, specifically by relying on fundamental rights to interpret the General Exceptions Clauses of Articles XX GATT and XIV GATS, and by improving access to information about GHG emissions embedded in certain products via eco-labelling schemes.

WTO rules define a range of conditions that should be complied with in order to avoid the risk of GHG reduction-related conditionalities being misused for

55 Ibid, [330]–[331].

protectionist purposes. These norms also allow some room for human rights protection, ensuring greater consistency between the climate change and trade agendas, with specific regard to the intertwined rights to environment and health.

It may be in the interest of developing countries themselves to support such an integration. They are the ones that will be most severely affected by climate disruptions and may benefit from an improved coordination between the two agendas. Along these lines, coordination should preferably form part of a package that, consistently with the right to development, would also include transfer of clean technologies and capacity-building for adapting to climate change.

Part IV

Institutional prospects

13 Systemic integration between climate change and human rights at the United Nations?

Spyridon Aktypis, Emmanuel Decaux and Bronwen Leroy

Introduction

The UN has been active in the field of climate change for a long time through several organs, specifically UNEP. Its action led to the adoption of fundamental instruments in the field. The key convention in the matter is the UNFCCC, which was adopted in 1992 at the end of the Rio Earth Summit. The UNFCCC entered into force on 21 March 1994 and currently includes 196 parties.

Initially, the UN took an environmental approach to climate change, excluding human rights. However, things changed with the adoption of the Malé Declaration on the Human Dimension of Global Climate Change, which was endorsed by Small Island States in 2007. Indeed, the Declaration prompted the UN to integrate human rights in the climate change discourse, on the one hand, within the context of the UNFCCC, and, on the other, through classical UN human rights bodies, that is, Charter-based bodies and treaty bodies, particularly the HRCte. Only recently therefore has the UN taken a stand on the relationship between climate change and human rights and work is still in progress.

This chapter critically presents the different UN initiatives involving climate change and human rights, particularly in light of the UNFCCC regime and classical UN human rights protection mechanisms, especially the HRC and its subsidiary organs. The stand of UN human rights treaty bodies exposed in works such as the general comments of the HRCte and the CESCR is also seminal. Nevertheless, the HRC and the UNFCCC are the general UN institutional frameworks respectively dealing with human rights and climate change and naturally tend to polarise the debate.

The ultimate aim of this research is to highlight the impact of human rights on the UNFCCC regime and, vice-versa, that of climate change on the UN human rights protection system, as well as their reciprocal interactions. The chapter concludes that an effective approximation of the two regimes is complex, but politically and normatively feasible. This approximation would be facilitated by a human rights approach to the right to a sustainable environment.

1. The Malé Declaration: an entry point for human rights to climate change

After the Inuit petition to the IAComHR raised the issue of the relationship between climate change and human rights in 2005,[1] the question was seriously taken up by the Republic of the Maldives. The Maldives is an island State in the Indian Ocean–Arabian Sea area, consisting of a double chain of 26 atolls, whose highest point is less than 2.5 metres above sea level. Being a low-lying State, the Maldives is seriously threatened with disappearance by rising sea level.[2]

In 2007, the Maldives convened a Small Island States Conference to address the human rights implications of climate change, which produced the Malé Declaration on the Human Dimension of Global Climate Change. The Malé Declaration explicitly highlights the link between climate change and human rights, stressing that 'climate change has clear and immediate implications for the full enjoyment of human rights'.[3]

The Declaration also points out 'the *fundamental right* to an environment capable of supporting human society and the full enjoyment of human rights', which is recognised in 'the constitutions of over one hundred States and directly or indirectly in several international instruments'.[4] This statement is particularly important, because it seems to assume the human rights nature of the right to a sustainable environment and consequently links climate change, negative environmental effects, and human rights.

Hence, the Declaration called on the UNFCCC institutions to address the relationship between climate change and human rights. More specifically, the Malé Declaration invited States to join a formal process ensuring a 'post-2012 consensus' on the stabilisation of temperatures below 2°C above pre-industrial levels and GHG concentrations at less than 450 ppm.[5] Furthermore, the Declaration invited the UNFCCC COP on Climate Change to 'seek the cooperation of the Office of the United Nations High Commissioner for Human Rights and the United Nations Human Rights Council in assessing the human rights implications of climate change'.[6] It has thus been stated that the Declaration sets out a 'road map for bringing human rights into the climate change negotiations'.[7]

The Declaration was submitted to the attention of COP 13 in Bali, with a clear emphasis on the necessity of regarding the UNFCCC negotiation process not only

1 *Inuit Petition to the Inter-American Commission on Human Rights, Seeking Relief from Violations Resulting from Global Warming Caused by Acts and Omissions of the United States* (7 December 2005) 103–4.

2 For further details, see John Knox, 'Linking Climate Change and Human Rights at the UN' (2009) 33 *Harvard Environmental Law Review* 477, 480–1.

3 *Malé Declaration on the Human Dimension of Global Climate Change* (14 November 2007) Preamble, <www.ciel.org/Publications/Male_Declaration_Nov07.pdf >.

4 Ibid, Preamble (emphasis added).

5 Ibid, [1].

6 Ibid, [3].

7 Knox (2009) 483.

from a trade perspective, but also through a human rights lens.[8] It prompted the 'OHCHR to conduct a detailed study into the effects of climate change on the full enjoyment of human rights', including 'relevant conclusions and recommendations' for submission 'prior to the tenth session of the Human Rights Council'.[9] Furthermore, the Declaration invited the HRC to convene a debate on climate change and human rights.[10]

2. UN human rights protection mechanisms and climate change

2.1. Acknowledging the relevance of climate change: UN HRC Resolution 7/23

Before the relationship between climate change and human rights was clearly brought to the attention of the UN by the Maldives, some preliminary UN studies on human rights tackled the issue of climate change on a case-by-case basis.

The late Human Rights Commission developed preliminary work on the link between human rights and climate change, requesting the Secretariat to present and update a study on 'Human Rights and the Environment as part of Sustainable Development' in 2004 and 2005.[11] The Sub-Commission on the Promotion and Protection of Human Rights was even more specific and developed a debate on the challenge of small islands and the potential issue of '*climate refugees*', which had been raised for the first time by Françoise Hampson.[12] This issue was also discussed by the HRC Advisory Committee in the context of the studies by Jean Ziegler on the right to food, including the new concept of '*hunger refugees*'.[13] However, the UN HCR always took care to stick to a restrictive legal wording in defining the notion of 'refugee' under the 1951 Geneva Convention.[14]

In March 2007, following the adoption of the Malé Declaration, Bolivia, Bhutan, Greece, the Maldives, Nigeria, Indonesia and the Philippines raised the serious consequences of climate change for the full enjoyment of human rights at the UN HRC's seventh regular session, and called on the Council to address them.[15]

8 *Report of the Conference of the Parties on Its Thirteenth Session*, Bali, 3–15 December 2007, UN Doc FCCC/CP/2007/6 (14 March 2008) 24, [122].

9 *Malé Declaration* (2007) [4].

10 Ibid, [5].

11 UN Secretary-General, *Human Rights and the Environment as Part of Sustainable Development*, E/CN.4/2004/87 (6 February 2004) and E/CN.4/2005/96 (19 January 2005).

12 Sub-Commission on Promotion and Protection of Human Rights, Resolution 2004/10 (9 August 2004), in *Report of the Sub-Commission on the Promotion and Protection of Human Rights on Its Fifty-Sixth Session*, E/CN.4/2005/2 and E/CN.4/Sub.2/2004/48 (21 October 2004) 11–12, 33.

13 See, in particular, Jean Ziegler, *Report to the Human Rights Council on the Right to Food*, Doc A/HRC/4/30 (19 January 2007) 17, [47].

14 See Hugh Massey, *UNHCR and De Facto Statelessness* (UNHCR, 2010).

15 See, for instance, Abdulla Shahid, Minister of Foreign Affairs of the Maldives, *Statement at High-Level Segment of the Seventh Session of the Human Rights Council* (4 March 2008) <www.foreign.gov.mv/v3/?p=speech&view=all>.

In March 2008, the Maldives submitted Draft Resolution A/HRC/7/L.21/Rev.1 to the attention of the HRC, with the support of 78 co-sponsors,[16] which prompted the Committee to unanimously adopt Resolution 7/23 on 'Human Rights and Climate Change'.[17]

The Preamble to Resolution 7/23 acknowledges the impact of climate change on 'the full enjoyment of human rights', based on the UN Charter and the International Bill of Rights. This statement is crucial and has been effectively defined as an 'entry point' for human rights to climate change.[18] The Preamble also recognises the role of the UNFCCC as the reference general framework for addressing global warming. Furthermore, although it does not explicitly mention a right to a sustainable environment, the Preamble points out the crucial importance of the right to a sustainable development, including the environmental needs of the present and future generations, in light of HRC Resolution 2005/60. Moreover, the Preamble stresses the link between the enjoyment of human rights, environmental protection, the right to health and climate change.

The Resolution finally recommended that the UN HCHR conduct a study and produce a report on the relationship between climate change and human rights, in consultation with other governmental and non-governmental bodies, including the IPCC and the Secretariat of the UNFCCC, to be made available to the Conference of the States Parties to the UNFCCC.

Resolution 7/23, and, in particular, the recognition of the impact of climate change on the enjoyment of human rights, was approved despite the scepticism of major oil producing and consuming countries.[19] For instance, the US described the current human rights protection system, based on first and second generation human rights, as 'ill-equipped' with respect to the phonemenon of climate change, pointing out a discrepancy between individually oriented and State-based human rights on the one hand, and aggregate anthropogenic GHG emissions on the other.[20]

2.2. The OHCHR Report: outlining a framework for climate change and human rights

In the follow-up to HRC Resolution 7/23, on 15 January 2009 the OHCHR released an analytical report on human rights affected by climate change, which

16 HRC, *Report on Its Seventh Session*, UN Doc A/HRC/7/78 (14 July 2008) 149, [276].

17 HRC (2008) 65. The Preamble to the Resolution recognises the particular impact of climate change on the 'world's poor' and '*low-lying and other small island countries, countries with low-lying coastal, arid and semi-arid areas* or areas liable to floods, drought and desertification, and developing countries with fragile mountainous ecosystems' (emphasis added).

18 Marc Limon, 'Human Rights and Climate Change: Constructing a Case for Political Action' (2009) 33 *Harvard Environmental Law Review* 439, 467.

19 Knox (2009) 490; Limon (2009) 445.

20 *Observations by the United States of America on the Relationship between Climate Change and Human Rights* (2008) [22]–[26] <www.ohchr.org/Documents/Issues/ClimateChange/Submissions/USA.pdf>.

was then presented a year later.[21] The report was conducted in consultation with States and relevant international organisations, including the IPCC and the Secretariat of the UNFCCC. This is a first comprehensive evaluation of the relationship between climate change and fundamental rights, against the background of environmental protection. The report includes, first, an analysis of the negative effects of climate change on human rights, and, second, general inferences on the role that human rights can play in fighting climate change.[22]

In light of scientific data provided in the fourth IPCC assessment,[23] the OHCHR report does not explicitly state that anthropogenic GHG emissions violate human rights,[24] but recognises the impact of climate change for the 'full range of human rights'.[25] This impact is global, long-lasting, intergenerational, and indirect, to the extent that it is mediated by environmental deterioration caused by global warming.[26] The environment thus emerges as the key factor in linking climate change and human rights, and is particularly threatened by shrinking ice, rising sea level and extreme weather events.[27] Environmental protection is clearly regarded as a fundamental right, not as a simple principle of international law. Indeed, although the 'right to a safe and healthy environment' is not explicitly considered to have a human rights nature, the report does not hesitate to refer to human rights instruments embedding a right to live in a safe and healthy environment. In addition to the Stockholm and Rio Declarations, of particular relevance are the ACHPR and the San Salvador Protocol to the ACHR.[28]

In the chain of causation linking climate change and human rights, displacement and increased conflicts are indicated as major causes of human rights violations.[29] The effects of climate change are expected to have a negative impact, in particular, on the rights to life, adequate food, water, health, adequate housing and self-determination.[30] Specific groups, such as women, children and indigenous people are indicated as particularly affected by climate change.[31] In addition to substantive rights, procedural rights to access information and participation are seen as crucial.[32]

21 OHCHR, *Report on the Relationship between Climate Change and Human Rights*, A/HRC/10/61 (15 January 2009).
22 For an analytical presentation of the report, see Knox (2009) 484; Marc Limon, 'Human Rights Obligations and Accountability in the Face of Climate Change' (2009–10) 38 *Ga J Int'l & Comp L* 543, 548–569; Limon (2009) 446–7.
23 IPCC, *Fourth Assessment Report*, AR4 (2007).
24 See Knox (2009) 488.
25 OHCHR (2009) [20].
26 Ibid, [16]–[21], and [89]–[91].
27 Ibid, [8].
28 Ibid,[17], footnote 17.
29 Ibid, [55] ff.
30 Ibid, [21] ff.
31 Ibid, [48] ff.
32 Ibid, [78]–[79].

From the standpoint of remedies, both mitigation and adaptation measures are indicated as possible means to protect human rights from climate change, particularly within the context of international cooperation.[33] This has been interpreted as possibly overriding the territorial application of fundamental rights.[34] However, the report warns against possible threats posed to human rights by adaptation and mitigation measures, such as the negative impact of agro-fuel production on food availability.[35]

Finally, general inferences highlight that, although useful, human rights are probably not perfectly suitable to fight climate change, to the extent that they are individual and jurisdiction-based.[36] In particular, according to the report, it would be difficult to prove the responsibility of specific legal persons for specific human rights violations, especially in light of the complex causal link between climate change and fundamental rights. In this respect, the report includes the reservations expressed as to Resolution 7/23, with regard to the possibility of relying on human rights against climate change,[37] although it is finally assumed that these impediments should be overcome by the obligation to fulfil human rights.[38]

Overall, the report sketches a general outline of the relationship between climate change and human rights, and, to date, is the cornerstone for any further study and action. Politically, it can be considered a milestone with regard to potential subsequent engagements. Scientifically, the value of this document is quite limited, since analysis is confined to a general overview of basic issues. In particular, the report does not say anything as to substantive and procedural consequences of human rights violations committed by means of GHG emissions.[39]

The content of the report was embedded in further HRC instruments, which spawned new HRC discussions and decisions on the relationship between climate change and human rights. In particular, Resolution 10/4 of 25 March 2009 convened a 'panel' on this topic in June 2009.[40] Resolution 18/22 of 30 September 2011 went a step further and organised an official 'seminar' with eminent keynote speakers, such as Mary Robinson, in February 2012.[41] The results of these undertakings were made available to subsequent COPs to the UNFCCC. In 2014, by means of Resolution 26/L.33 the HRC decided to further engage State

33 Ibid, [12] ff, and [84] ff.

34 Knox (2009) 478.

35 OHCHR (2009) [65] ff. See also the chapter by Bridget Lewis in this volume.

36 Ibid, [69] ff.

37 See section 2.1 above.

38 OHCHR (2009) [86]–[87].

39 See also Margaretha Wewerinke, *Climate Change, Human Rights and the International Legal Order: the Role of the Human Rights Council* (University of Cambridge, 4CMR P Series, 2013) 3.

40 HRC, Resolution 10/4, *Human Rights and Climate Change* (25 March 2009) [1]. For a detailed analysis of Resolution 10/4 and the following debates, see Limon (2009–10) 559–82.

41 HRC, Resolution 18/22, *Human Rights and Climate Change* (17 October 2011) [2(a)]. See the speeches of the seminar at: <www.ohchr.org/EN/Issues/HRAndClimateChange/Pages/HRClimateChangeIndex.aspx>; also: <www.mrfcj.org>.

commitment and discussion as to the relationship between climate change and human rights.[42]

Moreover, it has been proposed that the HRC appoint a Special Rapporteur on the issue of climate change and human rights.[43] Although this has not happened, the HRC encouraged the holders of special procedures mandates to take the relationship between climate change and human rights into account.[44] This has taken place, in particular, with regard to special procedures on the rights to food and on adequate housing as a component of the right to an adequate standard of living and on the right to non-discrimination.[45] Practical assessments on the issues have followed,[46] and might eventually lead to addressing recommendations to States in the case of ascertained violations.[47]

The possibility has also been envisaged of relying upon the HRC Universal Periodic Review, in order to assess the impact of climate change on human rights globally and address recommendations to major GHG emitters.[48] To a certain extent, this has already taken place, since, for instance, the 2008 Universal Report by Tuvalu, significantly a small island State, highlights the effects of climate change on human rights.[49]

On 3 March 2014, the 25th session of the HRC opened with high-level statements expressing concern about the impact of climate change on human rights, particularly in light of the devastation caused by Typhoon Haiyan in the Philippines, and discussion is intensifying.[50]

3. UNFCCC mechanisms and human rights

The UNFCCC provides a framework for negotiating specific protocols setting binding limits on GHG emissions, in view of the stabilisation of 'greenhouse gas

42 HRC, *Draft Resolution on Human Rights and Climate Change*, A/HRC/26/L.33 (23 June 2014) <http://ieenvironment.org/wp-content/uploads/2013/05/A-HRC-26-L33.pdf>.

43 See, for instance, CIEL, *Human Rights and Climate Change: Practical Steps for Implementation* (2009) 17.

44 HRC (2009) [3].

45 HRC, Resolution 6/27, *Adequate Housing as a Component of the Right to an Adequate Standard of Living* (14 December 2007) [7]; Resolution 7/14, *The Right to Food* (27 March 2008).

46 *Report of the Special Rapporteur on Adequate Housing as a Component of the Right to an Adequate Standard of Living, and on the Right to Non-discrimination in this Context, Raquel Rolnik: Addendum: Mission to Maldives*, UN Doc HRC/13/20/Add.3 (11 January 2010) [68]; Olivier de Schutter, *Press Release: Climate Change, a Ticking Time Bomb Food Security*, Geneva (16 December 2009) <www.srfood.org/index.php/en/areas-of-work/food-production-and-resources/climate-change>.

47 For a detailed analysis of the possible integration of climate change into human rights special procedures, see CIEL (2009) 13 ff.

48 CIEL (2009) 22–3.

49 HRC, Working Group on the Universal Periodic Review, *National Report Submitted in Accordance with Paragraph 15(A) of the Annex to Human Rights Council Resolution 5/1*, UN Doc A/HRC/WG.6/3/TUV/1 (12 September 2008) (Tuvalu) [2], [42]–[43], [56]–[57], [75].

50 *Human Rights Council Hears Statements on the Impacts of Climate Change* (15 March 2014) <www.regionalinterfaith.org.au/home/worldwide-news/1140-human-rights-council-hears-statements-on-the-impacts-of-climate-change.html>; HRC, *Human Rights and Climate Change*, Concept note (6 March 2015).

concentrations in the atmosphere at a level that would prevent dangerous anthropogenic interference with the climate system'.[51] This aim is underpinned by the principle of common but differentiated responsibility,[52] which is at least corroborated by the idea of a right to a sustainable environment.[53]

UNFCCC parties have met annually from 1995 in Conferences (COPs) to assess progress as to climate change. In 1997, COP-3 adopted the Kyoto Protocol, establishing legally binding obligations for developed countries to reduce their GHGs. The Protocol entered into force in 2005 and its first commitment period covered the time-span 2008–12, with an overall reduction target of 5 per cent compared to 1990 GHG emission levels.[54]

Following the Malé Declaration and the OHCHR Report on Climate Change and Human Rights, at the end of COP-15 in Copenhagen the Ad hoc Working Group on Long-term Cooperative Action under the UNFCCC acknowledged the relevance of human rights to climate change.[55] Furthermore, in its 2010 Report on the Cancun Conference on climate change COP-16 recognised the necessity of 'fully respecting human rights' in all climate change-related actions.[56] This is an 'entry point' for human rights to the UNFCCC, which follows and parallels HRC Resolution 7/23.[57]

Along these lines, procedural mechanisms have been envisaged in order to facilitate the integration of human rights issues within the UNFCCC regime. In particular, the possibility of creating a UNFCCC subsidiary body or expert group on climate change and human rights has been put forward, so as to 'inject' a human rights-based approach into UNFCCC negotiations.[58]

In 2012, Doha COP-18 outlined a second commitment period under the Kyoto Protocol for the time-span 2013–20, including an extended number of countries and more compelling targets, aiming to reduce GHG emissions by 18 per cent compared to 1990 levels.[59] Amendments concerning the second commitment period have not yet entered into force, owing to an insufficient number of ratifications, so that, for the time being, the new targets are voluntary.

51 UNFCCC, Art 2.
52 Ibid, Preamble.
53 See also Limon (2009) 474.
54 Kyoto Protocol, Article 3(1).
55 Ad hoc Working Group on Long-term Cooperative Action under the UNFCCC, *Report on Its Eight Session*, FCCC/AWGLCA/2009/17 (5 February 2010) Preamble.
56 UNFCCC COP, *Decision 1/CP.16, The Cancun Agreements: Outcome of the Work of the Ad Hoc Working Group on Long-term Cooperative Action under the Convention*, UN Doc FCCC/CP/2010/7/Add.1 (15 March 2011) 4, [8] (emphasis added). For further references to human rights in UNFCCC negotiations see Limon (2009–10) 582–6.
57 See section 2.1 above.
58 CIEL (2009) 31–2.
59 UNFCCC-COP (Serving as the Meeting of the Parties to the Kyoto Protocol), Report, Addendum, Doha, 26 November–8 December 2012, FCCC/KP/CMP/2012/12Add.1 (2013) Decision 1/CMP8, *Doha Amendment to the Kyoto Protocol Pursuant to Its Article 3, Annex I*, 10, [c], inserting new para 3bis into the Kyoto Protocol.

The effectiveness of the Kyoto Protocol is also limited by the fact that the US is not a party to it, Canada withdrew in 2012, and Japan and Russia have indicated that they do not intend to assume GHG reduction targets under the second commitment period.[60] Furthermore, the Protocol does not embed effective monitoring mechanisms for compelling States to comply with their reduction targets, let alone with their human rights obligations.

Currently, within the UNFCCC regime governments envisage the negotiation of a new universal climate change agreement, which is set to be adopted in Paris in December 2015 and enter into effect in 2020. Preliminary material includes enhanced action in order to achieve the 'highest possible mitigation efforts under the Convention by all Parties' and limit 'global average temperature increase to below 2°C or 1.5°C'.[61] However, no explicit reference to human rights has so far been incorporated. Therefore, the 2010 'entry point' has not 'kick-started' the expected 'organic assessment' of the 'application of human rights principles in the context of climate change policy'.[62]

4. The stand of other UN organs

In addition to the general UN human rights protection mechanisms, which naturally play a key role, specialised UN bodies might address or have effectively addressed the relationship between climate change and human rights. Owing to the 'horizontal' nature of human rights, different UN institutions are potentially involved.

For instance, in its 2007/2008 Human Development Report, the UN Development Programme argued that climate change represents 'a systematic violation of the human rights of the world's poor and future generations, and a step back from universal values'.[63] In June 2009, after the Security Council started discussing the impact of GHG emissions on international peace and security the UN General Assembly adopted Resolution 63/281, addressing the security implications of climate change.[64] In the outcome document of the Rio+20 Conference, UN Member States committed to addressing the negative impact of climate change on human rights.[65]

UN human rights treaty bodies can play a particularly important role in compelling States to address the relationship between climate change and human rights. These organs range from the HRCte to the CESCR and specialised treaty bodies dealing with vulnerable groups. Although the effectiveness of their

60 UNFCCC-COP, *Doha Amendment* (2013) *Annex I*, 9–10.

61 Ad Hoc Working Group on the Durban Platform for Enhanced Action, *Accelerating the Implementation of Enhanced Pre-2020 Climate Action* (ADP.2014.8.DraftText) [3]–[10].

62 Limon (2009) 467.

63 UNDP, *Human Development Report 2007/2008: Fighting Climate Change in a Divided World* (2008) 4.

64 UN GA, *Climate Change and Its Possible Security Implications*, A/RES/63/281 (11 June 2001).

65 Rio+20, UN Conference on Sustainable Development, *Outcome of the Conference – The Future We Want*, A/CONF.216/L.1 (19 June 2012).

procedures is hampered by the fact that they can only result in recommendations, that is, non-compulsory measures, these bodies can assess the impact of climate change on the effective enjoyment of human rights in concrete situations, by means of the periodic reporting process or individual communications. Further insights can be developed by these organs via instruments such as general comments and observations.

Significantly, the CESCR has both explicitly and implicitly developed climate change considerations based on different ICESCR provisions. In its General Comment No 15, focusing on Articles 11 and 12 ICESCR, the Committee highlighted the necessity that States 'assess the impacts of actions that may impinge upon water availability and natural-ecosystems watersheds, such as *climate change*, desertifications and increased soil salinity, deforestation and loss of biodiversity'.[66] In General Comment No 12, considering Article 11 ICESCR, the Committee noted that the 'precise meaning of "adequacy" [of food availability and access] is to a large extent determined by prevailing social, economic, cultural, *climatic*, ecological and other conditions'.[67]

In its 2009 Concluding Observations on Australia, the CESCR expressed concerns about climate change affecting the right to an adequate standard of living, with particular regard to indigenous people, including the rights to food and water. As a consequence, the Committee encouraged Australia to 'intensify its efforts to address issues of climate change, including through carbon reduction schemes' and to 'reduce its GHG emissions and take necessary and adequate measures to mitigate the adverse consequences of climate change, impacting the right to food and the right to water for indigenous people'.[68]

In one of its statements on the World Food Crisis, the Committee invited States parties to implement 'strategies to combat global climate change that do not negatively affect the right to adequate food and freedom from hunger, but rather promote sustainable agriculture, as required by Article 2 of the United Nations Framework Convention on Climate Change'.[69]

It is noteworthy that explicit references to climate change increase in later observations of the CESCR.[70]

66 CESCR, General Comment No 15, *The Right to Water (Arts 11 and 12 of the Covenant)* (20 January 2003) [28] (emphasis added).

67 CESCR, General Comment No 12, *The Right to Adequate Food (Art 11)* (12 May 1999) [7] (emphasis added).

68 CESCR, *Concluding Observations: Australia*, E/C.12/AUS/CO/4 (12 June 2009) [27].

69 CESCR, *The World Food Crisis*, E/C.12/2008/1 (20 May 2008) [13].

70 For an overview of the stands taken by the CESCR on the relationship between climate change and second generation human rights, see Marcos A Orellana, Miloon Kothari and Shivani Chaudry, *Climate Change in the Work of the Committee on Economic, Social and Cultural Rights* (CIEL, 2010) 20 ff.

5. Bridging the gap between the UNFCCC and human rights

The above analysis shows that there is a relatively recent, but increasing, recognition of the relationship between climate change and human rights within the context of the classical UN human rights protection mechanisms, especially via the HRC. The concrete results of this recognition are still unclear.[71] For the time being, it has produced explorative studies on the relationship between climate change and human rights, but it has proved difficult to concretely integrate the outcome of these studies within the UNFCCC framework.

From a general standpoint, the whole UNFCCC regime can be interpreted as a response to climate change, and thus to the obligations to respect, protect and fulfil human rights. In other words, the international community has already positively answered the question as to whether or not it is necessary to act against anthropogenic GHG emissions.[72] The fact that this question was not initially framed as a human rights one is not particularly relevant, to the extent that it is also a human rights issue. Of course, this does not exhaust human rights questions raised by climate change, with particular regard to the issue as to whether or not the current response is sufficient in light of the existing international human rights protection system.

From a normative standpoint, it has been correctly noted that the UNFCCC permits a human rights interpretation of climate change obligations.[73] Indeed, the objective of avoiding dangerous GHG interference with the climate system under Article 2 UNFCCC must be read within the context of other treaty provisions. Namely, the Preamble to the Convention acknowledges that climate change and its adverse effects are ‘a common concern of humankind’. ‘Adverse effects’ are defined in Article 1 as environmental changes affecting not only ‘resilience or productivity of natural and managed ecosystems’, but also the ‘operation of socio-economic systems . . . human health and welfare’.[74] Furthermore, all States must take precautionary measures to achieve these objectives, based on the principle of ‘common but differentiated responsibilities and respective capabilities’.[75] This principle entails for developed countries an obligation to take the lead in combating climate change, considering the needs of developing countries, ‘especially those that are particularly vulnerable to the adverse effects of climate change, and of those Parties, especially developing country Parties, that would have to bear a disproportionate or abnormal burden under the Convention’.[76] Of course, these aims are reinforced by the Kyoto Protocol, which pursues the ‘ultimate objective

71 See HRC (2015).
72 Limon (2009) 448.
73 Wewerinke (2013) 6.
74 UNFCCC, Article 1(1).
75 Ibid, Article 3(3).
76 Ibid, Article 3.

of the Convention'.[77] Such a regulatory approach facilitates a mutually reinforcing interpretation of the UNFCCC regime and human rights, according to the principle of systemic integration, thus reading human rights instruments and the UNFCCC as compatible.[78] This view integrates the 'horizontal' nature of human rights and complements a strict application of the principles of general and special laws, which requires a case-by-case comparison between, on the one hand, human rights treaties, and, on the other, the UNFCCC regime.[79]

Concretely, with regard to the impact of human rights on the UNFCCC regime, a systemically integrated approach should lead to results such as assessing the level of dangerousness of anthropogenic GHG emissions under Article 2 of the UNFCCC, determining States' mitigation targets under the Kyoto Protocol, in particular, in light of the right to development, and taking adaptation measures based on most affected human rights.[80] More specifically, these achievements can be realised through such elements as considering basic individual and collective claims alongside scientific evidence, as well as enhanced accountability and respect for the principle of non-discrimination and equity. Procedurally, human rights would entail stronger international cooperation, transparency and participation in decision-making, involving a level playing field in negotiations otherwise driven by economically powerful States.[81] These achievements might be assessed via the UN human rights implementation mechanisms, which could thus fill the procedural gap currently affecting the UNFCCC regime and lead to the potential imposition of classical reparation measures under the law of State responsibility.[82] From the point of view of mutual supportiveness, these accomplishments would be essential.[83] In fact, the current UNFCCC regime has proved largely insufficient for ensuring environmental sustainability. Therefore, strengthening mitigation and adaptation targets within the UNFCCC system based on human rights obligations would be not only systemically correct, but also necessary.[84]

With regard to the impact of climate change on different human rights treaties, several implications can be envisaged. For instance, taking into account the issue of climate change within the context of the ICESCR and ICCPR might lead to

77 Kyoto Protocol, Preamble.

78 International Law Commission, *Fragmentation of International Law: Difficulties Arising from the Diversification and Expansion of International Law*, A/CN.4/L.682 (2006) 25 ff, [37] ff; Wewerinke (2013) 7.

79 See, for instance, Ottavio Quirico, 'Amidst Fragmentation and Coherence: a Systemic Interpretation of the World Heritage Convention and the UNFCCC Regime' (2012) 10 *New Zealand Yearbook of International Law* 33, 66 ff.

80 Simon Caney, 'Climate Change, Human Rights and Moral Thresholds', in Stephen Humphreys (ed), *Human Rights and Climate Change* (CUP, 2009) 69 ff.

81 Limon (2009) 450 ff; CIEL (2009) 31.

82 See the *Draft Articles on Responsibility of States for Internationally Wrongful Acts*, in ILC, *Report to the General Assembly*, UN Doc A/56/10 (2001) Arts 28–39; Wewerinke (2013) 9; Susanna Willcox, 'A Rising Tide: the Implications of Climate Change Inundation for Human Rights and State Sovereignty' (2012) 9 *Essex Human Rights Review* 1, 13.

83 Limon (2009) 466–7.

84 Ibid, 450.

an environmentally oriented interpretation of the rights to life and health. Most significantly, the impact of climate change on human rights should facilitate the definitive establishment of environmental protection as a human right, which is only an emerging issue in international law, but certainly not an established principle. Although it is not yet possible to maintain that climate change has led to creating a human right to a sustainable environment,[85] the impact of climate change on the environment is so invasive that it has the potential to ultimately determine the definitive international establishment of environmental protection as a human right. This is all the more possible in light of the fact that a sustainable environment is necessary for the further enjoyment of all human rights.[86] In turn, setting out a human right to a sustainable environment would facilitate the harmonisation of fundamental rights within the UNFCCC regime, integrating the principle of common but differentiated responsibility.[87]

Therefore, it can be concluded that the major achievement of the studies so far developed on the relationship between climate change and human rights is the identification of a relevant inconsistency in the human rights regulatory framework, that is, the absence of recognition of the human rights nature of the right to a sustainable environment.[88] Indeed, the existence of a right to a sustainable environment can be posited, especially in light of the acknowledgment of a correlative obligation not to emit transboundary pollution.[89] However, its human rights characterisation is far from established in international law.[90] This is all the more true if one thinks that the human rights nature of the claim to a sustainable environment is largely recognised by domestic constitutions,[91] and might thus already be a general principle of law, but is not yet regarded as a general principle of international law.[92] The identification of this normative inconsistency can be considered a major achievement, at least within the context of positivistic legal

85 See the contribution by Francesco Francioni and Ottavio Quirico in this volume.

86 See Fatma Zohra Ksentini, *Human Rights and the Environment, Review of Further Developments in Fields with which the Sub-Commission Has Been Concerned*, Final Report of the UN Sub-Commission on Prevention of Discrimination and Protection of Minorities, UN ECOSOC, E.CN.4/Sub.2/1994/9 (6 July 1994) [172] ff.

87 See section 3 above.

88 Limon (2009) 469, speaks of a regulatory 'gap'. However, it is unclear whether or not the author assumes the existence of a right to a sustainable environment having a non-human rights nature in international law.

89 See, in particular, *Trail Smelter (USA v Canada)*, Award of 16 April 1938 and 11 March 1941 (2006) III *RIAA* 1905. See also Siobhán McInerney-Lankford, Mac Darrow and Lavanya Rajamani, *Human Rights and Climate Change: a Review of the International Legal Dimensions* (World Bank Study, 2011) 46. On the fundamental correlation between rights and duties see Wesley Newcomb Hohfeld, *Fundamental Legal Conceptions as Applied in Judicial Reasoning and Other Legal Essays* (YUP, 1920) 65.

90 HRC Resolution 2005/60, adopted on 20 April 2005, only recognises the importance of the environment for the enjoyment of human rights.

91 David R Boyd, *The Environmental Rights Revolution: a Global Study of Constitutions, Human Rights, and the Environment* (University of British Columbia Press, 2012).

92 See Karen E MacDonald, 'A Right to a Healthful Environment – Human Rights Habitats: Re-thinking Rights in an Age of Climate Change' (2008) 17 *EEELR* 213.

theory.[93] Of course, this calls for further investigation on the scope and extension in time of a possible human rights obligation to environmental protection, in light of the global and intergenerational nature of climate change.[94]

In this respect, it is noteworthy that building on resolutions related to human rights, the environment and climate change, particularly Resolutions 7/23, 10/4 and 18/22, the HRC established the mandate for the Independent Expert on Human Rights and the Environment via Resolution 19/10 of 22 March 2012, adopted by consensus as a revised draft just before the UN Conference on Sustainable Development was held in Rio in June 2012.[95] Anticipating the principles of the Rio Declaration on Environment and Development and reaffirming the Millenium Development Goals, the Resolution is based on the central tenet of environmental sustainability.[96] Inter alia, Resolution 19/10 requests the Independent Expert to study the human rights obligations relating to the enjoyment of a safe, clean, healthy and sustainable environment, adding a fundamental rights perspective to processes following Rio+20.[97] The independent expert, John Knox, was appointed in June 2012, with a three-year mandate renewed in March 2015 as a Special Rapporteur. So far, his work has resulted in three reports, which introduce and map the relationship between a sustainable environment and human rights, based on procedural and substantive obligations.[98] The reports highlight, in particular, the relationship between environmental degradation and the rights to life and health, and include a 'substantive obligation to adopt legal and institutional frameworks that protect against environmental harm'.[99] Within this framework, along the lines of Resolution A/HRC/26/L.33 the work of the Rapporteur led to drafting a Mapping Report pointing out existing relationships between climate change and human rights in basic regional and international human rights instruments.[100] Moreover, together with HRC

93 Norberto Bobbio, *Teoria Generale del Diritto* (Giappichelli, 1993).

94 Limon (2009) 472–3.

95 HRC, *Resolution on Human Rights and the Environment*, A/HRC/19/L.8/Rev.1 (22 March 2012). Although the HRC established a broad mandate, the perspective of global warming has been included via a specific reference to the HRC Resolutions on climate change and human rights embedded in the opening statement.

96 Ibid, Preamble.

97 Ibid, [2].

98 John Knox, *Report to the HRC of the Independent Expert on the Issue of Human Rights Obligations relating to the Enjoyment of a Safe, Clean, Healthy and Sustainable Environment*, UN Doc A/HRC/22/43 (24 December 2012); Id, UN Doc A/HRC/25/53 (30 December 2013) [47]; Id, UN Doc A/HRC/28/61 (3 February 2015) and Add.1 (Mission to France).

99 Knox (2013) [47]. Via consultations and visiting, the Rapporteur has so far studied the protection afforded to the right to environment in different legal systems, particularly Switzerland, France, Thailand, South Africa, Denmark, Panama and Kenya (<http://ieenvironment.org/consultations>).

100 HRC, *Mapping Human Rights Obligations relating to the Enjoyment of a Safe, Clean, Healthy and Sustainable Environment: Focus Report on Human Rights and Climate Change* (June 2014). See also John Knox et al, *The Effects of Climate Change on the Full Enjoyment of Human Rights: Report to the Climate Vulnerable Forum* (30 April 2015).

special procedures mandate-holders Knox commendably recommended that the new climate agreement to be negotiated in Paris at COP-21 in December 2015 take human rights into account, from both a substantive and procedural standpoint.[101]

Conclusion

The UN is at the forefront in the fight against climate change and has tackled the issue via two different channels, that is, the UNFCCC and classical human rights protection mechanisms, in particular, the HRC. Despite some attempts at integration, these avenues have so far proceeded separately, without merging with one another.

Political reasons, in particular, opposition by oil producing and consuming countries, can partially explain the substantive and institutional separation between human rights and climate change within the UN system. However, this argument is not decisive, because a formal recognition of the relationship between climate change and human rights has already taken place within the context of both the UNFCCC and HRC regimes, in addition to other UN institutional frameworks.

The separation also has normative roots, in particular, owing to the difficulty of establishing a causal link between global climate change and specific first and second generation human rights. Therefore, by shortening the chain of causation, the recognition of a fundamental right to a sustainable environment would facilitate a human rights-based approach to climate change. This would simplify the integration of human rights within the UNFCCC regime and, consequently, or even independently, trigger general and treaty-based UN monitoring mechanisms.

101 *A New Climate Change Agreement Must Include Human Rights for All: Open Letter to the States Parties to the UNFCCC* (17 October 2014).

14 Climate change and human rights in the Asia-Pacific

A fragmented approach

Ben Boer

Introduction

The Asia-Pacific stretches west to east from South Asia to Southeast Asia, to Northeast Asia and to the Pacific Islands. It is one of the world's regions most vulnerable to climate change, owing to long coastlines, high concentrations of population engaged in intense economic activity in low-lying coastal areas and river basins, a heavy reliance on agriculture, fisheries, forestry and exploitation of other natural resources.[1] The Fifth Report of the Working Group on Climate Change and Development states that 'the human drama of climate change will largely be played out in Asia, where over 60 per cent of the world's population, around 4 billion people, live'.[2]

The UN HRC has confirmed these observations through a series of resolutions over the past five years. These have underlined the negative impacts of climate change on vulnerable populations, and have drawn attention to the unequal impacts of climate change around the world, reiterating that 'the adverse effects of climate change will be felt most acutely by those segments of the population that are already in vulnerable situations owing to factors such as geography, poverty, gender, age, indigenous or minority status and disability'.[3]

1 See IPCC *Fifth Assessment Report* (2014) Synthesis; *AR4 Synthesis* (2007) 48. See also Stephen Tully, 'The Contribution of Human Rights as an Additional Perspective on Climate Change: Impacts within the Pacific' (2007) 5 *New Zealand Journal of Public International Law* 175 ff.

2 Working Group on Climate Change, *Up in Smoke – Asia and the Pacific* (2007) 3, available at <www.iied.org/pubs/pdfs/10020IIED.pdf>. The UN Economic and Social Commission for Asia and the Pacific stated in a 2013 report: 'Currently, there are 1.8 billion people in South and South-West Asia, 1.6 billion in North and North-East Asia, 0.6 billion in South-East Asia, 0.2 billion in North and Central Asia and 38 million in the Pacific. The region hosts the two most populous countries in the world: China with 1.4 billion people and India with 1.25 billion people' (*Population Trends in Asia and the Pacific*, available at <www.unescapsdd.org/files/documents/SPPS-Factsheet-Population-Trends-v3.pdf>).

3 *UN HRC Resolutions on Human Rights and Climate Change*, UN Docs A/HRC/26/L.33/Rev.1 (2014), A/HRC/RES/10/4 (25 March 2009) and A/HRC/RES/7/23 (28 March 2008). See also *Human Rights Council Open Letter from Special Procedures Mandate-holders of the Human Rights Council to the State Parties to the UN Framework Convention on Climate Change on the Occasion of the Meeting of the Ad Hoc Working Group on the Durban Platform for Enhanced Action in Bonn* (20–25 October 2014), available at <www.ohchr.org/Documents/HRBodies/SP/SP_To_UNFCCC.pdf>.

The most direct effects are due to sea-level rise (resulting from expansion of the oceans through warming). They include submergence, flooding, erosion and reduction of agricultural and residential land, with consequent involuntary internal and transboundary displacement, as well as significant reduction in the capacity of some countries to produce their own food. This causes increased food insecurity, loss of cultural identity and depletion of biological diversity, threats to territorial integrity and to the viability of States, and in the case of low-lying island States, threats to physical integrity as well. [4] All of these effects have human rights dimensions.

Human rights and human security in the context of climate change in Asia and the Pacific have gained more focus in recent years, with publications and reports being prepared on a regular basis.[5] This chapter addresses the issue in part within the framework of the right to development, which is closely intertwined with all other fundamental rights. Clearly though, the 'third generation' human right to development, set out in a 1986 UN General Assembly Declaration,[6] while characterized as 'an inalienable human right', is nuanced by the legal principles set out in the Brundtland Report,[7] and the ensuing 1992, 2002 and 2012 UN Conferences on environment and development, underpinned by the increasing constraints imposed by international environmental law regimes. Thus, the right to development, considered in conjunction with the concept of sustainability,[8] is now generally understood to incorporate social, economic, cultural and environmental elements as set out in various official statements on sustainable development.[9]

4 See IPCC (2014) Chapter 24, Asia, Chapter 25, Australasia, Chapter 29, Small Islands and Chapter 30, The Ocean. It is also the case that some of the major GHG emitters are located in Asia, with three countries being in the top 6 emitters. China is currently the world's highest emitter at around 29 per cent of total GHGs, followed by the United States at 15 per cent, the European Union at 11 per cent, India at 6 per cent and Japan at 4 per cent. China and India both have low per capita emission levels.

5 For example, see Malcolm Macintosh and Tapan Sakar, 'Climate Change in Human Security Issues in the Asia-Pacific region', in Moazzem Hossain and Anthony Selvanatha (eds), *Climate Change and Growth in Asia* (Edward Elgar, 2011) 214. See also International Bar Association, *Achieving Justice and Human Rights in an Era of Climate Disruption*, Climate Change Justice and Human Rights Task Force Report (2014) <www.ibanet.org/PresidentialTaskForceCCJHR2014.aspx>.

6 UN GA, *Declaration on the Right to Development*, A/RES/41/128 (1986).

7 World Commission on Environment and Development, *Our Common Future* (Brundtland Report, 1987) Annex on Legal Principles: 'All human beings have the fundamental right to an environment adequate for their health and well-being.'

8 See Klaus Bosselmann, *The Principle of Sustainability: Transforming Law and Governance* (Ashgate, 2008).

9 This was confirmed in the 2002 Johannesburg Declaration, characterizing economic development, social development and environmental protection as 'interdependent and mutually reinforcing pillars of sustainable development' (Annex, [5]), as well as in the Outcome Document of the Rio+20 Conference on Environment and Development (*The Future We Want* (2012)), and in the Open Working Group Proposal for Sustainable Development Goals (see <http://sustainabledevelopment.un.org/focussdgs.html>).

Institutionally, unlike the European, Inter-American and African human rights regimes, there is as yet no comprehensive human rights protection system in the Asia-Pacific. However, there is a slow movement across the region from classical first and second generation human rights claims towards third generation claims, particularly in Southeast Asia and, to an extent, in the Pacific Island region.[10] The signs of progress, though still largely incoherent, are apparent both in terms of institutional development as well as with regard to the constitutional interpretation of environment-related human rights provisions in the courts, particularly in India, Pakistan and the Philippines.[11] The climate change implications of human rights in this chapter are thus primarily considered through the eyes of the regional organizations, and particularly those that focus on environmental issues.

1. Developing a coherent human rights approach to climate change: initiatives in Asia and the Pacific

While a robust approach to climate change and human rights has yet to emerge in the Asia-Pacific region, some human rights soft law declarations and informal statements over the past 25 years have included provisions relevant to climate change, and give some indication of regional trends in thinking on these issues.[12]

Among these initiatives, the Cairo Declaration on Human Rights in Islam,[13] adopted by the Organisation of the Islamic Conference and signed by several Asian countries in addition to Middle-Eastern countries, provides that 'everyone shall have the right to live in a clean environment', which entails a correlative duty to be fulfilled by the 'State and society in general' (Article 17(a)). This is established in connection with the right of individuals to a decent living that may enable them to meet their requirements and those of their dependents, including food, clothing, housing, education, medical care and all other basic needs (Article 17(c)).

Further, the 1993 Final Declaration of the Regional Meeting for Asia of the World Conference on Human Rights, in line with the Declaration on the Right to Development, affirms the 'right to development' as a 'universal and inalienable right and an integral part of fundamental human rights', to be realized via a

10 Australia and New Zealand, although forming part of the Asia-Pacific region, are not canvassed in this chapter from an institutional viewpoint; however, it should be noted that they do have well-developed human rights institutions, namely the Australian Human Rights Commission and the New Zealand Human Rights Commission.

11 See Ben Boer, 'Environmental Law and Human Rights in the Asia-Pacific', in Ben Boer (ed), *The Environmental Law Dimensions of Human Rights* (Oxford, 2015); Parvez Hassan and Azim Azfar, 'Human Rights and the Environment: a South Asian Perspective' (2014) 5(2) *Journal of Human Rights and the Environment* 192, 199 ff.

12 A number of these statements are available at <www.hurights.or.jp/archives/other_documents> (*Human Rights Declarations in the Asia-Pacific*).

13 *Cairo Declaration on Human Rights in Islam* (5 August 1990).

cooperative framework (Principle 17). This right encompasses the 'need to develop the right of humankind regarding a clean, safe and healthy environment'.[14]

In light of the threats posed by sea-level rise to small island States, in November 2007 the Republic of Maldives took the lead in the proclamation of the Malé Declaration on the Human Dimension of Global Climate Change. The Declaration was adopted by a wide range of small island developing States, recalling the 'fundamental right to an environment capable of supporting human society and the full enjoyment of human rights', as recognized 'in the constitutions of over one hundred States and directly or indirectly in several international instruments'. On this basis, the Declaration called for the inclusion of human rights in UNFCCC negotiations and became a crucial entry point for human rights in the climate change discourse, although the two areas have been integrated at the UN level only to a limited extent.[15]

The Tehran Declaration on Human Rights and the Environment, drafted as a result of a conference involving a number of international participants from the Asia-Pacific and elsewhere, immediately prior to the 2009 Copenhagen UNFCCC Conference of the Parties, sought to draw the link between climate change and human rights. The Declaration entreated the UNFCCC parties to 'ensure that appropriate measures are taken and requisite financing is provided for mitigation and adaptation to address the impacts that climate change has on the full realization of human rights and the well-being of all people'.[16]

2. Regional evolution of human rights and climate change concerns in Asia

For the purposes of addressing the institutional aspects of the development of policies and law relating to climate change and human rights through the prism of the regional environment programmes, we can divide the Asian region into South Asia, Northeast Asia and Southeast Asia.[17]

14 This is an initiative of the Ministers and representatives of Asian States, who met in Bangkok from 29 March to 2 April 1993, pursuant to UN General Assembly Resolution 46/116 of 17 December 1991, in the context of preparations for the World Conference on Human Rights.

15 On the Malé Declaration and its impact on the UNFCCC see the chapter by Spyridon Aktypis, Emmanuel Decaux and Bronwen Leroy in this volume.

16 The Tehran Declaration was distributed at the 2010 Copenhagen Conference of the Parties and published in 39/6 *Environmental Policy and Law* (2009) 327. See also 16(4) *International Journal of Cultural Property* (2009) 379. For comment see Janet Blake and Ben Boer, 'Human Rights, the Environment and the Tehran Declaration' (2009) 39(6) *Environmental Policy and Law* 302, 302–307.

17 UNEP also conducts region-wide initiatives supportive of the regional cooperative environment programmes, <www.unep.org/roap/Activities/MainstreamingEnvironmentthroughRegionalForums/tabid/1036898/Default.aspx>.

2.1. South Asia

The South-Asian Association for Regional Cooperation (SAARC), comprising Afghanistan, Bangladesh, Bhutan, India, Maldives, Nepal, Pakistan and Sri Lanka, hosts the South-Asian Co-operative Environment Programme, known as SACEP.[18] SACEP's mission is:

> to promote regional co-operation in South Asia in the field of environment, both natural and human, in the context of sustainable development and on issues of economic and social development which also impinge on the environment and vice versa; to support conservation and management of natural resources of the region and to work closely with all national, regional, and international institutions, governmental and non-governmental, as well as experts and groups engaged in such co-operation and conservation efforts.[19]

A Convention on Cooperation on Environment was concluded in 2004 by SAARC with the encouragement of UNEP, but it has not been ratified by all countries, and consequently is yet to have any substantive effect in generating significant action with regard to regional environmental management.

In 2007, the SAARC Council of Ministers stated that 'given the vulnerabilities, inadequate means and limited capacities, there was a need to ensure rapid social and economic development to make SAARC climate change-resilient.'[20] The results were the Dhaka Declaration on Climate Change and the SAARC Action Plan on Climate Change of 2008.[21] The Declaration envisages a proactive approach by States to the issue of global warming, indirectly invoking the fundamental right to life in its provisions.[22]

The most recent State of the Environment Report for SAARC encompasses development, water, health and life protection. Although these issues are not directly considered from a human rights standpoint, they all have obvious human rights implications. The report states:

> South Asia is very vulnerable to climate change . . . This will exacerbate the challenges of poverty reduction and improving access to safe drinking water, two of the Millennium Development Goals . . . Poor air quality threatens human health and causes other forms of environmental damage.[23]

18 SACEP was established by the 1981 Colombo Declaration on the South Asia Co-Operative Environment Programme.

19 See <www.sacep.org/html/about_visionmission.htm>.

20 Ibid.

21 See SAARC, *Summary, Dhaka Declaration and SAARC Action Plan on Climate Change* (2008) <http://saarc-sec.org/areaofcooperation/cat-detail.php?cat_id=54>.

22 'Member States . . . initiate and implement programmes and measures as per SAARC practice for adaptation for dealing with the onslaught of climate change to protect the lives and livelihood of the people [. . .]' (*Dhaka Declaration on Climate Change*).

23 SAARC, *State of the Environment Report* (2009) Foreword, iii, xiii <www.saarc-sec.org/userfiles/SAEO%202009.pdf>.

Further, the 2010 Thimphu Statement on Climate Change refers to the protection of 'archaeological and historical infrastructure of South Asia from the adverse effects of Climate Change'[24] and to the 'conservation of bio-diversity and natural resources', which are essential components of the right to culture.[25]

In addition, the 2013 meeting of the SACEP Governing Council agreed that regional cooperation is necessary in order to address climate change issues.[26] However, there appears to be no official discussion by SACEP on the links between climate change and human rights, despite the fact that a number of SACEP members, including Bangladesh and Maldives, already seriously suffer the effects of sea-level rise in terms of inundation of their coastal lands and major social and economic disruption as a consequence.[27]

The most recent summit meeting of the SAARC Council issued the Kathmandu Declaration, in which national leaders, inter alia, directed relevant bodies and mechanisms to implement the SAARC Agreement on Rapid Response to Natural Disasters, the SAARC Convention on Cooperation on Environment and the Thimphu Statement on Climate Change, including taking into account the continuing threats posed by climate change to some SAARC Member States.[28]

2.2. Northeast Asia

In Northeast Asia, the Northeast Asian Subregional Programme for Environmental Cooperation (NEASPEC) is coordinated by the UN Economic and Social Commission for Asia and the Pacific (ESCAP). NEASPEC comprises China, Japan, Mongolia, the Republic of Korea, the Democratic People's Republic of Korea and the Russian Federation. The Programme, initiated in 1993, aims to 'redress the dichotomy between economic growth and environmental protection and improve eco-efficiency and environmental sustainability for the economic development of the sub-region'.[29] Its Senior Officials meet once a year, and a

24 *Thimphu Statement* (2000) [VIII] <www.saarc-sec.org/userfiles/ThimphuStatementonClimateChange-29April2010.pdf>.

25 Ibid, [XI].

26 'Regional Cooperation Imperative to Solve Effects of Climate Change', *Dawn* (3 December 2013) <www.dawn.com/news/1060826>.

27 For example, see Ademola K Braimoh and He Qing Huang, *Vulnerability of Land Systems in Asia* (Wiley, 2014) passim; Sumudu Atapattu, *Climate Change in South Asia: towards an Equitable Legal Response within a Framework of Sustainable Development and Human Security* (IDLO, 2011) 21. With respect to Bangladesh, Moazzem Hossain, AHM Ali and Eliyathamby Selvanathan state: 'It is now certain that within the next two to three decades, Bangladesh will face a major catastrophe in both economic and social terms as it witnesses millions of people moving north to search for higher grounds as climate change refugees . . .' ('Analyses of Livelihoods in the Bay of Bengal Delta', in Moazzem Hossain and Eliyathamby Selvanathan (eds), *Climate Change and Growth in Asia* (Edward Elgar, 2011) 78, 94).

28 *Kathmandu Declaration* (November 2014) <www.onlinekhabar.com/2014/11/211832>.

29 *About NEASPEC*, at <www.neaspec.org/sites/default/files/About%20NEASPEC.pdf>.

number of environmental cooperation programmes have been initiated, including on biodiversity and on transboundary air pollution.

Notwithstanding the fact that regional countries host about one quarter of the world's population and are responsible for around one third of global carbon emissions,[30] there is no specific NEASPEC programme focused on any aspect of mitigation of or adaptation to climate change, although there have been tentative steps in this direction with the discussion of carbon footprint labelling and low-carbon cities under an eco-efficiency partnership initiative.[31] Given the history of interaction between its members, as well as their diverging national political philosophies, the fact that there is no evidence of any discussion of a regional human rights scheme within the NEASPEC region is not unexpected.

2.3. Southeast Asia: ASEAN initiatives

Southeast Asia comprises Brunei, Cambodia, Indonesia, Laos, Malaysia, Myanmar, Singapore, the Philippines, Thailand and Vietnam, which together constitute the Association of Southeast Asian Nations (ASEAN). Timor is yet to be accepted as a member.

ASEAN hosts a relatively substantial environment programme established in the late 1970s, with a number of major initiatives and regular meetings of the ASEAN Senior Officials on the Environment.[32] It has also issued a wide range of environmental declarations and charters, several of which include specific provisions on climate change policy and the need for member countries to mitigate and adapt to the effects of climate change. Among these is the 2007 ASEAN Declaration on Sustainability, which devotes a number of paragraphs to climate change and its mitigation. ASEAN also has a specific programme on cooperation concerning climate change[33] and concluded a Haze Agreement to address transboundary air pollution.[34] Thus, in contrast to other countries in the Asia-Pacific, where human rights issues associated with climate change have made very limited impact either politically or institutionally, ASEAN has taken a clearer stance.

In the past decade, ASEAN member countries have committed themselves more particularly to the promotion of human rights. The ASEAN Charter was negotiated and agreed upon in 2007, investing ASEAN with permanent legal personality, and committing ASEAN Member States to the principles of democracy, the rule of law and good governance, and respect for and protection of human rights and fundamental freedoms.[35] Part of this process was the decision

30 *NEASPEC's Milestones*, <www.neaspec.org/sites/default/files/About%20NEASPEC.pdf>.
31 See <www.neaspec.org/our-work/eco-efficiency-partnership>.
32 *ASEAN Cooperation on the Environment*, <http://environment.asean.org>.
33 *ASEAN Cooperation on Climate Change*, <http://environment.asean.org/asean-working-group-on-climate-change>.
34 *ASEAN Agreement on Transboundary Haze Pollution*, <http://haze.asean.org/?page_id=185>.
35 ASEAN Charter (2007), Preamble and Article 1(7).

to establish an 'ASEAN Community' by 2020. The ASEAN Community is intended to comprise three mutually reinforcing pillars: the political and security community, the economic community, and the socio-cultural community, for 'the purpose of ensuring durable peace, stability, and shared prosperity in the region.' A 'blueprint' has been generated for each of these pillars. The Political and Security Community Blueprint goes beyond the requirements of traditional security and takes into account non-traditional aspects vital to regional and national resilience, such as the 'economic, socio-cultural, and environmental dimensions of development.'[36] This Blueprint also indicates support for the promotion and protection of human rights and fundamental freedoms as found in the ASEAN Charter,[37] but does not link it directly to climate change or the environment. The only consideration of climate change in the three blueprints is embedded in the Socio-Cultural Community Blueprint, which provides for cooperative and differentiated GHG mitigation and adaptation measures for ASEAN Member States in addressing the 'issue of climate change and its impacts on socio-economic development, health and the environment'.[38]

In 2007, ASEAN members along with Australia, China, the Republic of India, Japan, South Korea and New Zealand adopted the Singapore Declaration on Climate Change, Energy and the Environment.[39] The Declaration recognizes the impact of climate change on socio-economic development, health, and the environment, and envisages mitigation and adaptation measures, for example, enhancing access to safe drinking water, particularly in developing countries.[40] It also emphasizes the need to enhance adaptive capacities and for the international community to urgently act to address the growth of global GHG emissions, taking into account diverse national and regional circumstances on the basis of the principle of common but differentiated responsibilities and the capabilities of individual countries. While the Declaration does not directly mention the issue of human rights, this is nevertheless implied in the reference to vulnerability to climate change in the region and the need to implement appropriate mitigation and adaptation measures to minimize the impact of climate change.

36 ASEAN Political–Security Community Blueprint (2007) Article 17, <www.asean.org/archive/5187-18.pdf>.

37 ASEAN Charter, Article 7.

38 ASEAN Socio-Cultural Community Blueprint (2007) D10, [40], Responding to Climate Change and Addressing Its Impacts. This Strategic Objective aims to:

> Enhance regional and international cooperation to address the issue of climate change and its impacts on socio-economic development, health and the environment, in ASEAN Member States through implementation of mitigation and adaptation measures, based on the principles of equity, flexibility, effectiveness, common but differentiated responsibilities, respective capabilities, as well as reflecting on different social and economic conditions.

Available at <www.asean.org/archive/5187-19.pdf>.

39 The Declaration was published on the occasion of the Third East Asia Summit (EAS) in Singapore in November 2007.

40 Singapore Declaration, [15].

Most significantly, following the establishment of the ASEAN Intergovernmental Commission on Human Rights (ASEAN ICHR) in 2009, the ASEAN Human Rights Declaration (HRD) was agreed by ASEAN Member States in 2012. The Declaration confirms all of the fundamental rights found in the 1948 UDHR, together with the 1966 Covenants on Civil and Political Rights and on Economic, Social and Cultural Rights. While it does not mention human rights in connection with climate change as such, the provisions of Article 28 are sufficiently broad to cover it:

> Every person has the right to an adequate standard of living for himself or herself and his or her family including:
>
> a. The right to adequate and affordable food, freedom from hunger and access to safe and nutritious food;
> b. The right to clothing;
> c. The right to adequate and affordable housing;
> d. The right to medical care and necessary social services;
> e. The right to safe drinking water and sanitation;
> f. The right to a safe, clean and sustainable environment.

With regard to Article 28(a), a great deal of attention has been focused on the maintenance of food security as a fundamental human right.[41] In fact, with changing long-term weather patterns, the viability of vast tracts of previously highly productive arable land is being diminished in many areas of Asia and in the long term will inevitably result in the reduction in the reliable supply of food.[42]

Concerning Article 28(c), a high percentage of the land in Bangladesh and other river deltas in Asia are regularly flooded, a phenomenon that is increasingly exacerbated by the effects of climate change, resulting, among other consequences, in whole communities being displaced. Furthermore, in low-lying areas of Asia, access to safe drinking water and sanitation is being affected by salt-water intrusion into water tables, as well as into fresh water lenses that are the main source of drinking water in some States; these issues are directly addressed through Article 28(e).[43]

More comprehensively, along the lines of the African Charter on Human and Peoples' Rights,[44] the San Salvador Protocol to the American Convention on Human Rights[45] and the 2004 Arab Charter on Human Rights,[46] Article 28(f)

41 See for example, FAO, *The Right to Adequate Food,* Fact Sheet No 34, <www.fao.org/righttofood/en>.

42 See the contributions by Franca and Lenzerini and Piergentili in this volume.

43 See the *Symposium on Water, Sanitation and Hygiene in Asia* (2014) <www.esdfocus.org/water-sanitation-hygiene-symposium>.

44 1520 UNTS 128 (1981), entered into force 1986.

45 OASTS 69, reprinted in 28 ILM 156 (1989), entered into force 1999.

46 *Arab Charter on Human Rights* (22 May 2004), reprinted in 12 *Int'l Hum Rts Rep* (2005) 893.

HRD expressly recognizes a right to a 'safe, clean and sustainable environment'. However, unlike the African Charter, the ASEAN Declaration takes an individualistic stance on this right (by the use of '[E]very person' at the beginning of Article 28, as opposed to '[A]ll peoples' in Article 24 of the African Charter). Furthermore, unlike the San Salvador Protocol and the Arab Charter, the right to environment is not linked to health, but is conceived of as part of the right to an adequate standard of living, and thus of the right to life. In many respects, the ASEAN Declaration is similar to the Arab Charter. This approach is also in line with the view of environmental protection taken by the ECtHR.[47]

The use of the descriptors 'safe, clean and sustainable' environment in Article 28(f) HRD is consistent with the reports emerging from the United Nations Special Rapporteur on Human Rights and the Environment,[48] and is particularly meaningful with respect to climate change policies. This provision can be seen to correspond to the various effects of climate change. Rising sea levels, extreme weather events and increase in vector-borne diseases, for example, relate to the issue of environmental *safety*. Pollution caused by flooding of storm water drains and inundation of coastal infrastructure relates to the issue of *cleanliness*. Finally, climate change can be said to affect almost every aspect of human endeavour, generally in a detrimental fashion, resulting in *unsustainable* living conditions for hundreds of millions of people.

It is also noteworthy that the recognition of a right to environment in the ASEAN HRD is an important step for those States in the region that do not yet recognize such a right within their constitutions, such as Malaysia, Laos, Cambodia and Brunei.[49]

Overall, while the provisions of the ASEAN HRD can be regarded as strong in their language, in the short term their actual effect is likely to be relatively weak as a result of the fact that the ASEAN ICHR has not been accorded specific investigation or enforcement powers. Article 6.1 of its Terms of Reference specifies that '[D]ecision-making in the AICHR shall be based on consultation and consensus in accordance with Article 20 of the ASEAN Charter'. Despite the fact that there is no specific provision relating to dispute resolution, the Commission may in due course become stronger as more experience is gained in dealing with the issue of human rights around the region. For example, it is significant that the

47 See the contributions by Christine Bakker and Ottavio Quirico in this volume. See also Braimoh and Qing Huang (2014).

48 See John Knox, *Mapping Report of the Independent Expert on the Issue of Human Rights Obligations relating to the Enjoyment of a Safe, Clean, Healthy and Sustainable Environment*, A/HRC/25/53 (30 December 2013); Id, *Preliminary Report of the Independent Expert on the Issue of Human Rights Obligations Relating to the Enjoyment of a Safe, Clean, Healthy and Sustainable Environment*, UN Doc. A/HRC/22/43 (24 December 2012).

49 David R Boyd, *The Environmental Rights Revolution: a Global Study of Constitutions, Human Rights, and the Environment* (2012) 164 ff. See also Boer (2015).

Commission conducted an initial workshop on Human Rights, Environment and Climate Change in September 2012, which, although not issuing any specific recommendations or conclusions, was nevertheless an important symbolic step, because it explicitly recognized the links between human rights and climate change.[50]

Finally, it can be noted that ASEAN is considering taking action to protect the rights of indigenous peoples with regard to the effects of climate change, within the context of Vision 2020, which aims to ensure environmental protection, sustainable use of natural resources, and high quality of life for people.[51]

3. Human rights and climate change concerns in the Pacific Islands region

Only a few Pacific Island jurisdictions are signatory to the core human rights treaties such as the 1966 ICESCR and the ICCPR. In these circumstances, it is understandable that there are currently no specific regional human rights instruments applicable in the Pacific Islands. Nevertheless, some countries are actively engaged in human rights discourses, particularly on the issue of the effects of climate change.[52] The establishment of a region-wide Human Rights Commission has also been discussed, possibly leading to a regional Pacific Human Rights Charter,[53] following a process similar to the establishment of the ASEAN ICHR.

Three relevant regional organizations are active in the context of climate change and human rights. The first is the Secretariat of the Pacific Community (SPC), which includes 22 Pacific island countries and territories, as well as the four founding countries of Australia, France, New Zealand and the United States.[54] It has a wide remit that encompasses human rights, public health, geoscience, natural resources, disaster management, gender, youth and culture.[55] SPC is actively engaged in debates on human rights, especially concerning climate change in small-island developing States.[56]

The second organization is the Secretariat of the Pacific Regional Environment Programme (SPREP),[57] which has the same membership as the SPC. SPREP's

50 ASEAN ICHR, *Workshop on Human Rights, Environment and Climate Change* (18 September 2014).

51 *ASEAN Vision* (2012); *ASEAN, Climate Change, REDD+ and Indigenous Peoples* (AIPP Printing Press, 2011) 7.

52 Pacific Islands Forum Secretariat and the Secretariat of the Pacific Communities, *Human Rights and Governance: Role of Human Rights and Governance in Addressing SIDS Vulnerabilities Including Climate Change* (19 April 2013).

53 See SPC, *Regional Human Rights Mechanisms: Pathways for the Pacific* (2012).

54 SPC, *Members of the Pacific Community*, available at <www.spc.int/en/about-spc/members.html>.

55 See *Welcome to SPC*, available at <www.spc.int/en/about-spc/welcome.html>.

56 Pacific Islands Forum Secretariat and the Secretariat of the Pacific Communities (2013).

57 Established in the 1980s, it is now a vibrant body addressing the full panoply of environmental issues in the Pacific region, including climate change: <www.sprep.org>.

focus is on the 'protection and sustainable development of the region's environment'. This body has an extensive climate change programme across the Pacific region. Its website notes:

> Climate change is already disproportionally affecting the islands of the Pacific. Although islanders have done little to contribute to the cause – less than 0.03% of current global GHG emissions – they are among the first to be affected. Most islands are experiencing climate change impacts on communities, infrastructure, water supply, coastal and forest ecosystems, fisheries, agriculture, and human health. The consequences of sea level rise, sea temperature increases, ocean acidification, altered rainfall patterns, and overall temperature rise will be increasingly felt. SPREP has been designated by Pacific heads of government as the lead agency to coordinate the region's response to this challenge.[58]

In September 2009, SPREP established the Pacific Adaptation to Climate Change (PACC) programme. PACC is a partnership between 14 Pacific Island countries, and includes regional and national agencies and communities aimed at making island communities more resilient to the effects of climate change. It is 'working to enhance adaptive capacity on the ground' and is 'driving the mainstreaming of climate risks into national development planning and activities.'[59] The fourteen PACC countries have now developed best-practice adaptation strategies. These focus on coastal zone management, food security, food production, and water resources management, with each country hosting a pilot project on adaptation in order to enhance climate change resilience. The emphasis is placed on mainstreaming adaptation at all levels, from communities to governmental policy-making.[60] A small but significant example of adaptation support work is the building of a 'climate-proof' road on the island of Kosrae in the Federated States of Micronesia.[61] While the PACC programme does not directly mention human rights, several are clearly relevant to coastal zone management, food security and water resources management. The programme is also motivated by the knowledge that climate change threatens sustainable development, and thus needs to be taken into account in decision-making on development activities.

Further, the SPREP's Pacific Islands Framework for Regional Action on Climate Change Variability and Sea Level Rise (2006–15) seeks to address 'the impact of

58 The goal under SPREP's Climate Change Strategic Priority is that by 2015 all SPREP members 'will have strengthened capacity to respond to climate change through policy improvement, implementation of practical adaptation measures, enhancing ecosystem resilience to the impacts of climate change, and implementing initiatives aimed at achieving low-carbon development' (*Climate Change Overview*, <www.sprep.org/climate-change-home>). For a comprehensive overview of the impacts of climate change in the Pacific see Tully (2007) 182 ff.

59 *Pacific Adaptation to Climate Change*, <www.sprep.org/pacc>.

60 Ibid.

61 *Climate-proof Road Opens in Kosrae, FSM*, <www.sprep.org/climate-change/climate-proof-road-opens-in-kosrae-fsm>.

climate change through an integrated, multi-stakeholder approach that links specific sectors such as agriculture; energy; forestry and land use; health; coastal zone management; marine ecosystems; ocean management; tourism and transport.'[62] Some references in the document, such as the necessity to 'protect human health from climate change related diseases' and to 'integrate traditional knowledge' into risk management policies,[63] are directly relevant to human rights considerations.

Despite detailed policy development and practical adaptation initiatives by SPREP, these can only be taken so far. Desirably, adequate law and policy frameworks should back them at the national level. However, many countries are by no means vigorous when it comes to implementing adaptation strategies.[64]

The third organization is the Pacific Islands Forum, which in 2013 adopted the Majuro Declaration for Climate Leadership,[65] a document that pays particular attention to the situation of low-lying islands. The Declaration presents climate change as the 'greatest threat to the livelihoods, security and well-being of the peoples of the Pacific.'[66] It also stresses the necessity to 'respond urgently and sufficiently to the social, economic and security impacts of climate change to ensure the survival and viability of all Pacific small island developing States, in particular low-lying atoll States, and other vulnerable countries and regions worldwide' (Article 4). This wording is implicitly underpinned by basic claims, such as the rights to life and self-determination. Within this framework, States are considered 'responsible' in the name of everyone, and an annex includes binding GHG reduction targets for Pacific countries (Article 7).

4. Displacement issues

4.1. Climate change and human displacement in the Asia-Pacific region

One of the most acute and demonstrable effects of climate change on human rights in the Asia-Pacific region is involuntary displacement, both internal and transboundary, which has garnered the attention of a number of analysts.[67]

62 Asia Pacific Forum, *Human Rights and the Environment*, Final Report and Recommendations (The Asia Pacific Forum of National Human Rights Institutions, 2007).

63 *Pacific Islands Framework for Action on Climate Change 2006–2015*, 7 and 22.

64 See Ben Boer and Pepe Clarke, *Ecosystem-Based Adaptation to Climate Change in the Pacific Islands* (SPREP, 2011).

65 *Majuro Declaration*, available at <www.daghammarskjold.se/wp-content/uploads/2014/12/44th-PIFS-Majuro-Outcome.pdf>.

66 *Majuro Declaration*, Article 1. On the link between climate change and the right to life in the Pacific area see Owen Cordes-Holland, 'The Sinking of the Strait: the Implications of Climate Change for Torres Strait Islanders' Human Rights Protected by the ICCPR' (2008) *Melbourne J Int'l L* 406, 408, 419–20 and 426–7; Tully (2007) 184 ff.

67 See, for example, Davor Vidas, 'Sea-Level Rise and International Law: at the Convergence of Two Epochs' (2014) 4 *Climate Law* 70; Jane McAdam, *Climate Change, Forced Migration, and International Law* (OUP, 2012).

Internationally, despite the fact that climate migration has been on the international agenda for over 20 years, as McAdam and Saul point out, 'there remains no international organization charged with official responsibility for the issue.'[68] The fifth Climate Change Assessment Report 2014 briefly focuses on this matter, stating:

> Displacement risk increases when populations that lack the resources for planned migration experience higher exposure to extreme weather events, such as floods and droughts. Expanding opportunities for mobility can reduce vulnerability for such populations. Changes in migration patterns can be responses to both extreme weather events and longer-term climate variability and change, and migration can also be an effective adaptation strategy.[69]

In particular, the Pacific Island region is by its nature a marine and coastal environment, with a myriad of small island low-lying States, some of which are already being inundated by sea-level rise, when combined with high tides and storm surges.[70] Those islands which are only a few hundred metres across are especially affected. The few high-island States (for example, Papua New Guinea, Vanuatu and Solomon Islands) of course also have low-lying areas and low islands within their jurisdictions.[71] The highest point in some island countries is just a few metres above sea level.

Severe impacts, including disappearance of some islands, are predicted for parts of the Marshall Islands, Kiribati, Micronesia, Nauru and Tonga, Fiji and the Solomon Islands, Vanuatu and Samoa. Some of these countries are among the least developed and most of them are generally inhabited by indigenous people, many of which have subsistence or semi-subsistence lifestyles.[72] One of the most well-known examples of involuntary displacement concerns the Carteret Islanders in Papua New Guinea, many of whom have now moved to the island of Bougainville, with consequent losses of traditional lifestyles, fishing resources and cultural identity.[73]

68 Jane McAdam and Ben Saul, *An Insecure Climate for Human Security? Climate-Induced Displacement and International Law*, Sydney Centre for International Law, available at <www.peacepalacelibrary.nl/ebooks/files/357399072.pdf>.

69 IPCC (2014) 73.

70 Alisa Ceri Warnock, 'Small Island Developing States of the Pacific and Climate Change: Adaptation and Alternatives' (2007) 4 *NZYIL* 247, 249 ff. Reports from small island States stress problems of inundation within the UNFCCC context (UNFCCC, *Coastal Adaptation Technologies*, FCCC/TP/1999/1 (1999)). See also Cordes-Holland (2008) 408 ff; Tully (2007) 170.

71 Although this section focuses primarily on displacement issues in the Pacific, many issues are directly applicable to Asian countries, and ASEAN has made substantial efforts to deal with the problems, as noted earlier (see *ASEAN Cooperation on Climate Change* (2013)).

72 Warnock (2007) 251; Tully (2007) 173.

73 See Eric Kwa, 'Climate Change and Indigenous People in the South Pacific: the Need for Regional and Local Strategies', in Benjamin J Richardson, Yves Le Bouthillier, Heather McLeod-Kilmurray and Stepan Wood (eds), *Climate Law and Developing Countries: Legal and Policy Challenges for the World Economy* (Edward Elgar, 2009) 102–24.

As stated by Boer and Clarke:

> Sea-level rise, on the scale predicted by the Intergovernmental Panel on Climate Change, will lead to significant displacement of local populations in many parts of the Pacific, including low-lying island States such as Kiribati and Tuvalu, as well as small islands and coastal areas within larger Pacific Island countries, such as Solomon Islands and Papua New Guinea.[74]

In light of the prohibitive costs of adaptation measures such as sea walls, abandonment of settlements and displacement of communities seem inevitable in some countries.[75] This will generate migration flows and affect the right to self-determination.[76] However, other fundamental rights, such as the rights to food and water, are also significantly affected.[77] In the Pacific, displaced people are at risk of being deprived of both land and traditional livelihoods and they will face special challenges in meeting basic needs. Displacement is likely to increase the risk of conflicts over land and resources, political instability, and, in particular instances, discrimination and perhaps violence, where old and new communities fail to integrate adequately.

4.2. The inadequacy of traditional human rights law in the face of climate-induced displacement

In recent years there has been a good deal of research into issues of internal and transboundary displacement of people affected by various aspects of climate change. While it is not the intention to review the complex legal and policy issues involved here,[78] several New Zealand cases involving questions of displacement are used to illustrate some of the complex problems faced by communities and decision-makers.

In the case of *Ioane Teitiota v Chief Executive of Ministry of Business, Innovation and Employment*,[79] the applicant was a national of the Pacific atoll State of Kiribati, who

74 See Boer and Clarke (2011) 23.

75 Warnock (2007) 259 ff.

76 See the chapter by Cameron Moore in this volume. See also Cordes-Holland (2008) 426.

77 See Marc Overmars and Sasha Beth Gottlieb, *Adapting to Climate Change in Water Resources and Water Services in Caribbean and Pacific Small Island Countries* (Contribution to the 5th World Water Forum, 2009).

78 See, for example, McAdam and Saul, *An Insecure Climate*; McAdam (2012); Emily Crawford and Rosemary Rayfuse, 'Climate Change and Statehood', in Rosemary Rayfuse and Shirley V Scott (eds), *International Law in the Era of Climate Change* (Edward Elgar, 2012) 243; Susin Park, *Climate Change and the Risk of Statelessness: the Situation of Low-lying Island States* (UNHCR, 2011).

79 *Ioane Teitiota v The Chief Executive of Ministry of Business, Innovation and Employment* [2014] NZCA 173 (New Zealand Court of Appeal, 8 May 2014); *Ioane Teitiota v The Chief Executive of Ministry of Business, Innovation and Employment* [2013] NZHC 3125 (High Court of New Zealand, 26 November 2013); *AF (Kiribati)* [2013] NZIPT 800413 (Immigration and Protection Tribunal of New Zealand, 25 June 2013).

had travelled to New Zealand with his wife on a temporary permit in 2007 and remained there after their permits expired in 2010. Three children of the couple were born in New Zealand, but were not legally regarded as being entitled to New Zealand citizenship. Mr Teitiota applied for refugee and/or protected person status to a Refugee and Protection Officer under the New Zealand Immigration Act 2009, but his application was declined. Mr Teitiota then appealed this decision to the Immigration and Protection Tribunal. The primary basis for his application was the effect of climate change:

> The appellant claims an entitlement to be recognized as a refugee on the basis of changes to his environment in Kiribati caused by sea-level-rise associated with climate change. The issue for determination is whether the appellant is able to bring himself within the Refugee Convention or New Zealand's protected person jurisdiction on this basis.[80]

The Immigration Act refers to the definition of 'refugee' under the 1951 United Nations Convention Relating to the Status of Refugees.[81] Article 1(A)(2) of the Convention defines a 'refugee' as:

> any person who . . . owing to well-founded fear of being persecuted for reasons of race, religion, nationality, membership of a particular social group or political opinion, is outside the country of his nationality and is unable, or, owing to such fear, is unwilling to avail himself of the protection of that country; or who, not having a nationality and being outside the country of his former habitual residence as a result of such events, is unable or, owing to such fear, is unwilling to return to it.

Evidence was led in the Immigration and Protection Tribunal concerning sea-level rise around Kiribati, so as to show that the effects of population growth, urbanization and limited infrastructure development were being exacerbated by the sudden onset of environmental events such as storms, and the slow-onset processes of sea-level rise. Because of his concerns in relation to coastal erosion, increasing intrusion of salt into freshwater and in the light of his awareness of the climate change debate, the appellant and his wife travelled to New Zealand in 2007. Evidence was led that Mr Teitiota did not wish to return to Kiribati because of the combined pressures of over-population and sea-level rise.[82]

The Tribunal nonetheless rejected the claim. With regard to the status of Mr Teitiota as a refugee, it was held that 'underlying environmental events and processes favour no civil or political status' under s 129 of the New Zealand

80 *AF (Kiribati)* (25 June 2013) [2].

81 New Zealand Immigration Act 2009, s 129(1) (Recognition as Refugee): 'A person must be recognised as a refugee in accordance with this Act if he or she is a refugee within the meaning of the Refugee Convention.'

82 *AF (Kiribati)* (2013) [5] ff.

Immigration Act 2009 interpreted in light of the Refugee Convention.[83] Furthermore, the Tribunal denied that the applicant was 'in danger of being subjected to cruel treatment' under s 131 of the Act.[84]

After rejection of the application in the first instance, Mr Teitiota applied for leave to appeal to the New Zealand High Court. The essence of the application was again whether the word 'refugee' under the New Zealand Immigration Act included those who are refugees because of climate change and its effects.[85] The Court rejected the appeal by endorsing the stand of the Immigration and Protection Tribunal, considering 'impermissible' the 'attempt to expand dramatically the scope of the Refugee Convention and particularly Article 1A(2)'.[86]

The High Court specified:

> The appellant raised an argument that the international community itself was tantamount to the 'persecutor' for the purposes of the Refugee Convention. It completely reverses the traditional refugee paradigm. Traditionally a refugee is fleeing from his own government or a non-State actor from whom the government is unwilling or unable to protect him.[87]

The decision of the Immigration and Protection Tribunal was further appealed to the New Zealand Court of Appeal. On behalf of Mr Teitiota it was argued that:

> 'being persecuted' does not require human agency and the word 'refugee' is capable of encompassing persons having to flee irrespective of the cause. In Mr Teitiota's case, it is the 'act of fleeing climate change' because of the serious harm it will do to Mr Teitiota and his family coupled with the Kiribati Government's unwillingness or inability to deal with the factors 'instituted by climate change' that constitute Mr Teitiota a refugee.[88]

The Court endorsed the reasoning of the judge of the High Court who had heard the original leave to appeal and found that it was not appropriate to grant leave to appeal on any of the questions of law posed.[89] Significantly, the Court went on to say:

> No-one should read this judgment as downplaying the importance of climate change. It is a major and growing concern for the international community.

83 Ibid, [75]–[76],

84 Ibid, [88] and [95]–[96].

85 *Ioane Teitiota* (2013) [40].

86 Ibid, [63].

87 Ibid, [55]; see also Steve Trent, 'New Zealand High Court Rejects Kiribati Climate Refugee Appeal' (Environmental Justice Foundation, November 2013) <www.ejfoundation.org/node/997>.

88 *Ioane Teitiota* (2014) [16].

89 Ibid, [40].

> The point this judgment makes is that climate change and its effect on countries like Kiribati is not appropriately addressed under the Refugee Convention.[90]

Accordingly, the findings of the New Zealand Immigration and Protection Tribunal were confirmed and leave to appeal was again dismissed.[91]

The position taken by the New Zealand Immigration and Protection Tribunal in the *Teitiota* case was confirmed in two linked cases, which constituted an appeal against a deportation order for a family from the atoll country of Tuvalu (the fourth smallest country in the world, with a total land area over nine islands of 25.9 square kilometres).

In the first case, the appellants made submissions concerning the applicability of the 1951 Refugee Convention, and the 1984 Convention against Torture. However, in the event, the appellants abandoned these claims. The Tribunal then considered the applicability of the ICCPR concerning arbitrary deprivation of life (Article 6) and torture, cruel, inhuman or degrading treatment or punishment (Article 7). However, it declined to grant refugee status to the appellants on the basis of the effects of climate change, on the ground of the right to life or with respect to cruel, inhuman or degrading treatment, concluding, with regard to the latter, that:

> [O]n the evidence before the Tribunal, there is no basis for finding that there is a danger of the appellants being subjected to cruel treatment by the State failing to discharge its obligations to protect its population and territory from the adverse impacts of natural disasters and climate change.[92]

In a subsequent case involving the same appellants, the Tribunal slightly softened its stance. It upheld a request for protection under s 207 of the New Zealand Immigration Act, with regard to humanitarian issues concerning close family ties, as well as climate change issues, finding that taking these 'into account on a cumulative basis', there were 'exceptional circumstances of a humanitarian nature, which would make it unjust or unduly harsh for the appellants to be removed from New Zealand'.

The Tribunal accepted that:

> exposure to the impacts of natural disasters can, in general terms, be a humanitarian circumstance, nevertheless, the evidence in appeals such as this must establish not simply the existence of a matter of broad humanitarian concern, but that there are exceptional circumstances of a humanitarian

90 Ibid, [41].
91 Ibid, [42]–[43].
92 *AC (Tuvalu)* [2014] NZIPT 800517–520 (Immigration and Protection Tribunal of New Zealand, 4 June 2014) [120]–[121].

> nature such that it would be unjust or unduly harsh to deport the *particular appellant* from New Zealand.[93]

However, no particular finding was reached with respect to the submissions based on climate change. Indeed, the Tribunal held it unnecessary to reach any specific conclusions on the issue, since family ties were considered sufficient to justify exceptional humanitarian circumstances, making deportation unjust.[94]

The result of these cases is not surprising,[95] but points to serious gaps in international and national law concerning the plight of the millions of people potentially and actually affected by anthropogenic climate change. In addition to the other impacts of global warming, the phenomenon of transnational and internal displacement of people is one of the clearest manifestations that the current human rights framework, both in the Asia-Pacific and globally, is not adequate to respond to climate change-induced environmental phenomena.[96] Classical first and second generation human rights are insufficient to address the problem of climate migrants.[97]

Internationally, a solution put forward by a number of analysts is either to amend the 1951 Refugee Convention in order to address environmental refugees (an unlikely and potentially dangerous manoeuvre), to include additional provisions in the successor instrument to the Kyoto Protocol to the UNFCCC. Another solution is to establish a new international agreement specifically aimed at the problem of climate migrants.[98] These issues have become part of a lively debate in discussion forums such as *Refworld*, hosted by the United Nations Refugee Agency.[99]

93 *AD (Tuvalu)* [2014] NZIPT 501370–371 (Immigration and Protection Tribunal of New Zealand, 4 June 2014) [30] and [32]–[33] (emphasis in original).

94 Ibid, [33].

95 Australian courts have taken a similar position (see Case No 0907346 [2009] RRTA 1168 (10 December 2009)).

96 Given the gaps that exist in the development of human rights regimes in the Asia-Pacific region, various non-governmental organizations have been active over some years in putting forward human rights declarations, some including provisions on the right to environment (a *Directory of Asian Human Rights Organizations* is found at <http://asiapacific.anu.edu.au/blogs/asiarights/new-directory-of-asian-human-rights-organizations>). As early as 1989, the Human Rights Committee of the Law Association of Asia and the Pacific (LAWASIA) elaborated the 1989 *Draft Pacific Charter of Human Rights* (available at <http://pacific.ohchr.org/docs/Draft_Pacific_Charter_Memoranda.pdf>). The Asia Pacific Forum on Human Rights supports the 'establishment and strengthening of national human rights institutions' across the Asia-Pacific region (<www.asiapacificforum.net>). It is presently comprised of 19 countries and produced a report in 2007 advocating 'the adoption and implementation of a specific right to an environment conducive to the realisation of fundamental human rights' (*Human Rights and the Environment: Final Report and Recommendations* (2007) <www.asiapacificforum.net/support/issues/acj/references/right-to-environment>).

97 See Warnock (2007) 273 ff, Tully (2007) 187 ff.

98 See the contribution by Benoît Mayer and Christel Cournil in this volume. See also Stefan Gruber, 'Human Displacement and Climate Change in the Asia Pacific Region', in Boer (2015).

99 If predictions that up to 250 million people could be displaced by climate change by 2050 are correct (*Should International Refugee Law Accommodate Climate Change?*, on website of the United Nations

Regionally, it is essential that Asia-Pacific countries develop long-term strategies to reduce internal displacement and adaptation measures to protect and assist internally displaced persons, via international cooperation.[100] In order to manage future conflicts over land and resources, each country should, inter alia: 'develop a national plan for minimizing displacement' and 'review national legislation to ensure that the human rights of internally displaced persons are adequately protected by law'.[101]

Conclusion

Despite the fact that the Asia-Pacific is one of the world's most threatened regions with regard to climate change-induced environmental effects, a clear human rights approach to these effects has not yet emerged. One of the reasons may be related to the fact that, unlike the African, European and Inter-American human rights regimes, no cohesive substantive and procedural human rights system has yet developed in the Asia-Pacific. However, the establishment of the sub-regional ASEAN Intergovernmental Commission on Human Rights and the acceptance of the ASEAN Human Rights Declaration, while institutionally weak and slow in implementation at this stage, are significant developments that could be followed by the other Asia-Pacific sub-regions. Over time, this could enhance the capacity of individual countries to address the human rights implications of the effects of climate change, to the benefit of people in the region as a whole.

Refugee Agency, <www.un.org/apps/news/story.asp?NewsID=48201#.VKDeLAPCV>), then legal solutions must be found in order to address these fast-developing human catastrophes.

100 Note that the then Minister for Foreign Affairs in New Zealand, Phil Goff, stated in 2000 that, should law-lying islands become uninhabitable 'New Zealand, Australia, the Pacific and other countries would come to their assistance' (Phil Goff, Minister for Foreign Affairs, New Zealand (21 June 2000) Reuters News Service, online: <www.heatisonline.org>).

101 Boer and Clarke (2011) 24.

15 A wider human rights spectrum to fight climate change in Africa?

Faustin Ntoubandi and Roland Adjovi

Introduction

It is predicted that by 2080 the impacts of climate change, such as sea-level rise, droughts, heat waves, floods and rainfall variations, could drive 600 million people into malnutrition and increase the number of people facing water scarcity by 1.8 billion.[1] Within such a worrying context, Africa is usually described as the continent most vulnerable to climate change and climate variability, owing to multiple stresses and to low mitigation and response capacities.[2]

Predictions are very alarming, since the African continent's inhabitants are assumed at 1.033 billion people and account for about 15 per cent of the world's population.[3] For instance, it is estimated that by 2020, between 75 and 250 million Africans will be exposed to water stress caused by climate change. However, Africa only accounts for 3 per cent of the world's energy consumption, and contributes approximately 3.8 per cent of total GHG emissions,[4] which are considered to be the main drivers of climate change and climate variability.[5] Nonetheless, the cost of adapting to or mitigating the impact of climate change is very high: at least 5 to 10 per cent of the GDP.[6]

Therefore, climate change poses multifaceted challenges to Africa. One well-studied aspect is the impact of climate change on the natural resources and conflicts in the continent.[7] However, the broader question of the consequences of

1 UNDP, *Human Development Report – Sustaining Human Progress: Reducing Vulnerability and Building Resilience* (2014).

2 IPCC, *Climate Change: Impact, Adaptation and Vulnerability* (2014) Chapter 22, Africa.

3 World Population Statistics, *Africa Population* (20 May 2013) <www.worldpopulationstatistics.com/africa-population-2013>.

4 IPCC, *Mitigation of Climate Change* (2014) Technical Summary, 43.

5 Asaf Kumssa and John F Jones, 'Climate Change and Human Security in Africa' (2010) 17 *International Journal of Sustainable Development & World Ecology* 453, 456; Donald Anthony Mwiturubani and Sanita Jo-Ansie van Wyk (eds), *Climate Change and Natural Resources Conflicts in Africa* (Institute of Security Studies, Monograph No 170, 2010) <www.issafrica.org/uploads/Mono170.pdf> 5.

6 IPCC, *Synthesis Report* (2007) 56 ff.

7 Mwiturubani and van Wyk (2010); Oli Brown and Alec Crawford, *Climate Change and Security in Africa. A Study for the Nordic-African Foreign Ministers Meeting, 2009* (International Institute for Sustainable Development, Ministry of Foreign Affairs of Denmark, March 2009).

global warming for human rights in Africa, and more generally their reciprocal relationships, has been less debated. This chapter aims to systemically explore such a relationship by considering substantive human rights provisions that create tools for fighting climate change (1); and how these tools are implemented by African institutions from the viewpoint of both primary norms on rule-making and secondary rules on enforcement (2).

1. The substantive context

1.1. Effects of climate change on the African continent

Climate change has particularly adverse effects on the environment and health of African peoples.[8] For instance, it has the potential to affect the regular cycle of rainfall. This can result in unpredictable storm surges. In fact, heavy rainfall variability constitutes a threat to the survival of small islands and coastal regions in Africa, which are subjected to erosion, floods and sea-level rise. Together with global warming caused by climate change, heavy rainfall is a major cause of environmental and land degradation.[9] Such phenomena constitute a serious threat to human settlements, life and property. From this perspective, the destruction caused by El Niño in Dar es Salaam in December 2011 is considered to be the worst disaster Tanzania has ever witnessed.[10] Acute droughts, on the other hand, can cause significant depletion of water resources, particularly freshwater resources and reliability of water,[11] increased desertification and decreased average staple food and agricultural productivity, leading to food insecurity.[12] Thus, the levels of many lakes and basins in Africa, for instance, Lake Chad and Lake Victoria, are decreasing as a result of droughts and water evaporation that have drastically affected many ecosystems, wetlands, groundwater aquifers and rivers in several regions in the continent.[13] Furthermore, severe drought spells reduce water availability and pastures for rearing livestock, thus threatening the traditional lifestyle of nomadic herders whose survival is tied to livestock activities.[14] Sedentary farmers are affected too, to the extent that droughts negatively impact on cropland,

8 Siobhán McInerney-Lanford, Mac Darrow and Lavanya Rajamani, *Human Rights and Climate Change – A Review of the International Legal Dimensions* (World Bank, 2011) 11. See also Friedrich Ebert Stiftung (Ghana Office), *Climate Change*, <www.fesghana.org/index.php?page=climate-change>.

9 African Union and European Union, *Africa–EU Declaration on Climate Change* (2008) [13]–[14] <http://ec.europa.eu/clima/events/0043/africa_eu_declaration_en.pdf>.

10 Aidan G Msafiri, 'Ethics for Climate Justice and Sustainability through Valued-based Approaches: a New Tanzanian Model and Paradigm Shift', in Oliver C Ruppel and Others (eds), *Climate Change: International Law and Global Governance* (Nomos, 2013) Volume II: Policy, Diplomacy and Governance in a Changing Environment 661, 666.

11 See IPCC, *Climate Change: Impact, Adaptation and Vulnerability* (2014) Chapter 3, Freshwater Resources, Chapter 22, Africa, 1215 ff.

12 Msafiri (2013) 664; African Union and European Union (2008) [13]–[14].

13 Msafiri (2013) 665.

14 Ibid, 666.

biodiversity and the seasonal predictability they depend on for food, shelter, security, health, culture and life.[15]

As far as health hazards and risks are concerned, fluctuating temperatures have a direct influence on the survival and resilience of lethal vectors and pathogens that pose serious threats to human health. The spread and transmission of vector-borne infectious diseases, such as malaria, meningitis, dysentery, cholera, plague, Rift Valley fever, and schistosomiasis, are facilitated by extreme weather conditions.[16] Climatic changes that can affect the transmission of vector-borne infectious diseases include high temperature, humidity, altered rainfall, soil moisture and sea-level rise. With unpredictable increases in temperatures, killer diseases, such as malaria, can expand easily and affect a large number of peoples.[17] Transmission increases in camps where displaced populations are gathered as a result of floods. Droughts may also adversely affect nutrition and cause infectious diseases to spread.[18] In addition, natural disasters such as floods, droughts and storms can seriously affect health in a variety of ways, including physical injuries or death by drowning.[19] Africa is particularly vulnerable to climate change threats to health security because the majority of poor people live in rural areas, where access to basic necessities such as clean water, sufficient food and health services is not guaranteed. It appears, therefore, that, by reducing access to water, food, health and shelter, climate change undermines human security in the African continent.[20]

Another area of concern is the climate change-induced movement of persons, which constitutes a serious threat to the right to shelter. In this respect, since its First Assessment Report the IPCC indicated that climate change might severely increase human mobility.[21] Indeed, accelerated degradation of the environment and of natural resources is expected to aggravate the number of migrants.[22]

15 IPCC, *Climate Change: Impact, Adaptation and Vulnerability* (2014) Chapter 7, Food Security and Food Production Systems, Chapter 22, Africa, 1218 ff; Friedrich Ebert Stiftung (Ghana Office), *Climate Change and Its Impact on the Livelihood of Farmers and Workers in Ghana* (2012); Kumssa and Jones (2010) 456; African Union and European Union (2008) [13]–[14].

16 Kumssa and Jones (2010) 456–7.

17 Oliver C Ruppel and Sanita van Wyk, 'Climate Change-induced Movement of Persons in Africa: Human Rights Responses to Aspects of Human Security', in Oliver C Ruppel and Others (2013) 799, 803.

18 Andy Haines and Others, 'Climate Change and Human Health: Impacts, Vulnerability and Public Health' (2006) 120 *Public Health* 585; Msafiri (2013) 666.

19 Kumssa and Jones (2010) 457.

20 IPCC, *Climate Change: Impact, Adaptation and Vulnerability* (2014) Chapter 11, Human Health: Impact, Adaptation and Benefits, 715 ff, Chapter 22, Africa, 1221 ff; Jon Barnett and Neil Adger, 'Climate Change, Human Security and Violent Conflict' (2007) 26 *Political Geography* 639, 640. As to the definition of the concept of 'human security', see Ruppel and van Wyk (2013) 800–1; Kumssa and Jones (2010) 454; UNDP, *Human Development Report – New Dimensions of Human Security* (1994) 22–3.

21 IPCC, *Climate Change: Impact, Adaptation and Vulnerability* (2014) Chapter 22, Africa, 1240; *First Assessment Report* (1990) Chapter 5, 9.

22 Ruppel and van Wyk (2013) 805.

Populations of low islands, coastal and deltaic regions of Africa are expected to flee their homes and traditional living environments and spaces as a result of climate change. Although estimates on migration flows as a result of climate change are speculative, Africa's reduced adaptive capacity makes it particularly vulnerable and the number of climate migrants has considerably increased from 700,000 in 2008 to 1.7 million in 2010.[23]

1.2. Legal framework

A number of international and regional human rights instruments have implications for the protection of African peoples against the adverse effects of climate change.

Article 3 of the Universal Declaration of Human Rights, which constitutes the 'common standard of achievement for all peoples and all nations', provides that 'everyone has the right to life, liberty and security of person.'[24] This provision is further reinforced by various articles of the two 1966 International Human Rights Covenants, such as Article 6(1) ICCPR and Article 12(1) ICESCR.

Within the African context, by means of the Charter on Human and Peoples' Rights (ACHPR), African nations have granted every individual the 'right to respect for his life, liberty and security of his person',[25] and reaffirmed 'their adherence to the principles of human and peoples' rights and freedoms contained in the declarations, conventions and other instruments adopted by the Organization of African Unity . . . and the United Nations'.[26] In addition, the ACHPR protects the right of peoples to the 'best attainable state of physical and mental health'.[27] This claim should also encompass the rights to food and water, although these are not explicitly mentioned. [28]

Besides individual rights, the ACHPR encompasses some collective rights that are extremely relevant to climate change. Indeed, Article 20 recognises that people have a 'right to existence'. Article 21(1) provides that 'all peoples shall freely dispose of their wealth and natural resources' and '*in no case* shall a people be deprived of it'.[29] This rule is clearly linked to the basic tenet of self-determination and seems to be framed as peremptory, which should exclude derogations at all times, including emergency situations. Furthermore, Article 21(2) recognises a corresponding claim to compensation, since 'in case of spoliation the dispossessed people shall have the right to the lawful recovery of its property as well as to an

23 Ibid, 806.

24 UN GA, *Universal Declaration on Human Rights*, UN Doc A/RES/217 (III) (10 December 1948).

25 African Union, *African Charter on Human and Peoples' Rights* (adopted on 27 June 1981), OAU Doc CAB/LEG/67/3 rev. 5, 21 ILM 58 (1982) Articles 4 and 6.

26 Paragraph 10 of the Preamble.

27 Article 16.

28 See CESCR, General Comment No 14, *The Right to the Highest Attainable Standard of Health (Art 12 of the Covenant)*, E/C.12/2000/4 (2000).

29 Emphasis added.

adequate compensation.' Along these lines, the ACHPR acknowledges a progressive 'right to development' under Article 22, which is otherwise only considered soft regulation under general international law. Within this context, peoples are granted a collective right to a 'general satisfactory environment favourable to their development' (Article 24).

These rights are similarly stated in the African Charter on the Rights and Welfare of the Child (ACRWC)[30] and in the African Union (AU) Protocol to the African Charter on Human and Peoples' Rights on the Rights of Women in Africa (Maputo Protocol).[31] In addition, the Preamble to the ACRWC embeds an acknowledgement by the AU Member States that 'the situation of most African children remains critical due to the unique factors of their socio-economic, cultural, traditional and developmental circumstances, natural disasters, armed conflicts, exploitation and hunger [. . .]'.[32] On the other hand, the Maputo Protocol recognises a women's right to sustainable development and imposes on States parties an obligation to integrate women at all levels of development policies and programmes.[33]

Moreover, the 2009 AU Convention for the Protection and Assistance to Internally Displaced Persons in Africa (Kampala Convention) imposes on State parties a legal obligation to protect and assist internally displaced persons (IDPs). The Convention defines IDPs as 'persons or groups of persons who have been forced or obliged to flee or to leave their homes or places of habitual residence', particularly 'as a result of or in order to avoid the effects of armed conflicts, situations of generalized violence, violations of human rights or natural or human-made disasters', and who 'have not crossed an internationally recognized State border.'[34] The Kampala Convention explicitly recognises climate change as one of the drivers of internal displacement, when it provides that 'States Parties shall take measures to protect and assist persons who have been internally displaced due to natural or human-made disasters, including climate change'.[35] As a precautionary measure in support of this provision, under Articles 9(2)(j) and 10(3) the Convention also provides for an obligation for the Member States to study the environmental impact of any development project and to prevent any environmental degradation that could lead to internal displacement. Thus, State obligations under the Kampala Convention include, inter alia, the duty to prevent displacement from happening, for instance, by establishing early warning systems and by adopting disaster preparedness and management measures.[36] States parties must also assess the needs and vulnerabilities of displaced persons, as well as those

30 Articles 2(2)(g) and 14(2)(h).
31 Articles 16 and 18.
32 Para 4.
33 Article 19.
34 Article 1 (k).
35 Article 5 (4).
36 Article 4 (1) and (2).

of the host communities, and provide adequate assistance.[37] Finally, they are called upon to establish an effective legal framework to provide just and fair compensation, and other forms of reparation, for damages incurred as a result of displacement.[38]

This brief overview shows that the African human rights protection system departs, in many respects, from human rights protection provided for in other regional systems and in general international law.[39] In particular, the move from 'individualistic' first and second generations human rights to 'collective' human rights, some of which belong to the evolving category of third generation human rights, should provide more adequate tools in the fight against global warming from a fundamental rights perspective. The African continent thus represents a specific ground for testing the potential of human rights in fighting climate change.

2. Institutional approaches to climate change and human rights

2.1. Law and policy-making: responding to the adverse effects of climate change on human rights

African States seem to take seriously the global effort to combat climate change and mitigate its impacts on human security. In this regard, the vast majority of African States have signed or ratified the main international treaties on global warming, notably, the UNFCCC, the Kyoto Protocol and the 1994 Convention to Combat Desertification. However, these instruments have not been fully incorporated in domestic legislation.

The AU has extensively addressed the relationship between climate change and human rights in various declarations and decisions of its supreme organ, the Assembly of Heads of State and Government, beginning in 2007 with the Declaration on Climate Change and Development.[40] In this Declaration, African States officially recognised the potential of climate change to 'endanger the future well-being of the population, ecosystems and socio-economic progress of Africa'. As a logical consequence, they have decided to develop a number of policy initiatives and strategies at the AU level, aiming to mitigate the impacts of climate change on human security in Africa. Indeed, follow-up resolutions have expressed some guiding principles and established a framework for coordinating the climate change policies of the African States. In particular, led by Algeria, the 2012 AU Summit adopted a common position on climate change, which claims compensation for the fact that Africa suffers heavily from global warming, although it

37 Article 5 (5) and (6).

38 Article 12.

39 See the other contributions in part IV of this volume.

40 Adopted at the Eighth Ordinary Session of the Heads of State and Government of the AU, Addis Ababa, AU Doc Assembly/AU/Decl 4 (VIII) (30 January 2007).

contributes very little to it.[41] However, the position has no binding effect on States. Furthermore, if successful, this strategy would raise a set of questions that remain unresolved: specifically, as to how compensation funds would be managed and how corruption rampant on the continent would affect the effectiveness of these new financial resources.[42] Practically, the AU has taken initiatives such as the Great Green Wall, which combines mitigation and adaptation approaches, seeking to improve food security and livelihoods of the people in the Sahel and the Sahara.[43] This initiative brings together 20 countries from the Sahelo-Saharan region and organisations such as the African Forest Forum, Pan African Farmers Organization and FAO.[44]

At the administrative level, a Division on Environment, Climate Change, Water and Land Management has been established within the AU Commission, including a special unit on climate change and desertification, aiming to coordinate African States' policies on the subject.[45] Furthermore, in 2009 the Committee of African Heads of State and Government on Climate Change (CAHOSCC) was created with the aim of leading the African Group during conferences in climate change negotiations rounds, that is, so far Copenhagen, Cancun, Durban, Doha and Warsaw. As a result, African States have been represented in climate change negotiations not only individually, but also collectively by a delegation of the AU. The work of the Committee has been duly reported on and acknowledged by the AU Assembly of Heads of States and Governments.[46] However, it is legitimate to question the effectiveness of such a duplication of functions and the high risk of possible inconsistencies in the positions expressed, particularly because African States are not suffering from climate change in an equal manner.

In parallel to the CAHOSCC, the African Ministerial Conference on Environment (AMCEN), which was established in 1985, is also in charge of the coordination of the positions of the African States on climate change.[47] AMCEN has recently issued a Declaration stressing Africa's vulnerability to the effects of global warming, in particular its adverse impact on ecosystems, food security, and social and economic development.[48] The Declaration specifically notes the urgent need for Africa to adapt to the adverse impacts of climate change on human rights

41 *Decision on the African Common Position on Climate Change*, AU Doc Assembly/AU/Dec 236 (XII) (2009).

42 See Claire Furphy, *Corruption in Africa: a Crime against Development* (2010) <www.consultancy africa.com>.

43 *The African Wall*, <www.fao.org/docrep/016/ap603e/ap603e.pdf>.

44 Ibid, 5. For recent initiatives see the documents of the 23rd Ordinary Session of the AU Assembly (26–27 June 2014), particularly concerning agriculture and food security.

45 Webpage of the Division: <http://rea.au.int/en/division/ENR>.

46 See, for instance, *Decision on the UNFCCC and Preparations for the Global Climate Change Events in 2014*, AU Doc Assembly/AU/Dec 538 (XXIII) (2014); *Decision on the UNFCCC*, AU Doc Assembly/AU/Dec 399 (XVIII) (2012); *Decision on Climate Change Negotiations*, AU Doc Assembly/AU/Dec 308 (XV) (2010).

47 See the website dedicated to the Conference: <www.unep.org/roa/amcen>.

48 *Gaborone Declaration on Climate Change and Africa's Development* (18 October 2013).

and the correlative necessity to reduce GHGs released into the atmosphere by all countries. Moreover, the Declaration emphasises that climate change represents an urgent and irreversible threat to human societies and the planet, that the window of opportunity to avoid dangerous climate change is closing, and that there is a growing risk of runaway climate change and catastrophic impacts for Africa. In particular, the Declaration stresses the importance of a comprehensive work programme covering finance, technology transfer and capacity-building for sustainable agriculture within the context of the right to development under the Cancun Adaptation Framework. Indeed, agriculture is recognised as a 'means of livelihood and the backbone of the African economy' and, in the end, 'a matter of survival'.[49]

Another continental effort integrating environmental and climate change concerns is the New Partnership for Africa's Development (NEPAD) initiative. NEPAD was initiated in 2001 by the Conference of Heads of State and Government of the Organization of African Unity (now the AU), as a pledge to eradicate poverty in Africa and achieve sustainable growth and development. A few focus areas of NEPAD activities are of particular relevance to climate change and human rights, namely, agriculture and food security, health and agriculture, water sanitation and the environment, and natural resource management. NEPAD's strategy aims to assist African countries to include climate change responses in their national development processes, and to strengthen skills in adaptation, mitigation, technology and finance in order to combat environmental changes.[50] The AU has endorsed NEPAD's plans as an official AU programme, and an Action Plan for the Environment Initiative was adopted in 2003 to achieve NEPAD's objectives.[51] This continental dynamic nonetheless remains partially ineffective, because African States have not adopted implementing legal norms and procedures at the domestic level.

The AComHPR has also issued two specific resolutions on climate change and its relevance to the continent. The first Resolution was adopted in 2009 and called upon the Assembly of the Heads of State and Government of the AU to protect human rights from climate change.[52] Substantively, Resolution 153/2009 highlights the fact that the lack of reference to human rights in international climate change negotiations particularly endangers the rights to life, physical integrity and livelihood, especially those of vulnerable populations, including indigenous peoples, whether collectively or individually. Interestingly, the Commission recalls and connects, on the one hand, the rights to development and to a satisfactory environment favourable to development under the Maputo Convention and, on

49 Article 16 of the *Declaration on Climate Change and Africa's Development*, Issued at the Fifth Special Session of the African Ministerial Conference on the Environment in Gaborone on 17–18 October 2013.

50 AU, *NEPAD's Action Plan for the Environment Initiative* (2003).

51 AU, *New Partnership for Africa's Development Base Document* (2001).

52 AU, *African Commission's Resolution on Climate Change and Human Rights and the Need to Study Its Impact in Africa*, Res 153 (XLVI) 09 (25 November 2009).

the other, the Convention on Biological Diversity.[53] Procedurally, the Resolution takes two main steps. First, it invites the Assembly of the Heads of State and Government of the AU to include human rights standards in climate change negotiations. This aims to particularly ensure that procedural guarantees, such as prior informed consent, are respected so as to avoid forced relocation, unfair dispossession of property and loss of livelihoods. Protection of vulnerable peoples' rights is also a fundamental objective.[54] Second, the Resolution prompts the AU Assembly to include the AComHPR in continental negotiations concerning human rights and climate change.[55] One can nonetheless question the need for any of the Commissioners to be part of negotiating teams when the Commission itself has not demonstrated its relevance to global warming. In fact, Resolution 153/2009 called for studies on the impact of climate change on human rights, but these are yet to be concluded. In this respect, a Working Group on Extractive Industries, Environment and Human Rights Violations in Africa was established in November 2009 during the session in which Resolution 153 was adopted, but no correlation was established with climate change.[56] Otherwise, between 2009 and 2014 the Commission has undertaken no steps to conduct the study it assigned to itself, even though the subject is highly compelling. In a follow-up Resolution adopted in May 2014, the Commission recognised that climate change studies have not been concluded yet and entrusted a Working Group with the task of completing them.[57] Hopefully, these studies will help to define and frame the role of the African Commission in climate change negotiations.

Overall, at the policy-making level, the relationship between climate change, environment and human rights has been pointed out. Policy initiatives are in line with the 2009 UN Human Rights Council's pronouncement on the effects of climate change on the enjoyment of human rights, according to which 'human rights obligations and commitments have the potential to inform and strengthen international and national policy-making in the area of climate change, promoting policy coherence, legitimacy and sustainable outcomes'.[58] However, documents embedding Africa's policy linking climate change and human rights are short texts that do not develop any organic and articulated approach to the matter. Furthermore, no binding effect has been attached to policy initiatives in the field, which thus remain 'soft' legal instruments.[59] The possibility that African States take concrete action on these bases is therefore quite limited.[60]

53 Ibid, [1].

54 Ibid, [2].

55 Ibid, [3].

56 *Resolution 148 on the Establishment of a Working Group on Extractive Industries, Environment and Human Rights Violations in Africa* (25 November 2009).

57 *Resolution 271 on Climate Change in Africa* (14 May 2014).

58 UN GA, HRC, Resolution 10/4, UN Doc A/HRC/RES/26/27 (25 March 2009).

59 See Megan Chapman, 'Climate Change and the Regional Human Rights Systems' (2010) 10(2) *Sustainable Development Law and Policy* 37.

60 IPCC, *Climate Change: Impact, Adaptation and Vulnerability* (2014) Chapter 22, Africa, 1242.

2.2. Adjudicating climate change based on human rights?

Responsibility for the impact of GHG emissions on human rights may be secondarily reinforced by jurisdictional organs. In this respect, the environmental case law of the AComHPR and the ACtHPR is particularly relevant. The Commission and the Court have indeed protected the environment through either first and second generation human rights or an independent third generation human right to environment.

The recent decision of the AComHPR in *Centre for Minority Rights Development on behalf of Endorois Community v Kenya*, commonly known as the *Endorois* case, is crucial. In this case, the Commission adjudicated upon the removal of an indigenous community from its ancestral land. The Endorois traditionally used to graze their animals in the lowlands around Lake Bogoria in the rainy season, turning to the Monchongoi Forest during the dry period. In 1973, the government of Kenya created the Lake Hannington Game Reserve, forcibly removed the Endorois pastoralist community from its lands and relocated it to an area unsuitable to their pastoral way of life. Furthermore, in 2002 the government granted mining concessions on Endorois traditional lands, which led to the construction of a road and mining operations, creating pollution risks. The plaintiffs alleged that the forced eviction of the Endorois took place without proper consultation and prevented them from freely practising their culture without adequate compensation, in breach of fundamental rights protected under the ACHPR.

The AComHPR held the government of Kenya responsible for breaching first and second generation fundamental rights, specifically under Articles 1 (Obligations of Member States), 8 (Right to Freedom of Conscience), 14 (Right to Property), 17 (Right to Education), 21 (Right to Freely Dispose of Wealth and Natural Resources) and 22 (Right to Development) ACHPR. In particular, the cultural rights of the Endorois were understood in connection with the protection of forests and natural resources.[61] The Commission also held that polluting traditional environments infringes upon the right to culture of indigenous communities.[62] This is further implicit in the recognition of a breach of Article 22 ACHPR, given that under Article 24 the right to development is strictly linked to the right to a healthy environment. On this basis, the Commission determined that the government of Kenya breached its obligations to respect and fulfil fundamental rights of the Endorois community, including procedural claims to information, participation and access to judicial remedies in environmental matters.[63] This decision clearly follows the lines of the approach to climate change and human rights outlined by the AComHPR in Resolution 153/2009.[64]

61 *Centre for Minority Rights Development (Kenya) and Minority Rights Group (on behalf of Endorois Welfare Council) v Kenya*, Comm 276/03 (4 February 2010) [245] ff.

62 Ibid, [293].

63 Ibid.

64 See above section 2.1.

The ACtHPR[65] took the same stance in a case brought before it by the African Commission against the government of Kenya. The case is very similar to *Endorois*, but relates to the *Ogiek*, a community of approximately 20,000 members, largely dependent on the greater Mau Forest. In fact, without previous consultation, in October 2009 the Kenyan government issued an eviction notice to the Ogiek people living in the Mau Forest.[66] After exhaustion of domestic remedies,[67] the Ogiek, supported by several NGOs, filed a complaint with the AComHPR, which referred the case to the ACtHPR.[68] Allegedly, although the eviction apparently aimed to protect the environment, its ultimate purpose was to initiate large-scale logging activities.

Finding a situation of 'extreme gravity and urgency',[69] on 13 March 2013 the ACtHPR issued an order for provisional measures, recognising the immediate and irreparable danger for the survival of the Ogiek group. As a result, the Court ordered that the Kenyan government reinstate a protective framework against land transactions in the Mau Forest and refrain from any act that could jeopardise litigation rights. In particular, the Court acted based on Articles 2 (Entitlement to Protection under the Banjul Charter), 3 (Non-discrimination), 4 (Right to Life and Integrity of the Person), 14 (Right to Property), 17(2) and (3) (Right to Culture) and 22 (Right to Development) ACHPR.[70] Environmental protection was thus indirectly granted via different human rights, particularly the right to development, which is intrinsically linked to the right to environment under Article 24 ACHPR.

Although the *Endorois* and *Ogiek* cases do not directly concern environmental pollution, they have relevant implications for situations in which GHG emissions affecting indigenous communities are at stake, such as those of the Inuit and Athabaskan peoples in the Arctic.[71]

In other cases, the AComHPR received petitions directly based on violations of the right to environment enshrined in Article 24 ACHPR. The most notable case was brought by an NGO against Nigeria, because of oil exploitation in the Niger Delta and its negative impact on the social and economic conditions of

65 See the website of the African Court, operative since 2006: <www.african-court.org>. With the Protocol on the Statute of the African Court of Justice and Human Rights (2008), the ACtHPR merged with the African Court of Justice. However, this protocol is not yet in force.

66 See Joseph Sang, *The Ogiek in Mau Forest* (2000) <www.forestpeoples.org/sites/fpp/files/publication/2010/10/kenyaeng.pdf>.

67 See, in particular, *Kemai and Nine Others v Attorney General and Three Others*, Civil Case 238 of 1999 (High Court, Nairobi, 23 March 2000).

68 *AComHPR v Kenya*, Appl No 06/2012 (12 July 2012).

69 *AComHPR v Kenya*, Appl No 06/2012, Order of Provisional Measures (5 March 2013) [22].

70 Ibid, [20].

71 Inuit, *Petition to the Inter-American Commission on Human Rights, Seeking Relief from Violations Resulting from Global Warming Caused by Acts and Omissions of the United States* (7 December 2005); Arctic Athabaskan Council on behalf of All Arctic Athabaskan Peoples of the Arctic Regions of Canada and the United States, *Petition to the Inter-American Commission on Human Rights, Seeking Relief from Violations of the Rights of Arctic Athabaskan Peoples Resulting from Rapid Arctic Warming and Melting Caused by Emissions of Black Carbon by Canada* (23 April 2013).

the local population.[72] In the *Ogoniland* case, the Commission adjudicated upon the involvement of the military government of Nigeria in oil production through the Nigerian National Petroleum Development Company, a State oil company operating in consortium with Shell Petroleum Development Corporation. Oil extraction operations disregarded health and environmental issues of local communities. Allegedly, toxic wastes were released into the environment and local waterways and avoidable spills resulted from the consortium's failure to properly maintain its facilities. The Nigerian government did not adequately monitor oil operations, did not require adequate standard safety measures and failed to conduct health and environmental impact studies, also preventing scientists from accessing Ogoniland and undertaking necessary investigations. Oil operations thus caused environmental degradation and health problems, including skin infections, gastrointestinal and respiratory ailments, increased cancer risks, and neurological and reproductive problems affecting the Ogoni people. Furthermore, State security forces made available to the oil companies by the Nigerian government attacked, burned and destroyed several Ogoni villages and homes in response to a non-violent campaign organised against oil exploitation by the Movement for the Survival of Ogoni People. Additional attacks were conducted by unidentified gunmen.

The AComHPR held the Federal Republic of Nigeria in breach of Articles 2 (Right to Freedom from Discrimination), 4 (Right to Life), 14 (Right to Property), 16 (Right to Health), 18(1) (Protection of the Family and Vulnerable Groups), 21 (Right to Free Disposal of Wealth and Natural Resources) and 24 (Right to a General Satisfactory Environment) ACHPR. As a consequence, the Commission requested that the government of Nigeria allow proper investigation of the facts in issue and provide adequate compensation. Procedurally, the Commission asked the State of Nigeria to make appropriate environmental and social impact assessments for any future oil development, to establish effective and independent oversight bodies for the petroleum industry, and to give affected communities access to regulatory and decision-making procedures.

With particular regard to the right to environment, the African Commission stated:

> (t)he right to a general satisfactory environment, as guaranteed under Article 24 of the African Charter or the right to a healthy environment, as it is widely known . . . imposes clear obligations upon a government. It requires the State to take reasonable and other measures to prevent pollution and ecological degradation, to promote conservation, and to secure an ecologically sustainable development and use of natural resources.[73]

72 *The Social and Economic Rights Action Center and the Center for Economic and Social Rights/Nigeria*, Comm No 155/96, Decision ACHPR/COMM/A044/1 (AComHPR, 2001).

73 Ibid, [52].

Significantly, the Commission established a direct connection between the right to environment and the right to health, holding that 'Article 12 of the International Covenant on Economic, Social and Cultural Rights (ICESCR), to which Nigeria is a party, requires governments to take necessary steps for the improvement of all aspects of environmental and industrial hygiene'.[74] As a result, the Commission held:

> the right to enjoy the best attainable state of physical and mental health enunciated in Article 16(1) of the African Charter and the right to a general satisfactory environment favourable to development . . . obligate governments to desist from directly threatening the health and environment of their citizens. The State is under an obligation to respect the just noted rights and this entails largely non-interventionist conduct from the State for example, not from carrying out, sponsoring or tolerating any practice, policy or legal measures violating the integrity of the individual.[75]

Unlike the *Endorois* and *Ogiek* cases, *Ogoniland* directly concerns environmental pollution tackled by means of the combined rights to health and environment, which must be respected, protected and fulfilled. In particular, the nature of this case allowed the Commission to clarify that governments are under a positive duty to protect their citizens from harmful acts committed by private entities. In light of the fact that action was initiated by two non-governmental organisations, one based in Nigeria and another based in the US, the Commission stressed the importance of *actio popularis* in the field of environmental protection.[76] This decision is thus a milestone in the effort of outlining a complete substantive and procedural framework for the right to (a (climatically) sustainable) environment, based on an *erga omnes* approach.[77] Arguably, it has inspired the text of Resolution 153/2009 and should exceptionally facilitate protection and action against GHG emissions before the AComHPR and the ACtHPR *de lege ferenda*. This should avoid a negative stance such as that taken by the IAComHR on the Inuit Petition, which is based, inter alia, on the right to a healthy environment regarded as a customary international claim.[78]

74 Ibid.

75 Ibid.

76 Ibid, [49].

77 This case has another dimension with the execution of Ken Saro Wiwa, one of the earliest members of the Movement for the Survival of the Ogoni People, and the subsequent litigation under the Alien Tort Statute in the United States of America (see US Supreme Court, *Kiobel v Royal Dutch Petroleum Co*, Case No 10–1491 (17 April 2013)). See also Ingrid B Wuerth, '*Kiobel v Royal Dutch Petroleum Co*: the Supreme Court and the Alien Tort Statute' (2013) 107 *AJIL* 601.

78 *Inuit Petition* (2005) 74.

Conclusion

Since many African States are 'developing', they are particularly vulnerable to climate change. The continent's regulatory framework offers relevant tools to fight the impact of climate change on human rights, and these tools are different from those provided for in other regional and international human rights systems. In particular, human rights protection in Africa is marked by a progressive shift from first and second generation human rights to third generation rights. This innovative perspective permits to approach climate change not only from the standpoint of classical individual rights, but also from that of collective rights, particularly the right to (a (climatically) sustainable) environment, comprehensively framed from the viewpoint of substance and consequent procedural implications.

In practice, the AComHPR and the ACtHPR paved the way for exploiting these instruments to provide direct and indirect protection against climate change, by adjudicating cases particularly based on the interconnected rights to environment and development, specifically those of indigenous peoples. This approach is nevertheless not paralleled by adequate initiatives as to decision-making. In fact, the AU has adopted progressive policies linking climate change and human rights, but these are superficially articulated from the substantive viewpoint and institutionally inconsistent, which ultimately results in limited effectiveness.

16 Missing opportunities to shed light on climate change in the Inter-American human rights protection system

Mouloud Boumghar

Introduction

The American Continent plays a pivotal role in the relationship between climate change and human rights. On the one hand, the US, which is historically one of the main GHG emitters,[1] is located in this part of the world. On the other, climate change has an invasive negative impact on fundamental rights in the region.

With regard to North America, it is considered that global warming affects water supply and distribution, ecosystems and biodiversity, food and agricultural security, and health.[2] Extreme coastal storms such as Hurricane Katrina, causing death and property destruction, have clearly brought the phenomenon to the attention of the public.

Similar trends have been registered in Central and South America, where further problems arise for human settlements as well as low-lying and coastal areas.[3] For instance, given that more than 20 per cent of the population in the region is concentrated in large cities, problems related to water availability are of great concern.[4]

Furthermore, climate change has a particular impact on the Polar Regions. First, the Antarctic is close to South America, and, second, part of North America is included in the Arctic Circle. Because of the ice-covered environment, global warming has a dramatic impact on these areas. For instance, ice-shrinking is causing a decline in polar bear population and a shift in evaporation and rainfall patterns, as well as impacting availability of freshwater resources. Ocean acidification is determining changes in the ranges of marine species. This affects the basic rights of indigenous peoples living in the Arctic, for example with regard to food security, which has further relevant cultural implications.[5]

Like Europe and Africa, but unlike the Asia-Pacific, in America human rights are protected via specific substantive and procedural mechanisms, particularly

1 See World Bank Group, *State and Trends of Carbon Pricing* (2014) 19.

2 IPCC, *Fifth Assessment Report* (2014) Chapter 26, North America, 1456 ff.

3 Ibid, Chapter 27, Central and South America, 1516 ff.

4 Ibid, 1531.

5 Ibid, Chapter 27, Polar Regions, 1570 ff.

centred on the IAComHR and the IACtHR, within the context of the OAS. Combining this legal framework and the geographic features of the region, which includes the Arctic area inhabited by indigenous peoples, makes the American continent interesting for exploring the relationship between climate change and human rights. Such is the focus of this chapter, which first investigates the relationship between the Inter-American human rights protection system and climate change from a theoretical perspective, and subsequently considers specific initiatives and cases on the issue.

1. Fundamental Inter-American human rights protection mechanisms and climate change

Several international and regional human rights instruments, mostly adopted by the OAS, have implications for the protection of peoples in the Americas against the adverse effects of climate change.

A few months before the adoption of the UDHR by the UN General Assembly on 10 December 1948, the OAS General Assembly adopted the ADRDM on 2 May 1948. More than 20 years later, on 22 November 1969 the OAS Member States adopted the ACHR (Pact of San José), which then entered into force on 18 July 1978. An Additional Protocol to the ACHR in the area of Economic, Social and Cultural Rights (Protocol of San Salvador) was adopted on 17 November 1988 and entered into force one year later. These texts embed claims that are interpreted in light of international human rights law[6] and are generally considered to be affected by the adverse effects of climate change.

Both Article 1 ADRDM and Article 4 ACHR protect the right to life. Article 13 ADRDM recognises that '[e]very person' has 'the right to the preservation of his health through sanitary and social measures relating to food, clothing, housing and medical care, to the extent permitted by public and community resources'. Article 10(1) of the San Salvador Protocol states that '[e]veryone shall have the right to health, understood to mean the enjoyment of the highest level of physical, mental and social well-being'. Article 12 of the Protocol provides for the right to food, that is, 'the right to adequate nutrition which guarantees the highest level of physical, emotional and intellectual development'. Moreover, the same provision expressly acknowledges that '[e]veryone shall have the right to live in a healthy environment and to have access to basic public services', which entails a State duty to promote, protect, preserve and improve environmental protection.

6 See, inter alia, IACtHR, *Mary and Carrie Dan v US*, Case 11.140, Report No 75/02 (27 December 2002) [96]–[97]; Cecilia Medina, 'The Role of International Tribunals: Law-Making or Creative Interpretation?', in Dinah Shelton (ed), *The Oxford Handbook of International Human Rights Law* (OUP, 2013) 649, 661 ff; Lucas Lixinski, 'Treaty Interpretation by the Inter-American Court of Human Rights: Expansion at the Service of the Unity of International Law' (2010) 21 *EJIL* 585.

This core regulation is complemented by other relevant human rights instruments.[7] Among those, the Inter-American Convention on the Elimination of Racism, Racial Discrimination and Related Forms of Intolerance (IACERDI) establishes a direct link between the principle of non-discrimination and environmental protection, setting out an obligation for States to 'prevent, eliminate, prohibit, and punish . . . all acts and manifestations of racism, racial discrimination, and related forms of intolerance', including '[t]he restriction or limitation . . . of the right of every person, to access and sustainably use water, natural resources, ecosystems, biodiversity, and ecological services that are part of each State's natural heritage, protected by the relevant international instruments and their own national laws' (Article 4(xiv)). The Convention has nevertheless not yet entered into force.

These substantive obligations are mainly enforced via the IAComHR and the IACtHR, which adjudicate upon breaches of the ACHR, IACERDI[8] and ADRDM, although the latter is not binding, despite its universal application. The effectiveness of the system is nonetheless limited by several factors. Substantively, since the ratification of the ACHR is not a condition for membership with the OAS, among the 34 members of the continental organisation only 23 States are parties to the Pact of San José.[9] Even if the most populated countries of Central and South America, such as Mexico, Brazil and Argentina, have ratified the Pact, both the US and Canada in the Northern part of the continent are not parties to this instrument.[10] The San Salvador Protocol only includes 16 States parties, that is, less than the ACHR. Procedurally, the IACtHR has contentious jurisdiction over State communications and individual petitions, but only subject to explicit acceptance by States Parties and prior examination by the IAComHR (Articles 61 and 62 ACHR). As to the San Salvador Protocol, under Article 19(6) individual petitions are exclusively admitted for breaches of trade union rights (Article 8(1)(a)) and the right to education (Article 13). Therefore, fundamental claims embedded in the Protocol, such as the right to environment, are not subject to the jurisdiction of the Commission and the Court.

7 These include the Inter-American Convention to Prevent and Punish Torture (adopted 9 December 1985, into force 28 February 1987); the Protocol to the American Convention of Human Rights to Abolish the Death Penalty (adopted 8 June 1990, into force among ratifying States); the Inter-American Convention on the Prevention, Punishment and Eradication of Violence against Women (also known as 'Convention of Belém do Pará', adopted 9 June 1994, into force 3 May 1995); the Inter-American Convention on Forced Disappearance of Persons (adopted 9 June 1994, into force 28 March 1996); the Inter-American Convention on the Elimination of All Forms of Discrimination against Persons with Disabilities (adopted 7 June 1999, into force 14 September 2001); the Inter-American Convention against All Forms of Discrimination and Intolerance (6 June 2013, not yet into force); and the Inter-American Convention against Racism, Racial Discrimination and Related Forms of Discrimination (6 June 2013, not yet in force).

8 By virtue of Article 15.

9 See <www.oas.org/dil/treaties_B-32_American_Convention_on_Human_Rights_sign.htm>.

10 As well as, inter alia, Trinidad and Tobago and Venezuela, which denunciated the Convention respectively in 1998 and 2012.

2. The stance of the Organisation of American States

Institutionally, the OAS has recognised the importance of studying climate change and taking mitigation and adaptation measures against global warming in the American Continent.[11] However, initiatives approaching climate change from the standpoint of fundamental rights are much more limited.

The main instrument of the OAS linking climate change and human rights is Resolution 2429, which was adopted on 3 June 2008.[12] This document was drafted within the context of the OAS action on sustainable development, in accordance with UN HRC Resolution 7/23 on Climate Change and Human Rights. It thus replicates on a regional scale the stance taken by the main UN human rights body on the issue.[13]

Resolution 2429 commits the OAS to explore 'possible links between climate change and human rights'[14] and prompts it to pursue GHG adaptation and mitigation policies, specifically aiming to increase the resilience of vulnerable States and populations.[15] More precisely, the Resolution instructs the IAComHR to explore possible connections between climate change and the full enjoyment of human rights.[16] This is supposed to take place in collaboration with other regional and international bodies, namely the OAS Department for Sustainable Development, the HRC, the OHCHR, and the IPCC. Procedurally, participation of States and civil society at large is envisaged in this process,[17] which nevertheless has only an 'investigative' aim. Indeed, although the Resolution invites the Permanent Council and the Permanent Executive Committee of the Inter-American Council for Integral Development (CEPCIDI)[18] to report to the OAS General Assembly on the issue, with support from the General Secretariat and the Executive Secretariat of the IACHR,[19] no mention is made of regulatory action.

Overall, despite the fact that it is only a short text, Resolution 2429 is meaningful, to the extent that it provides a comprehensive approach to the relationship between climate change and human rights, involving both law-making and jurisdictional bodies. Furthermore, even if it is a quite isolated document, the resolution can be read within the framework of broader OAS initiatives

11 See, in particular, the website of the OAS Department of Sustainable Development: <www.oas.org/en/sedi/dsd>. See also the Agreement Establishing the Inter-American Institute for Global Change Research promoted by the Department of International Law of the OAS.

12 OASGA, *Climate Change and Human Rights*, Res 2429, XXXVIII-O/08 (3 June 2008).

13 See Aktypis, Decaux and Leroy in this volume.

14 OASGA (2008) [3].

15 Ibid, [2].

16 Ibid, [4].

17 Ibid, [6].

18 According to Article 95(c)(1) of the OAS Charter, the Inter-American Council for Integral Development is a specialised inter-governmental body promoting and coordinating the execution of development programmes and projects assigned to subsidiary bodies and relevant organisations, on the basis of the priorities identified by the Member States, in areas such as economic and social development, including trade, tourism, integration and the environment.

19 OASGA (2008) [7].

acknowledging a relationship between human rights and the environment. In particular, on 5 July 2001 the Assembly of the OAS adopted Resolution 1819, which takes account of the action of the UN ECOSOC, and thus that of the Special Rapporteur Fatma Zohra Ksentini,[20] and recognises the importance of the relationship between environment and human rights. This is a short text that only recommends the Secretariat to undertake explorative studies, but it provides interesting substantive and procedural hints. On the one hand, it acknowledges that economic, social and cultural rights can constrain unsustainable environmental patterns. On the other, it highlights the importance of dissemination of information and political participation of affected groups in governing environmental issues infringing upon human rights. Along these lines, the Resolution commits the IAComHR and the Unit for Sustainable Development to investigate the relationship between fundamental rights and environmental protection, according to the comprehensive approach promoted by Resolution 2429/2009 on climate change and human rights. Follow-up studies have particularly highlighted the impact of environmental pollution on indigenous peoples.[21]

More broadly, initiatives linking climate change, environment and human rights are consistent with the approach of the OAS to fundamental rights and sustainable development stated in basic regulatory instruments. Thus, Article 30 of the 1948 Charter of the OAS affirms that '[t]he Member States . . . pledge themselves to a united effort to ensure international social justice in their relations and integral development for their peoples, as conditions essential to peace and security'.[22] Even more significantly, according to the Preamble to the 2001 Inter-American Democratic Charter of the OAS, a 'safe environment is essential to the integral development of the human being, which contributes to democracy and political stability'. Furthermore, under Article 15 of the Charter '[t]he exercise of democracy promotes the preservation and good stewardship of the environment', so that '[i]t is essential that the States of the Hemisphere implement policies and strategies to protect the environment, including application of various treaties and conventions, to achieve sustainable development for the benefit of future generations'. Along the same lines, Article 33 and the Preamble to the Social Charter of the Americas, adopted by the OAS General Assembly on 4 June 2012, explicitly recognise that 'a safe environment is essential to an integral development'. Article 17 also affirms that OAS Member States commit 'to promote healthy lifestyles and to strengthen their capacity to prevent, detect, and respond to chronic non-communicable diseases, current and emerging infectious diseases, and

20 OASGA, *Human Rights and the Environment,* Res 1819, XXXI-O/01 (5 June 2001); Fatma Zohra Ksentini, *Human Rights and the Environment, Review of Further Developments in Fields with which the Sub-Commission Has Been Concerned,* Final Report of the UN Sub-Commission on Prevention of Discrimination and Protection of Minorities, UN ECOSOC, E.CN.4/Sub.2/1994/9 (6 July 1994).

21 OAS, *Human Rights and the Environment,* Report of the General Secretariat pursuant to AG/RES 1819 (XXXI-O/01), OEA/Ser.G, CP/CAJP-1898/02 (4 April 2002).

22 'Integral development' encompasses the 'economic, social, educational, cultural, scientific, and technological fields through which the goals that each country sets for accomplishing it should be achieved'.

environmental health concerns'. Moreover, Article 18 states that 'water is fundamental for life and central to socioeconomic development and environmental sustainability'. The Social Charter specifically addresses climate change in Article 22, according to which '[n]atural and man-made disasters affect populations, economies, and the environment', so that Member States commit to 'improving regional cooperation and to strengthening their national, technical, and institutional capacity for disaster prevention, preparedness and response, rehabilitation, resilience, risk reduction, impact mitigation, and evaluation'. Within this context, States abide by the obligation to 'face the impact of climate variability, including the *El Niño* and *La Niña* phenomena, and the adverse effects of climate change that represent a risk increase in all countries of the Hemisphere, particularly for developing countries'.

3. Soliciting the Inter-American Commission on Human Rights

The IAComHR has received two petitions explicitly and directly linking human rights and climate change. Both have been filed on behalf of the Inuit people of the US and Canada and directed against a North American State Member of the OAS, the US in the first case and Canada in the second. The NGOs Centre for International Environmental Law and Earthjustice, along with Ecojustice Canada, drafted the petitions.

The first petition was lodged on 7 December 2005 with the IAComHR by Sheila Watt-Cloutier, then Chair of the Inuit Circumpolar Council (ICpC),[23] on behalf of herself and 62 other individuals, all Inuit of the Arctic regions of the US and Canada, including 48 Canadian residents.[24] The second petition was submitted by the Arctic Athabaskan Council (AAC)[25] to the IAComHR on 23 April 2013, on behalf of the Arctic Athabaskan people of the US and Canada.[26]

23 The ICpC is a pan-Inuit organisation and includes members from Canada, the US, Greenland and Russia. The Inuit consider the Circumpolar Region, including Arctic and sub-Arctic regions, as their ancestral homeland.

24 Inuit, *Petition to the Inter-American Commission on Human Rights, Seeking Relief from Violations Resulting from Global Warming Caused by Acts and Omissions of the United States* (7 December 2005) 103–4, <www.ciel.org/Publications/ICC_Petition_7Dec05.pdf> (first Inuit petition). See also Hari M Osofsky, 'The Inuit Petition as a Bridge? Beyond Dialectics of Climate Change and Indigenous Peoples' Rights', in William CG Burns and Hari M Osofsky (eds), *Adjudicating Climate Change: State, National, and International Approaches* (CUP, 2009) 272; Timo Koivurova, 'International Legal Avenues to Address the Plight of Victims of Climate Change: Problems and Prospects' (2007) 22(2) *J Envtl Law and Litigation* 267, 285–93.

25 The Arctic Athabaskan Council was created by several Athabaskan peoples' local governments of the Arctic regions of the US and Canada. Its main aim is to preserve and protect the Northern ecosystem (see <www.arcticathabaskancouncil.com>).

26 Arctic Athabaskan Council on Behalf of All Arctic Athabaskan Peoples of the Arctic Regions of Canada and the United States, *Petition to the Inter-American Commission on Human Rights Seeking Relief from Violations of the Rights of Arctic Athabaskan Peoples Resulting from Rapid Arctic Warming and Melting Caused by Emissions of Black Carbon by Canada* (23 April 2013) (second Inuit Petition) <http://earthjustice.org/sites/default/files/AAC_PETITION_13-04-23a.pdf>.

3.1. First Inuit petition: the United States, GHGs and human rights

The Inuit identify themselves as one of the numerous 'indigenous peoples' of the Hemisphere, to which the audacious case law of the IAComHR applies.[27] In fact, the Inuit have a characteristic link to their land, on which they rely for their economic survival, culture and identity.[28] In light of cases such as *Belize Maya* and *Saramaka*,[29] this is crucial to the first petition, which focuses on the heavy dependence of the Inuit on the specificities of the Arctic environment, despite the engagement of many of them in wage employment. Traditional knowledge of physical surroundings, including climate, which results from experience accumulated over several generations, is also vital to life and survival on ice.[30] Furthermore, the Inuit's lifestyle in the Arctic regions of the US and Canada relies on shared hunting and harvesting, that is, 'country food'.[31]

According to the ICpC, climate change caused by anthropogenic GHG emissions is objectively damaging the Arctic environment. Drawing on scientific studies, including the IPCC Third Assessment Report and US reports, the petitioner states that temperature is rising, causing the melting of sea ice, ice sheets, glaciers and permafrost, and sea level rise, with consequent alterations in species and habitats.[32] It is thus assumed that scientific consensus has emerged that global warming is caused by the increased concentration of GHGs in the atmosphere as a result of human activity, with the most severe destruction already taking place and progressively worsening in the Arctic according to robust projections.[33] By further relying on witnesses, the petition points out the adverse impact of climate change on traditional everyday life and survival in the Arctic.[34]

Overall, the petitioner asserts that climate change has made Inuit lands less accessible, more dangerous, unfamiliar and less valuable, harming the Inuit's property, subsistence harvest, travels, safety, health, education and culture. Allegedly, this is in breach of several human rights set forth in the ADRDM, dynamically interpreted in light of other regional and international instruments, which is compelled by the fact that the US has not ratified the ACHR.[35] Key to the petition is the collective nature of the right to property of indigenous peoples on their ancestral lands and resources, regarded as a pre-condition to the

27 Laurence Burgorgue-Larsen, 'The Rights of Indigenous Peoples', in Laurence Burgorgue-Larsen and Amaya Ubeda de Torres (eds), *The Inter-American Court of Human Rights: Case Law and Commentary* (Oxford University Press, 2011) 497.

28 *Inuit Petition* (2005) 17.

29 See section 4 below.

30 *Inuit Petition* (2005) 19.

31 Ibid, 15.

32 Ibid, 21 ff.

33 Ibid, 27 ff.

34 Ibid, 39 ff.

35 Ibid, 74 ff and 96 ff.

enjoyment of other human rights.[36] Such a claim would be affected by environmental degradation, which would in turn be covered by a customary right to environment.[37] This would entail a breach of individual Inuit's rights to the benefits of their culture,[38] to use and enjoy traditionally occupied lands as well as personal, intangible and intellectual property,[39] and the right to health.[40] Such violations are relevant to Article 23 ADRDM and would be in further breach of the rights to life, physical integrity, and security,[41] own means of subsistence,[42] residence, movement, and inviolability of the home.[43]

Subjectively, it is argued that these infringements are the result of acts and omissions by the US, which was the largest cumulative CO_2 emitter in the second half of the 20th century and the world's largest emitter of energy-related CO_2 at the time of the petition, accounting for nearly one quarter of global emissions.[44] Furthermore, with only 4.7 per cent of the world's population, in 2000 the US produced 24 per cent of total GHG emissions, that is, five times the global average, on a per-capita basis.[45] In this respect, US policy measures, such as the Global Climate Change Initiative, aiming to reduce GHG emissions by 18 per cent between 2002 and 2012, were deemed 'insufficient', 'ineffective' and even 'misleading'.[46] Because of this negative approach, the US is considered responsible

36 Ibid, 70 ff.

37 Ibid, 74.

38 Ibid, 74 ff. The right to culture is recognised in Article 13 ADRDM, whose first paragraph reads as follows:

> Every person has the right to take part in the cultural life of the community, to enjoy the arts, and to participate in the benefits that result from intellectual progress, especially scientific discoveries.

39 *Inuit Petition* (2005) 79 ff. Protection for intangible and intellectual property is facilitated by a broad interpretation of Articles 21 and 23 ADRDM.

Article 21 provides:

> 1. Everyone has the right to the use and enjoyment of his property. The law may subordinate such use and enjoyment to the interest of society; 2. No one shall be deprived of his property except upon payment of just compensation, for reasons of public utility or social interest, and in the cases and according to the forms established by law; 3. Usury and any other form of exploitation of man by man shall be prohibited by law.

Article 23 states:

> Every person has the right to own such private property as meets the essential needs of decent living and helps to maintain the dignity of the individual and of the home.

40 *Inuit Petition* (2005) 85 ff. See also Article 11 ADRDM.

41 *Inuit Petition* (2005) 89 ff. Under Article 1 ADRDM, 'every human being has the right to life, liberty and the security of his person.'

42 *Inuit Petition* (2005) 92 ff. This right can be inferred from an extensive interpretation of the right to property of indigenous peoples and is explicitly recognised in Article 1(2) of the ICCPR and ICESCR: 'In no case may a people be deprived of its own means of subsistence.'

43 *Inuit Petition* (2005) 92 ff. See also Articles 8 and 9 ADRDM.

44 *Inuit Petition* (2005) 69.

45 Ibid.

46 Ibid, 103 ff.

for failing to cooperate with international efforts to reduce GHGs and in breach of the obligation to provide effective remedies.[47]

Based on these premises, the petition invites the IAComHR to assess US acts and omissions not only under the ADRDM, but also according to other relevant international obligations, particularly under the ACHR, the Draft American Declaration on the Rights of Indigenous Peoples, the UDHR, the ICCPR, the ICESCR, as well as the UN Draft Declaration on the Rights of Indigenous Peoples.[48] In addition to regional and international human rights protection instruments, the US is considered to be in violation of general environmental rules, particularly the UNFCCC,[49] the principle of sustainable development, the precautionary principle, and the no-harm rule, protecting the Inuit living not only in the US, but also in Canada.[50] Thus, the petitioner requests the Commission to interpret the ADRDM in light of general international environmental law.

Finally, the ICpC asks the Commission to declare the responsibility of the US under the ADRDM, thus recommending that the US provide adequate relief; cooperatively adopt mandatory GHG mitigation measures; pre-emptively assess the impact of its policy on the Arctic environment; and cooperate with the Inuit in implementing specific GHG mitigation and adaptation plans.[51]

3.2. Second Inuit petition (Arctic Athabaskan people): Canada, black carbon and human rights

The second Inuit petition addresses black carbon emissions by Canada, which are roughly estimated at 98,000 annual tons and the State allegedly failed to regulate.[52] Diesel emissions, residential heating, and the burning of biomass in agriculture are identified as the major sources of black carbon.[53]

According to the AAC, the warming effect of black carbon on climate is only second to CO_2, and snow and ice are particularly sensitive to it.[54] The problem is thus not only the quantity of black carbon emissions, but also the proximity of its sources, since: '[a]lthough relatively smaller than emissions from lower latitudes, emissions from within or near the Arctic have a disproportionate effect because there is a greater likelihood they will deposit on Arctic snow and ice.'[55]

47 Ibid, 110–11.
48 Ibid, 74 and 96–7.
49 Into force as to the US in March 1994. To date, the US has nonetheless not ratified the Kyoto Protocol.
50 *Inuit Petition* (2005) 97–102.
51 Ibid, 118.
52 *Arctic Athabaskan Council Petition* (2013) 2.
53 Ibid.
54 Ibid, 2–3.
55 Ibid, 3.

Reduction of Canadian black carbon emissions is therefore invoked to slow down the melting of ice and snow in the Northern regions of the Hemisphere. Like the first Inuit petition, Athabaskan of Canada and Alaska see their economic, spiritual and cultural survival as an indigenous people threatened by severely rising temperatures.[56] As a consequence, the petitioner highlights the responsibility of Canada for failing to adequately regulate black carbon emissions that are partly causing Arctic warming and ice-melting.[57] For instance, it is claimed that the State of Alberta Energy and Utilities Board's Directive 060 on Upstream Petroleum Industry Flaring, Incinerating, and Venting does not significantly reduce black carbon emissions generated by flared, incinerated and vented gas.[58]

Rising temperatures caused by black carbon allegedly have a devastating effect on the lifestyle of the Arctic Athabaskan people, which is largely grounded in an ice-covered environment.[59] Thus, still based on cases such as *Belize Maya* and *Saramaka*, the petitioner argues that the negative conduct of Canada is in breach of several Arctic Athabaskan peoples' fundamental rights. In this respect, along the lines of the first Inuit petition the AAC invites the AComHR to interpret the ADRDM in the broader light of international law.[60] More specifically, the petition invokes the no-harm principle and the precautionary principle in environmental matters,[61] although environmental protection is only implicitly considered covered by a fundamental right itself, by means of a reference to Article 11 of the San Salvador Protocol and to OAS Resolution 2429/2008 on Climate Change and Human Rights.[62] Environmental protection is also regarded as central to the enjoyment of other fundamental individual claims,[63] so that Canada would ultimately be in breach of the rights to culture, property, health and means of subsistence.[64]

In light of these premises, the AAC requests the IAComHR to recommend that Canada adopt mandatory measures to limit emissions of black carbon from key sectors, consider the climate impacts of black carbon emissions on the Arctic and Arctic Athabaskan people before approving major government action, and implement, in coordination with Arctic Athabaskan people, a plan to protect their culture and resources from the effects of accelerated Arctic warming and melting.[65]

56 Ibid, 21 ff.
57 Ibid, 17–21 and 79 ff.
58 Ibid, 21.
59 Ibid, 21 ff.
60 Ibid, 49–50.
61 Ibid, 51–2.
62 Ibid, 54–5.
63 Ibid, 54–6 ff.
64 Ibid, 57 ff.
65 Ibid, 87.

4. A possible approach to human rights and climate change in light of the case law of the IAComHR and IACtHR

Both the first and second Inuit petitions require the IAComHR to make a site visit and hold a hearing to investigate the claims.[66] As to the first petition, the Commission did not make any site visit, but after dismissing all claims without issuing a report on motivations[67] held a hearing on the issue of global warming and human rights in the Americas on 1 March 2007.[68] At the time of writing, the Commission has not yet addressed the second petition.[69]

Thus, in light of prior Inter-American case law, how could the Commission have addressed the first Inuit petition? How could it address the second one? The IAComHR and IACtHR have developed an important jurisprudence, particularly in the matter of indigenous peoples, which is relevant to these questions. Some cases concern indigenous rights to property over their ancestral lands and natural resources, whereas others focus more on the impact of environmental pollution.[70] Such intertwined substantive obligations are complemented by correlative interrelated procedural rights, in terms of access to information, judicial remedies and participation in decision-making.

In *Yanomami*, the IAComHR dealt with the consequences of a Brazilian government's plan of exploitation of natural resources in the Amazonian region. This plan included the construction of a highway through the Yanomami indigenous people's territory, which compelled them to abandon their habitat and enabled massive works to begin, with devastating physical and socio-cultural effects, including contagious diseases. The Commission concluded that the failure of the Brazilian government to take timely and effective measures to protect Yanomami Indians resulted in the breach of their rights to life, liberty, personal security, residence, movement, and preservation of health and well-being provided for in Articles 1, 8 and 9 ADRDM.[71]

In *Maya Indigenous Communities of the Toledo District*, the IAComHR dealt with logging and oil concession by the State of Belize over traditional lands occupied by the Maya people. The Commission recognised that the rights of indigenous

66 *Inuit Petition* (2005) 118; *Arctic Athabaskan Petition* (2013) 86.

67 IAComHR, *Letter from Ariel E Dulitzky, Assistant Executive Secretary, OAS, to Paul Crowley, Legal Representative for Sheila Watt-Cloutier and Others, regarding Petition P-1413–05* (16 November 2006).

68 For a brief account, see Megan Chapman, 'Climate Change and the Regional Human Rights Systems' (2010) 10(2) *Sustainable Development Law and Policy* 37.

69 The petition is not mentioned in the Annual Reports of the IAComHR: <www.oas.org/en/iachr/decisions/cases_reports.asp>.

70 This makes the Inter-American human rights protection system richer than the European and African ones as to environmental litigation, despite the recognition of the right to a general satisfactory environment in Article 24 ACHPR (Riccardo Pavoni, 'Environmental Jurisprudence of the European and Inter-American Courts of Human Rights: Comparative Insights', in Ben Boer (ed), *Environmental Law Dimensions of Human Rights* (OUP, 2015); Lixinski (2010) 594).

71 *Indians Yanomami (Brazil)*, Case 7615, Resolution No 12/85 (IAComHR, 5 March 1985).

and tribal peoples over their ancestral territories are vital for their economic, social and cultural survival.[72] In this context, the individual and collective right to property provided for in Article 23 ADRDM becomes a pre-condition crucial to the exercise of other fundamental rights,[73] including the procedural rights to appropriate consultation and judicial remedies, as well as the substantive rights to life, health, cultural identity, religious freedom, labour rights, and psychological and moral integrity.[74]

This approach was confirmed by the IACtHR in *Yakye Axa Indigenous Community v Paraguay*, where the Court adjudicated upon the situation of an indigenous community evicted from its ancestral lands and living in a vulnerable socio-economic situation, which threatened individual and collective survival. Upholding the claims of the petitioners, the Court found a violation of the rights to property, life and fair trial under Articles 4, 8, 21, and 25 ACHR.[75] More specifically, the Court considered that access of indigenous peoples to their ancestral lands and related natural resources is basic to the enjoyment of the rights to food and water, which are further critical to the rights to health, a decent existence, and other fundamental claims, such as the rights to education and cultural identity.[76]

Significantly, in *Community of San Mateo de Huanchor v Peru* the petitioners complained of environmental pollution caused by a field of toxic waste sludge owned by Lizandro Proaño SA Mining Company, affecting a community including a majority of indigenous people. The IAComHR declared the petition admissible and held that, if proven, the effects of the environmental pollution of the sludge, which had created a public health crisis, could be characterised as a violation of the rights to personal security, property, fair trial, judicial protection, children's rights and the progressive development of economic, social and cultural rights

72 *Maya Indigenous Communities of the Toledo District (Belize)*, Report No 40/04, Case 12.053 (IAComHR, 12 Ocotber 2004) [114]. See also *Mayagna (Sumo) Awas Tingni Community v Nicaragua*, Merits, Reparations and Costs, Series C No 79 (IACtHR, 31 August 2001) [148]; *Indigenous and Tribal People's Rights over Their Ancestral Lands and Natural Resources. Norms and Jurisprudence of the Inter-American Human Rights System*, OEA/Ser.L/V/II, Doc 56/09 (IAComHR, 30 December 2009) 63.

73 *Maya Indigenous Communities* (2004) [113] and [116].

74 Ibid, [154]–[178] and [193]–[196].

75 *Yakye Axa Indigenous Community v Paraguay*, Merits, Reparations, Costs, Series C No 125 (IACtHR, 17 June 2005) [242].

76 Ibid, [167]. Similarly, in *Sawhoyamaxa Indigenous Community v Paraguay* an autochthonous community was deprived of ancestral lands and compelled to live alongside a road in vulnerable socio-economic conditions with limited access to health care, which resulted in the death of 19 individuals, including 18 children. Based on the rights to property, life and fair trial under Articles 4, 8, 21, and 25 ACHR, the IACtHR held the State of Paraguay responsible for these deaths under the obligation to fulfil human rights, ordering restitution of traditional lands as a means to get the Sawhoyamaxa people out of vulnerable conditions (*Sawhoyamaxa Indigenous Community v Paraguay*, Merits, Reparations, Costs, Series C No 146 (IACtHR, 29 March 2006) [164], [178] and [248]). See also *Afro-Descendant Communities Displaced from the Cararica River Basin (Operation Genesis) v Colombia*, Series C No 270 (IACtHR, 20 November 2013) [354].

embedded in Articles 4, 5, 8, 17, 19, 21, and 26 ACHR.[77] Procedurally, the Commission determined that domestic criminal proceedings not leading to definitive verdicts, as well as non-implemented administrative orders, are not effective remedies fulfilling the right of access to justice owned by individuals belonging to the Community of San Mateo de Huanchor, which had been breached as a result of the pollution stemming from mining activities, potentially in violation of Article 25 ACHR.[78]

Along these lines, in *Kichwa of Indigenous People of Sarayaku v Ecuador* the IACtHR dealt with the signature by Ecuador and a hydrocarbon company of a contract of exploration and exploitation of crude oil in a territory that was part of the ancestral lands of the Kichwas. Aware of the environmental damage caused by oil exploitation in the country, the Kichwas had always opposed oil drilling in their territory. However, despite the recognition of indigenous peoples' rights, Ecuador allowed oil exploration activities, leading to cutting trees and plants of environmental, cultural and nutritional value and placing powerful explosives on the surface and in the subsoil, affecting sources of drinking water for the Kichwas.[79] This prompted the IACtHR to hold Ecuador in breach of the rights to life, physical integrity, consultation and judicial protection under Articles 4, 5, 8, 21 and 25 ACHR.[80] Crucially, the Court stressed the State positive obligation to ensure effective participation of indigenous communities in development plans via prior, culturally appropriate and informed consultations carried out in good faith by independent competent bodies at early stages of a development or investment plan, in accordance with relevant international standards and best practices.[81]

The IAComHR took a step further in its Report on the Situation of Human Rights in Ecuador, dealing with the fundamental rights of indigenous peoples inhabiting Oriente, a region affected by development activities, particularly oil drilling.[82] In this case, the Commission held that environmental contamination and degradation threatens not only the rights to life, physical security and integrity, but also the 'right to live in an environment free from contamination'.[83] It therefore upheld a general State obligation to 'ensure that measures are in place to prevent and protect against the occurrence of environmental contamination which threatens the lives of the inhabitants' of affected territories.[84] Further to this, the

77 *Community of San Mateo de Huanchor and Its Members (Peru)*, Petition 504/03, Report No 69/04, Admissibility (IAComHR, 19 October 2005) [66].

78 Ibid, [44] ff.

79 *Kichwa Indigenous People of Sarayaku v Ecuador*, Merits and Reparations, Series C No 245 (IACtHR, 27 June 2012) [174].

80 Ibid, [341].

81 Ibid, [165] and [175] ff.

82 IAComHR, *Report on the Situation of Human Rights in Ecuador*, OEA/Ser.L/V/II.96, Doc 10 rev 1 (24 April 1997) Chapter VIII.

83 Ibid, The Situation in Oriente.

84 Ibid, Analysis. See also IAComHR, *Access to Justice and Social Inclusion: the Road towards the Strengthening of Democracy in Bolivia*, OEA/Ser.L/V/II, Doc 34 (28 June 2007) [253].

Commission determined that Ecuador has an obligation to implement the right to participation in decision-making and to an effective judicial recourse.[85]

These views also apply to non-indigenous peoples, although the jurisprudence of the IAComHR and the IACtHR is by far less developed in the field. Most significantly, in the case of the metallurgic complex of *La Oroya* in Peru, which according to the petitioners was a source of environmental pollution and contamination, the IAComHR held that the alleged death and health problems of victims resulting from actions and omissions by Peru in the face of environmental pollution could represent violations of the rights to life and physical integrity embedded in Articles 4 and 5 ACHR.[86] The Court also took a procedural stand on environmental pollution and concluded that the State lack or manipulation of information on pervasive environmental contamination and its effects on health was in breach of the rights to a fair trial, freedom of thought and expression, and judicial protection under Articles 8, 13 and 25 ACHR.[87] The IACtHR took a similar stand on procedural rights in *Claude Reyes et al v Chile*, dealing with a deforestation project and further linking access to information and participation in decision-making.[88]

However, as clearly pointed out by the IACtHR in *Salvador Chiriboga v Ecuador*, human rights are not absolute and must be balanced with other general interests, such as environmental protection itself.[89] With specific regard to indigenous groups, in *Saramaka People v Suriname* the Court determined that the protection of the right to property of indigenous and tribal communities over their lands and natural resources is limited. In fact, the Court recalled that the State may restrict the use and enjoyment of the right to property set forth in Article 21 ACHR, provided the restriction is established by prior legislation, necessary, proportionate, and aiming to achieve a legitimate objective in a democratic society.[90] The Court further specified that restrictions must not affect the survival of indigenous peoples.[91] Consequently, procedural rights were seen as a means to safeguard the substantive claims of indigenous peoples against unlawful restrictions. In fact, the State was requested to ensure the 'effective participation of the members of the Saramaka people, in conformity with their customs and traditions, regarding

85 IAComHR (1997) Conclusions and Recommendations.

86 *Community of La Oroya (Peru)*, Report No 76/09, Admissibility, Petition 1473-06 (IAComHR, 5 August 2009) [74]. For more details on this case, see FIDH, *Peru: Metallurgical Complex of La Oroya. When Investor's Protection Threatens Human Rights* (FIDH, 2012); Paula Spieler, 'The La Oroya Case: the Relationship between Environmental Degradation and Human Rights Violations' (2010) 18(1) *Human Rights Brief* 19, 21.

87 IAComHR (2009) [75].

88 *Claude Reyes et al v Chile*, Preliminary Objections, Merits, Reparations and Costs, Series C No 151 (IACtHR, 26 September 2006) [61]–[103] and [174].

89 *Salvador Chiriboga v Ecuador*, Preliminary Objection and Merits, Series C No 179 (IACtHR, 6 May 2008) [76] and [135].

90 *Saramaka People v Suriname*. Preliminary Objections, Merits, Reparations, Costs, Series C No 172 (IACtHR, 28 November 2007).

91 Ibid, [128].

any development, investment, exploration or extraction plan within Saramaka territory', so as to 'guarantee that the Saramakas receive a reasonable benefit from any such plan within their territory', and to ensure that 'independent and technically capable entities, with the State's supervision, perform a prior environmental and social impact assessment'.[92] Based on these premises, the IACtHR found a violation of the rights to property and judicial remedy of the Saramaka people under Articles 21 and 25 ACHR, because of logging and mining concessions in their territory granted by the State without appropriate consultations.[93] The same stance was taken by the IACtHR in *Xákmok Kásek v Paraguay*, a case concerning the establishment of a nature reserve in the ancestral lands of an indigenous community.[94]

In light of these cases, particularly the recognition of a right to environment in Ecuadorian Oriente, it is not impossible to uphold the Inuit claims that GHG emissions by the US and Canada are in breach of fundamental substantive rights, including the human right to environment, and related procedural claims. This stance is further supported by the case of *Kawas-Fernández v Honduras*, where, based on OAS and UN initiatives on climate change and human rights, the IACtHR recognised an 'undeniable link between the protection of the environment and the enjoyment of other human rights', and further recalled '[t]he ways in which the environmental degradation and the adverse effects of the climate change have impaired the effective enjoyment of human rights in the continent'.[95] Of course, the Inuit allegations cannot be considered unconditional, but must be balanced with other fundamental policy interests of the US and Canada within the limits of the best effort standard, along the lines of *Saramaka People v Suriname* and *Xákmok Kásek v Paraguay*.[96]

More specifically, the Athabaskan petition can be assessed based on the cases of *Community of San Mateo de Huanchor* and *La Oroya*, where a distinct source of environmental pollution leading to the breach of fundamental rights of indigenous and non-indigenous peoples was identified and a causality link could be established.

More complex are the circumstances of the first Inuit petition, where a link between anthropogenic GHGs and human rights is scientifically proven, but the emitter is not clearly identified.[97] Although these conditions are not addressed in the case law of the IAComHR and IACtHR, the problem could be resolved, especially at the State level, by resorting to the principle of statistical probability.

92 Ibid, [127]–[129].

93 Ibid, [214].

94 *Xákmok Kásek Indigenous Community v Paraguay*, Merits, Reparations, and Costs, Series C No 214 (IACtHR, 24 August 2010) [337].

95 *Kawas-Fernández v Honduras*, Merits, Reparations, and Costs, Series C No 196 (IACtHR, 3 April 2009) [148].

96 See also Siobhán McInerney-Lankford, Mac Darrow and Lavanya Rajamani, *Human Rights and Climate Change: a Review of the International Legal Dimensions* (World Bank, 2011) 35.

97 On this problem, see OHCHR, *Report on the Relationship between Climate Change and Human Rights*, A/HRC/10/61 (2009) [70].

Indeed, according to the International Law Association (ILA), this principle allows 'a finding of liability' when a party to a case can establish that the acts of others have resulted in a 'material increase in risk' of damage sufficient to determine causation.[98] As a consequence, fundamental procedural rights should be interpreted extensively in the matter of anthropogenic GHG emissions, so as to encompass basic political measures such as the US decision not to ratify the Kyoto Protocol. On this basis, the Inuit should have been pre-emptively informed of such a position and should have had a say in the matter.

However, the Inuit petitions are extraterritorial, and thus unprecedented, probably because of the unity of the Inuit and Athabaskan people and the need to put a spotlight on the Arctic environment as a whole. Therefore, the Commission undoubtedly has competence *ratione loci* over the petitions, to the extent that they relate to facts that occurred within the territory of the respondent States. Instead, in the absence of innovative jurisprudence it is difficult to maintain that the IAComHR could declare its jurisdiction over the extraterritorial effects of GHGs, including black carbon. Inuit residing in Canada (first petition) and Athabaskan people of Alaska (second petition) are not respectively subject to the jurisdiction of the US and Canada. Since, like the ECtHR, the IAComHR does not recognise the principle of extraterritorial 'cause–effect' responsibility upheld by respondent States in *Bankovic*,[99] it is difficult to assume that transboundary pollution creates a *de facto* or *de jure* control over foreign territory or brings victims under the control of OAS Member State agents operating extraterritorially.[100] The IAHComR should thus only be competent *ratione loci* to examine the Inuit petitions inasmuch as they allege violations of rights protected by the ADRDM that occurred respectively within US and Canadian territory.

Conclusion

Substantively, the Inter-American human rights protection system provides for interesting avenues to establish a link between climate change and human rights. These are nevertheless limited, in particular because, although it is explicitly acknowledged in the San Salvador Protocol to the ACHR, the right to a healthy environment is not justiciable before the IAComHR and the IACtHR.

Institutionally, the OAS has recognised a link between climate change and human rights in Resolution 2429/08, invoking comprehensive explorative studies by law-making and jurisdictional bodies. However, this non-binding instrument has not yet been followed by concrete regulatory initiatives in the field.

98 ILA, *Legal Principles Relating to Climate Change* (Sofia Conference, 2012) 40, <www.ila-hq.org/en/committees/index.cfm/cid/1029 >.

99 *Bankovic and Others v Belgium and Others*, Appl No 52207/99 (ECtHR, 12 December 2001).

100 See *Djamel Ameziane v US*, Petition P-900–08, Report No 17/12, Admissibility (IAComHR, 20 March 2012) [29]–[35]. But see John Knox, 'Climate Change and Human Rights Law' (2009) 50 *Va J Int'l L* 163, 203.

The IAComHR has been directly solicited by the Inuit and Arctic Athabaskhan petitions, which raise relevant theoretical and practical issues as to fundamental rights breached by anthropogenic GHGs, causation, imputation and extra-territoriality. The Commission dismissed the former petition without taking a stand on the questions it raises. The case law of the Commission in the matter of environmental protection, especially with regard to indigenous peoples, nevertheless shows that the IAComHR potentially has relevant substantive and procedural tools to assess the relationship between climate change and human rights. Hopefully, the Commission will shed light on these issues by dealing with the Arctic Athabaskan petition, which has recently been submitted and is still pending.

17 All in all it was all just bricks in the wall

European legal systems, climate change and human rights

Ottavio Quirico

Introduction

Climate change has an impact throughout Europe, causing environmental effects such as increased flooding, windstorms and glacier retreat.[1] This has key implications for a range of resources, for instance, water availability, and sectors, specifically transport, energy, tourism, agriculture, forestry, and health systems.[2] Repercussions depend not only on the direct effect of global warming on European States, but also on the way it affects foreign countries, particularly neighbouring regions in the Mediterranean area.[3]

From a human rights perspective, the European response to climate change is basically two-pronged. On the one hand, the EU has adopted major initiatives to regulate anthropogenic GHG emissions, based on its binding commitments under the Kyoto Protocol.[4] This action has taken place within the context of a progressive environmental policy, which is rooted in the case law of the ECJ, as developed since the 1970s. Such a process culminated in the recognition of advanced environmental protection in the primary sources of EU law, including Article 37 CFREU. The EU climate policy is also aware of human rights more generally, and thus 'respects the fundamental rights and observes the principles recognised in particular by the Charter of Fundamental Rights of the European Union'.[5] On the other hand, within the context of the Council of Europe, the ECtHR and the ECteSR have afforded environmental protection via 'classical' first and second generation human rights, specifically, the rights to private and family life and health. More generally, the Aarhus Convention provides a comprehensive umbrella for procedural environmental rights.

1 IPCC, Fifth Assessment Report, *Climate Change: Impacts, Adaptation and Vulnerability* (2014) Chapter 23, Europe, 1279 ff.

2 Ibid, 1300–2.

3 Ibid, 1303.

4 *Kyoto Protocol to the UNFCCC* (1998) Annex B; *Doha Amendment to the Kyoto Protocol*, C.N.718.2012. TREATIES-XXVII.7.c (8 December 2012) Annex B.

5 European Parliament and Council, *Directive 2009/29/EC Amending Directive 2003/87/EC so as to Improve and Extend the Greenhouse Gas Emission Allowance Trading Scheme of the Community* (23 April 2009) Preamble, [50].

Paradoxically, but only to a certain extent, this framework signals that the EU takes a human rights approach to environmental protection which is more advanced than the classical stand of the Council of Europe. Within such a context, this chapter explores possible links existing between climate change and human rights and the evolution of these ties in light of the future accession of the EU to the ECHR, which is supposed to bring together the two major European legal systems.

1. A comprehensive umbrella: climate change and procedural environmental rights under the Aarhus Convention

In order to understand climate change and environmental protection from a human rights perspective, it is proper to start with 'procedural' environmental rights, which help to shed light on substantive issues. Indeed, procedural claims are firmly established in the European framework because of the Aarhus Convention, which was adopted in 1998 at the Fourth Ministerial Conference of the 'Environment for Europe' process, under the aegis of UNECE.[6] Despite its general scope of application, this international treaty was meaningfully developed in the European context and so far, in addition to the EU, most Parties are European States.[7]

The Aarhus Convention particularly concerns 'Access to Information, Public Participation in Decision-making and Access to Justice in Environmental Matters'. 'Access to information' encompasses everyone's claim to receive environmental information held by public authorities on the state of the environment, affected human health, safety and related policies and measures (Article 4).[8] 'Participation' includes arrangements by public authorities enabling interested public and non-governmental organisations to comment on proposals, plans and programmes bearing on the environment (Article 6).[9] Finally, 'access to justice' entails a claim

6 See further Nicolas De Sadeleer, 'Enforcing EUCHR Principles and Fundamental Rights in Environmental Cases' (2012) 81 *Nordic Journal of International Law* 39, 49 ff.

7 UNECE, *Aarhus Convention, Status of Ratification*, <www.unece.org/env/pp/ratification.html>.

8 Aarhus Convention, Article 4 (Access to Environmental Information):

> Each Party shall ensure that, subject to the following paragraphs of this Article, public authorities, in response to a request for environmental information, make such information available to the public.

9 Aarhus Convention, Article 6 (Public Participation in Decisions on Specific Activities):

> The public concerned shall be informed, either by public notice or individually as appropriate, early in an environmental decision-making procedure, and in an adequate, timely and effective manner, inter alia, of: (a) The proposed activity and the application on which a decision will be taken; (b) The nature of possible decisions or the draft decision; (c) The public authority responsible for making the decision; (d) The envisaged procedure . . . (e) The fact that the activity is subject to a national or transboundary environmental impact assessment procedure.

to challenge public decisions that have been made without respecting the two aforementioned rights and environmental law more generally (Article 9).[10] Interestingly, these claims are ultimately based on Article 1 of the Convention, which provides that procedural rights are guaranteed 'in order to contribute to the protection of the right of every person of present and future generations to live in an environment adequate to his or her health and well-being'.[11]

Along the lines of provisions such as Articles 4 and 6 UNFCCC,[12] scholars have pointed out that access to information is crucial to ensure the transparency and effectiveness of GHG mitigation and adaptation action as to issues such as measurement, reporting and verification.[13] Furthermore, participation of civil society is regarded as essential to incorporate human rights standards and the voice of vulnerable groups in climate policies.[14] These rights are preliminary to the enjoyment of effective remedies, because proper legal action is hampered by a lack of information,[15] which is considered all the more important in light of the weakness of MEA enforcement mechanisms, with particular regard to the Kyoto Protocol.[16] Such a standpoint is upheld by the European Parliament. Indeed, the Subcommittee on Human Rights of the Parliamentary Committee on Environment, Public Health and Food Safety recently held a debate on the human rights implications of global warming and EU policy options, stressing the importance

10 Aarhus Convention, Article 9 (Access to Justice):

> Each Party shall, within the framework of its national legislation, ensure that any person who considers that his or her request for information under Article 4 has been ignored, wrongfully refused . . . inadequately answered, or otherwise not dealt with . . . has access to a review procedure before a court of law or another independent and impartial body established by law . . . Each Party shall, within the framework of its national legislation, ensure that members of the public concerned (a) Having a sufficient interest or, alternatively, (b) Maintaining impairment of a right, where the administrative procedural law of a Party requires this as a precondition, have access to a review procedure before a court of law and/or another independent and impartial body established by law, to challenge the substantive and procedural legality of any decision, act or omission.

11 Aarhus Convention, Article 1 (Objective). See also Stephen Stec and Susan Casey-Lefkowitz, *The Aarhus Convention: an Implementation Guide* (UNECE, 2000).

12 According to Article 4(1)(i) UNFCCC, States parties 'promote and cooperate in education, training and public awareness related to climate change and encourage the widest participation in this process, including that of non-governmental organisations'. Under Article 6 UNFCCC, States parties commit to promote '(ii) public access to information on climate change and its effects; (iii) public participation in addressing climate change and its effects and developing adequate responses'.

13 See Svitlana Kravchenko, 'Procedural Rights as a Crucial Tool to Combat Climate Change' (2010) 38 *Georgia J Int'l & Comp L* 613, 618 ff and 648.

14 Ibid, 635 ff and 648.

15 See *Ville de Lyon v Caisse des dépôts et consignations*, Case C-524/09 (ECJ, 22 December 2010); *Native Village of Kivalina and City of Kivalina v ExxonMobil Corp et al*, Complaint for Damages – Demand for Jury Trial, CV 08–1138 SBA (26 February 2008) [5]; Kravchenko (2010) 630.

16 Kravchenko (2010) 616.

of procedural environmental claims within the context of EU internal and external policies integrating climate change and fundamental rights.[17]

2. The EU, climate change and human rights

2.1. Direct and indirect environmental protection: towards the human right to (a sustainable) environment?

The EU is an international organisation comprised of 28 Member States, aiming to create a common market under the TEU and TFEU.[18] The system is completed by human rights protection outlined in the CFREU.[19] These are the primary sources of EU law and also embed the substantive principles of sustainable development and environmental protection. In particular, a high standard is outlined in Article 3 TEU for 'environmental protection' and 'improvement of the quality of the environment'. Such a commitment applies both internally and externally, within the context of sustainable development, in light of Article 21(2)(g) and (f) TEU.[20] More specifically, Article 21(2)(f) TEU sets out the obligation for the Union to cooperate in developing international measures to preserve and improve the quality of the environment and the sustainable management of global natural resources. Climate change action is part of this approach.[21]

Along these lines, Article 11 TFEU provides for the integration of environmental protection into EU policies. This norm basically summarises Articles 3 and 21 TEU as to environmental protection in internal and external EU action:

> *Environmental protection* requirements must be *integrated into the definition and implementation of the Union's policies and activities*, in particular with a view to promoting *sustainable development.*[22]

Such a provision is consistent with the TEU Preamble, which embeds the principles of sustainable economic progress and environmental protection in the establishment of the EU internal market. Furthermore, the necessity of achieving a 'high

17 European Parliament, Committee on Environment, Public Health and Food Safety, Subcommittee on Human Rights, *Joint Hearing on the Implications of Climate Change on Human Rights* (11 July 2012).

18 *Treaty of Lisbon Amending the Treaty on European Union and the Treaty Establishing the European Community*, 13 December 2007 OJ C 306 (17 December 2007). The EU was originally established as a European Economic Community under the EEC Treaty adopted in Rome in 1957.

19 Although the CFREU was initially adopted in 2000 as a non-binding instrument, solely aiming to make human rights known throughout the EU (OJ EC, 2000/C 364/01, 18 December 2000), in 2009 the Charter became a binding primary source of EU law with the entry into force of the Lisbon Treaty.

20 On the importance of this rule for a human rights-oriented external EU climate policy see Christel Cournil *et al*, *Climate Change and Human Rights: EU Policy Options*, EXPO/B/DROI/2011/20 (Directorate-general for External Policies, Policy Department, 2012) 28.

21 Ibid.

22 Emphasis added.

level of environmental protection', including the taking of climate change measures, is spelled out in detail in Article 191(2) TFEU, which is embodied in Title XX on 'EU Environmental Policies'. Article 191(1) also establishes a direct link between 'environmental protection' and 'human health'. This approach complements Article 1 of the Aarhus Convention, which provides for the right of everyone to 'an environment adequate to his or her health and well-being'. Significantly, Article 194, Title XXI TFEU (Energy) completes Title XX and specifies that energy supply must take place within the context of the need to 'preserve and improve the environment'.

Since they are included in the primary rules of the EU, these norms may be regarded as 'fundamental', or even '(quasi-)constitutional'.[23] Moreover, formal recognition of the human rights signification of environmental protection is embedded in Article 37 CFREU:

> A *high level of environmental protection* and the *improvement of the quality of the environment* must be *integrated into the policies of the Union* and ensured in accordance with the *principle of sustainable development.*[24]

This provision is almost a 'twin' of Article 11 TFEU. Its language is only slightly different, to the extent that it explicitly refers to a 'high level' of environmental protection and the 'improvement' of environmental quality.[25] Article 37 CFREU thus includes elements of Articles 3 and 21 TEU even more explicitly than Article 11 TFEU,[26] and is particularly important, because basically the CFREU currently occupies the same hierarchical level as the TEU and TFEU, and thus complements the founding treaties.[27] This definitely points out that the Union takes an overall human rights stand on the issue of environmental protection.

Albeit embedded in a Charter of 'Fundamental Rights', doubts have been expressed as to the nature of Article 37 CFREU, particularly as to whether it sets a justiciable claim.[28] It has indeed been observed that this norm is not formulated

23 Indeed, the TEU and TFEU were part of the rejected 'Constitution for Europe', together with the CFREU (*Treaty Establishing the Constitution for Europe*, OJ C 310 (16 December 2004)).

24 Emphasis added. Article 37 CFREU is embedded in Title IV (Solidarity), which includes 'solidarity rights'.

25 Furthermore, environmental protection is fostered under Article 37 'in accordance with' sustainable development, not 'in view of' its promotion.

26 See Gracia Marín Durán and Elisa Morgera, 'Commentary on Article 37 of the EU Charter of Fundamental Rights – Environmental Protection', in Steve Peers, Tamara Hervey, Jeff Kenner and Angela Ward (eds), *Commentary on the EU Charter of Fundamental Rights* (Hart, 2013) 983 ff.

27 Christel Cournil *et al* (2012) 41. On systemic issues concerning the ranking of the CFREU within the sources of EU law see Ottavio Quirico, 'Substantive and Procedural Issues Raised by the Accession of the European Union to the European Convention on Human Rights' (2010–11) 20 *Italian Yearbook of International Law* 31 ff.

28 For an in-depth discussion, see De Sadeleer (2012) 43 ff. On the interpretation of fundamental rights protected by EU law, see Damien Chalmers, Gareth Davies and Giorgio Monti, *European Union Law* (CUP, 3rd edn, 2014) 259 ff.

as a typical subjective right, since it does not include the expression 'everyone has the right to',[29] but outlines duties binding EU organs ('the policies of the Union must').[30] Therefore, based exclusively on Articles 51 and 52 CFREU, which address the 'institutions and bodies of the Union', and on the distinction that the Preamble to the Charter makes between 'rights, freedoms and principles', some scholars have concluded that Article 37 sets out 'principles', but not 'individually justiciable rights'.[31] Nonetheless, in light of the necessary correlation between rights and duties,[32] it is not easy to maintain that environmental protection, which 'must' be integrated into the policies of the EU, is not matched by a correlative 'right'. The rule is simply formulated based on a duty rather than a right, taking into account the obligatory side of a legal relationship, which has been explored within the context of the discussion on the nature and scope of a possible 'right to (a sustainable) environment'.[33] It can just be the case that the *erga omnes* nature of the duty in issue provides a general cause of action: uncertainty as to the holder of the right, which is universalised, does not necessarily exclude its existence. As to the content of the claim, even assuming that Article 37 CFREU does not set out a *general* right to environmental protection, it should at least be recognised that it creates a *specific* claim that *EU legislative acts* ought to be respectful of the environment. In this regard, the claim under Article 37 CFREU seems to match the 'right of every person of present and future generations' to environmental protection embedded in Article 1 of the Aarhus Convention. After all, 'procedural rights' under the Aarhus Convention are also framed in terms of *erga omnes* duties of European (and non-European) authorities ('each Party shall ensure'), but nobody questions their nature as rights rather than principles. Along these lines, it is interesting to note that the text of the Constitution for Europe, which subsequently became the text of the TEU, TFEU and CFREU, provided that 'the provisions of the Charter . . . [would have been] judicially cognisable' (Article II-112(5)).[34]

Besides, the compelling nature of Article 37 CFREU is ultimately not so crucial. Indeed, even assuming that Article 37 CFREU is not justiciable, the right to environment might be adjudicated upon by the ECJ, that is, the official authority

29 EU Network of Independent Experts on Fundamental Rights, *Commentary of the Charter of Fundamental Rights of the European Union* (2006) 315; Bruno Genevois, 'La Convention européenne des droits de l'homme et la Charte des droits fondamentaux de l'Union européenne: complémentarité ou concurrence?' (2010) 26 *Revue française de droit administratif* 437 ff; David Anderson and Cian Murphy, 'The Charter of Fundamental Rights', in Andrea Biondi and Piet Eeckhout, *EU Law after Lisbon* (Oxford University Press, 2012) 155, 166 ff.

30 De Sadeleer (2012) 48.

31 Marín Durán and Morgera (2013); EU Network of Independent Experts on Fundamental Rights (2006) 315.

32 See Wesley Newcomb Hohfeld, *Fundamental Legal Conceptions as Applied in Judicial Reasoning* (Yale UP, 1964).

33 See the contribution by Francioni and Quirico in this volume.

34 In favour of the justiciablity of current Article 37 CFREU, see De Sadeleer (2012) 46.

in the matter of EU law,[35] as a general principle of law inferred from domestic constitutions.[36] Indeed, although it is not a human rights Court, the ECJ has developed a human rights case law incidental to its decisions on the internal market,[37] based, in particular, on the constitutional traditions common to the Member States and international conventions, including the ECHR.[38] Along these lines, in a set of cases the ECJ has balanced market freedoms and environmental protection, with regard to both substantive and procedural issues.[39]

The principle of environmental protection has been made particularly clear by the ECJ in *Prosecutor v ADBHU*,[40] where the Court adjudicated upon the implementation in France of European Council Directive 75/439/EEC.[41] The Directive instructs EU Member States to take measures ensuring the safe collection and disposal of waste oils, preferably by means of recycling. In 1983, French legislation implementing Directive 75/439/EEC essentially prohibited the burning of waste oils and was contested by ADBHU, an association protecting manufacturers, dealers, and users of stoves and heating appliances designed to burn both fuel and waste oils, as an obstacle to free trade, movement of goods, and competition. Eventually, the ECJ held that 'environmental protection' constitutes 'one of the *Community's essential objectives*', and thus declared the prohibition of the burning of waste oils, in conditions other than those permitted under French legislation, consistent with Directive 75/439/EEC.[42] With specific regard to climate change, in *Ålands Vindkraft AB v Energimyndigheten* the ECJ adjudicated upon the lawfulness of the limited application of national support schemes for green electricity provided by energy companies operating in Sweden. Based on Directive

35 Article 19(1) TEU, former Article 164 TEC as revised by the Maastricht Treaty.

36 See, for instance, the French Environmental Charter (*Charte de l'environnement*), which is part of the '*bloc de constitutionnalité*' (loi constitutionnelle 2005–205), and provides: 'Chacun a le droit de vivre dans un environnement équilibré et respectueux de la santé – Everyone has the right to live in a sustainable and healthy environment' (Article 1). See also De Sadeleer (2012) 40; *Explanations relating to the Charter of Fundamental Rights (Explanatory Notes)*, OJ C 303/17 (2007) Article 37, providing that 'the principles set out in this Article [37 of the CFREU] have been based on Articles 2, 6 and 174 of the EC Treaty, which have now been replaced by Article 3(3) of the Treaty on European Union and Articles 11 and 191 of the Treaty on the Functioning of the European Union. It [Article 37] also draws on the provisions of some national constitutions.'

37 From *Internationale Handelsgesellschaft mbH v Einfuhr- und Vorratsstelle für Getreide und Futtermittel*, Case 11/70 (ECJ, 17 December 1970) onwards. See Sybe A De Vries, 'Balancing Fundamental Rights with Economic Freedoms According to the European Court of Justice' (2013) 9 *Utrecht Law Review* 169 ff.

38 See, for instance, *Elliniki Radiophonia Tiléorassi AE and Others v Dimotiki Etairia Pliroforissis and Others*, Case C-260/89 (ECJ, 18 June 1991) [41].

39 For an overview of the cases see Lidia Iancu and Timothy Ghilain, *The Role of the Court of Justice in Environmental Justice as a Third Generation Human Right* (Effectius, 2011); Mathew L Schemmel and Bas de Regt, 'The European Court of Justice and the Environmental Protection Policy of the European Community' (1994) 17 *Boston College Int'l & Comp L Rev* 53.

40 *Prosecutor v ADBHU*, Case 240/83 (ECJ, 7 February 1985).

41 European Council, *Directive 75/439/EEC on the Disposal of Waste Oils* (16 June 1975).

42 Ibid, [13], emphasis added.

2009/28 on the promotion of the use of energy from renewable sources,[43] the Court concluded that such a territorial restriction of interstate trade is consistent with the purpose of protecting the environment, in the absence of a harmonised supranational support scheme for green electricity.[44] These views strengthen the argument that environmental protection is a general principle of EU law, possibly ranking higher than the founding treaties.[45]

Whereas in *ADBHU* and *Ålands Vindkraft* the ECJ directly balanced basic market freedoms and environmental safeguards, in other cases the Court disclosed the possibility of providing environmental protection by balancing classical first and second generation human rights. Particularly, in *Schmidberger* the ECJ dealt with pollution and environmental damage caused by heavy trucks on Alpine motorways, which had long been a matter of public concern in Austria. In 1998, Transitforum Austria Tirol, an environmental Austrian organisation, gave notice of its intention to hold a protest demonstration that would block the Brenner motorway between Austria and Italy for approximately 30 hours. The Austrian authorities did not ban the demonstration, which duly took place on 12 and 13 June 1998. Schmidberger, a German transport company, claimed that the blockage of the motorway was in breach of the free movement of goods. By means of a preliminary ruling, the ECJ held that the conduct of Austrian authorities was justified in light of the fundamental freedoms of expression and assembly. Incidentally, the Court also held that 'the protection of the environment and public health . . . may, under certain conditions, constitute a legitimate objective in the public interest capable of justifying a restriction of the fundamental freedoms guaranteed by the Treaty [TEC], including the free movement of goods'.[46]

2.2. A human rights-based assessment of EU climate policies

Along the lines of its high standard of environmental protection, internally the EU has committed to ambitious GHG reduction targets under the UNFCCC,[47] as complemented by the Kyoto Protocol.[48] By 2020, the EU will cut its emissions to 20 per cent below 1990 levels, whereas by 2012 the target was 8 per cent.[49] The 2030 Framework for Climate and Energy Policy also envisages a 40 per cent

43 European Parliament and Council, *Directive 2009/28/EC on the Promotion of the Use of Energy from Renewable Sources and Amending and Subsequently Repealing Directives 2001/77/EC and 2003/30/EC* (23 April 2009).

44 *Ålands Vindkraft AB v Energimyndigheten*, Case C-573/12 (ECJ, 1 July 2014) [92]–[94].

45 Article 6(3) TEU. See also Quirico (2010) 35.

46 *Eugen Schmidberger, Internationale Transporte und Planzüge v Republik Österreich*, Case C-112/00 (12 June 2003) [66].

47 EU Council Decision, *Conclusion of the United Nations Framework Convention on Climate Change*, 94/69/EC (15 December 1993).

48 EC Council Decision, *Approval, on behalf of the European Community, of the Kyoto Protocol to the United Nations Framework Convention on Climate Change*, 2002/358/EC (25 April 2002).

49 *Kyoto Protocol* (1998) Annexe B; *Doha Amendment to the Kyoto Protocol* (2012) Annexe B.

reduction target.[50] Moreover, the Roadmap for Moving to a Low-Carbon Economy sets an 80–95 per cent reduction target by 2050.[51] Currently, Directive 2003/87 (EU Emission Trading System – ETS),[52] as amended by Directive 2009/29/EC, is the main regulatory instrument in the field. This Directive caps GHG emissions from the major EU industrial sectors via a system of tradable GHG allowances. Residually, the 2009 Efforts Sharing Decision defines GHG reduction targets for sectors that are not included in the EU ETS, such as building, agriculture and transport.[53] Since Directive 2009/29 explicitly acknowledges respect for fundamental rights, particularly as established in the CFREU,[54] be it recognised as a right or only as a principle, Article 37 CFREU has, among other fundamental 'principles/rights' an important role to play in the definition of the EU climate policy.[55]

Following the external projection of environmental protection under Article 21 TEU, the EU also pursues an ambitious international sustainable policy, with particular regard to climate change. This goes as far as to claim priority of sustainability over international commercial engagements.[56] Along these lines, climate change regulatory instruments establish a direct connection between global warming and third generation human rights, in particular, the right to development. This is the case, for instance, of the EU Action Plan on Climate Change and Development.[57] More generally, in a Common Position on Sustainable Development ahead of the Rio+20 Conference, the European Parliament stressed the negative impact of climate change on human rights.[58] Furthermore, even when international EU climate change-related regulation does not mention 'human rights', an implicit connection is established by means of references to

50 European Commission, *Climate Action, 2030 Framework for Climate and Energy Policy*, <http://ec.europa.eu/clima/policies/2030/index_en.htm>; European Council, *Conclusions on 2030 Climate and Energy Policy Framework*, SN 79/14 (23 October 2014).

51 European Commission, *Climate Action, Roadmap for Moving to a Low-Carbon Economy in 2050*, <http://ec.europa.eu/clima/policies/roadmap/index_en.htm>.

52 European Parliament and Council, *Directive 2003/87/EC Establishing a Scheme for Greenhouse Gas Emission Allowance Trading within the Community and Amending Council Directive 96/61/EC* (13 October 2003).

53 European Parliament and Council, *Decision No 406/2009/EC on the Effort of Member States to Reduce Their Greenhouse Gas Emissions to Meet the Community's Greenhouse Gas Emission Reduction Commitments up to 2020* (23 April 2009). For an exhaustive overview of EU climate change policy instruments see Johanna Cludius, Hannah Förster and Verena Graichen, *GHG Mitigation in the EU: an Overview of the Current Policy Landscape* (WRI, 2012).

54 Preamble, [50].

55 See Cournil *et al* (2012) 70.

56 See European Parliament, *Resolution on a New Trade Policy for Europe under the Europe 2020 Strategy*, 2010/2152(INI) (2011); *Resolution on International Trade Policy in the Context of Climate Change Imperatives*, 2010/2103(INI) (2010) [11].

57 Council of the EU, *Climate Change in the Context of Development Cooperation*, DEVGEN 241ENV 637 (24 November 2004); Cournil *et al* (2012) 52.

58 European Parliament, *Resolution on Developing a Common EU Position ahead of the United Nations Conference on Sustainable Development (Rio+20)* (29 September 2011) [E].

'sustainable development', which is often considered basic to the enjoyment of fundamental claims in human rights instruments.[59] Within this context, it is significant that the EU has tried to enforce its ETS extraterritorially. Indeed, EU Directive 2008/101 amended Chapter II of Directive 2003/87, and thus extended the EU ETS to the aviation sector from 1 January 2012, including 'all flights which arrive at or depart from an aerodrome situated in the territory of a Member State to which the Treaty [TEC, currently TFEU] applies'. The extension has nonetheless been temporarily suspended, pending ICAO negotiations on an international GHG reduction mechanism for aviation.[60]

With specific regard to adaptation, the EU seeks to integrate it in all its policies,[61] and some initiatives highlight relevant human rights implications.[62] In particular, the White Paper on a European Framework for Action in view of Adapting to Climate Change provides an overview of the impact of climate change on human rights.[63] The document stresses the negative effects of global warming on human and animal and plant health,[64] water resources, food production, and biodiversity.[65] Consequently, the Paper proposes action improving the resilience of these resources to climate change, especially via a regional approach under Article 4 UNFCCC.[66] Progress in the EU Health Strategy and the EU Forest Action Plan is thus seen as a possible avenue to enforce adequate adaptation.[67] Furthermore, a recent Communication of the European Commission highlights the importance of adaptation strategies other than relocation to achieve long-lasting solutions to the problem of climate-induced migration,[68] which in the view of the EU is a crucial global issue.[69]

59 For instance, Article 32 of the Cotonou Agreement Establishing a Partnership between the EC and African, Caribbean and Pacific (ACP) States (2000, revised in 2005) regulates multilateral cooperation in climate change matters concerning small island States. Article 9 of the Cotonou Agreement provides that 'respect for all fundamental rights and fundamental freedoms . . . [is] an integral part of sustainable development' (see Cournil *et al* (2012) 43–4).

60 ICAO, *Consolidated Statement of Continuing ICAO Policies and Practices Related to Environmental Protection – Climate Change*, Resolution A38-18 (2013) 90–9.

61 European Commission, White Paper, *Adapting to Climate Change: towards a European Framework for Action*, COM(2009) 147 final, Brussels (2009).

62 European Commission, Directorate-General Environment, Letter to the UNHCHR, *Information for Preparation of a Study on the Relationship between Climate Change and Human Rights*, DG ENV/C1/LA/as ARES (3 July 2008). See also Despina Sinou, 'Les changements climatiques et la diplomatie de l'Union européenne en matière de droits de l'homme', in Christel Cournil and Anne-Sophie Tabau (eds), *Politiques climatiques de l'UE et droits de l'homme* (Bruylant, 2013) 103, 120.

63 European Commission (2009).

64 See also European Centre for Disease Prevention and Control, *Climate Change and Communicable Diseases in the EU Member States*, Technical Document (2010).

65 European Commission (2009) 3–5.

66 Ibid, 5.

67 Ibid, 9–10.

68 European Commission, *Communication on Climate Change, Environmental Degradation, and Migration: an EU Strategy on Adaptation to Climate Change*, SWD 138 final (2013) 34.

69 Ibid, 6–7; Chloé Anne Vlassopoulos, 'Migrations et changements climatiques: un nouvel enjeux de politique publique pour l'Union européenne?', in Cournil and Tabau (2013) 217 ff; Cournil

From the standpoint of rule-making, internal and external EU initiatives on climate change measures have been deemed 'credible' by a recent Report on EU Policy Options as to Climate Change and Human Rights, adopted by the Directorate for External Policies.[70] This means that the environmental law framework outlined in primary EU law sources has effectively led to establishing an advanced climate policy. Internally, whether or not such a policy is effectively consistent with human rights must be determined in light of measures implementing basic climate Directives, along the lines of the climate change-human rights consistency invoked in Directive 2009/29/EC. Thus, for instance, since they were only 4 per cent lower than those distributed in 2005, overall CO_2 allowances for the year 2006 have been deemed insufficient under the obligation to protect fundamental rights.[71] A 1.74 per cent annual reduction in GHG allowances has also been criticised as inadequate to ensure human rights protection.[72] More generally, climate policy-making is already shaped by human rights, in particular by the right to development.[73] Indeed, the allocation of State GHG allowances under the Effort Sharing Decision takes place within the context of common but differentiated responsibility,[74] which is an intrinsic element of the right to development.[75] The same is true of GHG allowance-auctioning within the EU ETS, which takes into account, for instance, the heavy dependence of States on fossil fuel for energy production.[76] With specific regard to adaptation, the main risk of human rights infringements arises from the necessity of balancing different fundamental claims, especially owing to the fact that the EU comprehensively integrates adaptation to climate change in its regulatory initiatives.[77] Externally, the cooperative framework established for international European policies under the Lisbon Treaty is considered a privileged tool for defining a human rights approach to climate change. Indeed, cooperation is also an essential component of the UNFCCC regime.[78] However, the lack of explicit references to human rights in international regulatory instruments has been regarded as an important cause of the failure of the EU proposals on climate change in the global arena, where multilateral action is required in light of the horizontal structure of interstate society.[79] An ideal 'road map' has also been put forward to integrate

et al (2012) 60 ff. Displacement in Europe is mainly caused by the economic impact of climate change on basic economic activities, such as tourism and agriculture (IPCC (2014) 1300).

70 Cournil *et al* (2012) 28.

71 Ibid, 71.

72 Ibid, 71. See also Anne-Sophie Tabau, 'Les politiques européennes d'attenuation passéés au crible des droits de l'homme', in Cournil and Tabau (2013) 157, 163–4.

73 Sandrine Maljean-Dubois and Vanessa Richard, 'Les droits de l'hommme dans les politiques extérieures européennes de développement et d'adaptation aux changements climatiques', in Cournil and Tabau (2013) 129 ff.

74 European Parliament and Council (2009) Article 14.

75 Cournil *et al* (2012) 72.

76 Ibid, 73.

77 Ibid, 81; Tabau (2013) 189 ff.

78 Cournil *et al* (2012) 50 ff.

79 Ibid, 29.

climate change and human rights, including encouraging immediate action, adopting an official EU statement and guidelines, and embodying these concerns in the mandate of the EU Special Representative on Human Rights.[80] Although appreciable, this approach falls short of developing a precise legal stand on the matter.

From the viewpoint of secondary rules,[81] it is not easy to enforce human rights standards possibly breached by EU climate change policies.[82] A progressive assessment is likely to emerge in the future, particularly via the practice of the ECJ. Indeed, as we have seen, the ECJ has paved the way for direct environmental protection, which might assume a human rights characterisation in light of Article 37 CFREU, despite the arguably excessive doubts raised as to its justiciability. Furthermore, the Court might frame climate policies via specific first and second generation human rights.[83] So far, in the most interesting case where the ECJ dealt with GHG emissions, the Court adjudicated upon the validity of Directive 2003/87, extending extraterritorially the application of the EU ETS in the aviation sector.[84] In this dispute, the Court upheld the validity of the extraterritorial extension of the EU ETS, based on the concepts of 'unlimited jurisdiction' and 'high level of environmental protection', within the context of GHG reduction targets established under the UNFCCC and the Kyoto Protocol.[85] Unfortunately, the ECJ did not sufficiently elaborate on the '*erga omnes*' interpretation of environmental protection from GHG emissions and its

80 Ibid, 38–9, 47–8.

81 Following the Kelsenian categorisation: 'If A, B ought to be' (Hans Kelsen, *Pure Theory of Law* (Peter Smith, 1989) 77).

82 See the contribution by Quirico, Bröhmer and Szabó in this volume.

83 See above section 2.1.

84 *Air Transport Association of America and Others v Secretary of State for Energy and Climate Change*, Case C-366/10 (ECJ, 21 December 2011).

85 Ibid, [125] and [128]:

> In laying down a criterion for Directive 2008/101 to be applicable to operators of aircraft registered in a Member State or in a third State that is founded on the fact that *those aircraft perform a flight which departs from or arrives at an aerodrome situated in the territory of one of the Member States*, Directive 2008/101, inasmuch as it extends application of the scheme laid down by Directive 2003/87 to aviation, *does not infringe the principle of territoriality or the sovereignty* which the third States from or to which such flights are performed have over the airspace above their territory, *since those aircraft are physically in the territory of one of the Member States of the European Union* and are thus subject on that basis to the *unlimited jurisdiction* of the European Union . . . as European Union policy on the environment seeks to ensure *a high level of protection in accordance with Article 191(2) TFEU*, the European Union legislature may in principle choose to permit a commercial activity, in this instance air transport, to be carried out in the territory of the European Union only on condition that operators comply with the criteria that have been established by the European Union and are designed to fulfil the environmental protection objectives which it has set for itself, in particular where those objectives follow on from an international agreement to which the European Union is a signatory, such as the *Framework Convention* and the *Kyoto Protocol* (emphasis added).

nature.[86] Certainly, the Court did not intend to provide a human rights interpretation of environmental protection, since it grounded its normative reasoning in Article 191(2) TFEU, not in Article 11 TFEU, nor in Article 37 CFREU. However, systemically a 'high level of environmental protection' is required not only by Article 191(2) TFEU, but also by Article 37 CFREU. Therefore, following the guidelines established by the Vienna Conventions on the Law of Treaties,[87] an integrated interpretation of these two rules shows that the ECJ ultimately supported implicitly the extraterritorial extension of the EU ETS on a human rights basis.[88]

3. The Council of Europe system: first and second generation human rights and climate change

The Council of Europe is an international organisation aiming to promote State cooperation as to human rights, democratic development and the rule of law.[89] It currently comprises 47 Member States, including the 28 EU countries. Its main initiative is the 1950 ECHR, which has been signed by all 47 Member States and sets out a catalogue of fundamental rights subject to the jurisdiction of the ECtHR. The ECHR and its Additional Protocols currently do not provide for a right to environment, and therefore the issue of climate change and its impact on human rights is not covered *per se* either by the Convention or by the case law of the ECtHR. However, since the beginning of the 1990s the ECtHR has incorporated the parameter of environmental protection in its interpretation of classical first and second generation rights embedded in the ECHR. The Strasbourg Court has thus provided indirect environmental protection, particularly via the substantive rights to life (Article 2) and to private and family life

86 This option was nonetheless pointed out by the Advocate General (Juliane Kokott, Opinion in *Air Transport Association of America*, Case C 366/10 (2011) [152]–[154]). See Ottavio Quirico, 'Primary Sovereign Rights or Secondary Environmental Duties? Critical International Law Issues Raised by the Extraterritorial Application of the EU ETS in the Aviation Sector', in Christine Bakker and Francesco Francioni (eds), *The EU, the US and Global Climate Governance* (Ashgate, 2014) 143, 152 ff.

87 Vienna Conventions on the Law of Treaties (1969 and 1986) Article 31. See also ILC, *Fragmentation of International Law: Difficulties Arising from the Diversification and Expansion of International Law* A/CN.4/L.682 (2006) [220].

88 This confirms the potentially binding nature of Article 37 CFREU (Cournil *et al* (2012) 41). In the future, breaches of first and second generation human rights might arise in situations where, for instance, the EU and its Member States acquire CDM credits from clean development projects that are in violation of fundamental rights, such as the Aguan Biogas Project in Honduras (Carbon Market Watch, *Aguan Biogas Project*, <http://carbonmarketwatch.org>; Cournil *et al* (2012) 74). Furthermore, EU and Member States' measures reversing clean investment policies owing to negative financial conjunctures might infringe upon the fundamental right to property (Tabau (2013) 183–5; Cournil *et al* (2012) 77). On this issue, see also Vadi in this volume.

89 This body was established in 1949 via the Treaty of London (*Statute of the Council of Europe*, 5 March 1949).

(Article 8), as well as the procedural rights to a fair trial and to an effective remedy (Articles 6 and 13).

Significantly, in *Oneryildiz v Turkey* the Grand Chamber of the ECtHR adjudicated upon the case of people living in a slum below a rubbish tip in Istanbul, who died because of a methane gas explosion also entailing environmental consequences. The tip had been poorly managed by the Istanbul City Council, which had failed to address an expert report warning about the dangers of an explosion. The Court determined that Turkey had breached the rights to life, property and effective remedy under Articles 2 and 13 ECHR, and 1 Additional Protocol 1, for not effectively preventing the environmental accident resulting in the death of the applicant's relatives and damage to their house and other belongings.[90]

More systemically, the ECtHR has considered the issue of environmental protection under Article 8 ECHR, which provides for the right to private and family life.[91] In particular, in *Tatar v Romania* the Court adjudicated upon an accident that occurred at the gold mine of the municipality of Baia Mare, involving the release of approximately 100,000m^3 of cyanide-contaminated water into the environment.[92] The applicants had unsuccessfully filed various administrative and criminal complaints, fearing that the use of sodium cyanide by the gold mine company, and primarily water contamination via cyanide-infected tailings, could have an adverse impact on the health of their family, particularly aggravating the condition of a son suffering from asthma. The Court determined that Romania had breached the duty to assess the risks posed by the company's activity and to take suitable measures protecting the right to private life and home. Although the case in issue concerned more the health of the applicants than the enjoyment of the amenities of home, the Court examined the facts exclusively under Article 8 ECHR. In all likelihood, this interpretive shift ultimately led the ECtHR to extending the scope of the right to private and family life, so as to include the 'enjoyment of a healthy and protected environment'.[93]

In *Hatton and Others v the United Kingdom*, re-assessing the issue of noise disturbances at Heathrow Airport, the ECtHR considered procedural flaws in UK policies leading to increased noise from night flights, which had previously not been deemed in breach of Article 8 ECHR.[94] The Court held that the overly narrow scope of judicial review at the time of adoption of the policies was in breach of Article 13 ECHR.[95]

90 *Öneryildiz v Turkey*, Appl No 48939/99 (ECtHR, 30 November 2004) [69] ff, [119] ff.

91 See, for instance, *López Ostra v Spain*, Appl No 16798/90 (ECtHR, 9 December 1994); *Guerra and Others v Italy*, Appl No 14967/89 (ECtHR, 19 February 1998).

92 *Tatar v Romania*, Appl No 67021/01 (ECtHR, 27 January 2009) [25] ff.

93 Ibid, [107] and [112].

94 *Hatton and Others v the United Kingdom*, Appl No 36022/97 (ECtHR, 8 July 2003) [104]. See also *Hatton and Others v the United Kingdom*, Appl No 36022/97 (ECtHR, 2 October 2001).

95 Ibid, [148].

Overall, although the ECHR is a living instrument that should be 'interpreted in the light of present-day conditions', as the ECtHR has consistently declared,[96] the Strasbourg Court seems to be profoundly attached to its role as a pan-European guarantor of first and second generation human rights established in the aftermath of the Second World War. Similarly, the ECteSR, which adjudicates on breaches of the 1986 European Social Charter (ESC), another fundamental instrument adopted by the Council of Europe, has assessed the relationship between climate change and human rights based on the right to health. Particularly in *Marangopoulos*, the ECteSR was called to make a decision on the allegation that Greece had failed to comply with its obligation to protect public health against air pollution under Article 11 ESC.[97] Greece allowed the operation of lignite mines, that is, the main source of energy at the time of the complaint, and power stations fuelled by lignite, causing, inter alia, GHG emissions, without taking all necessary steps to reduce their environmental impact. Allegedly, total suspended particulates emitted by processing lignite exceeded tolerable limits set by the EU and an abnormally high prevalence of respiratory diseases was registered in affected areas. Greek energy policy on lignite was questioned, since it implied 'continued massive use of lignite as being quite incompatible with the Kyoto Protocol objectives'.[98] The 'European Commission's approval for Greece's [GHGs] first national allocation plan (NAP1), adopted in 2004' and consequent 'emissions allowance distribution' were deemed 'not significant' and inadequate to achieve internationally agreed GHG reduction targets.[99] The absence of a credible policy for renewable energy was also put forward.[100]

Based, inter alia, on the UNFCCC, the Kyoto Protocol, EC implementation measures (including the ETS), TEC provisions ensuring a high standard of environmental protection, the Aarhus Convention, and environmental provisions under the Greek Constitution,[101] the ECteSR upheld the plaintiff's claim both substantively and procedurally. More specifically, in light of the case law of the ECtHR, IACtHR, AComHPR, and UN CESCR, the ECteSR determined that the right to health embedded in Article 11 ESC includes the 'right to a healthy environment'.[102] A connection was also established between the right to health under the ESC and the right to life protected under Article 2 ECHR.[103] According to the Committee, even taking into account the margin of discretion granted to national authorities, Greece breached Article 11(1) ESC, setting out an obligation 'to remove as far as possible the causes of ill-health', because its lignite-based energy

96 See, for instance, *Airey v Ireland*, Appl No 6289/73 (ECtHR, 9 October 1979) [26].
97 *Marangopoulos Foundation for Human Rights (MFHR) v Greece*, Collective Complaint No 30/2005 (ECteSR, 6 December 2006).
98 Ibid, [34].
99 Ibid.
100 Ibid, [36].
101 Ibid, [113] ff, [130] ff, [151] ff, [160] ff.
102 Ibid, [160].
103 See also Bakker in this volume.

policy exceeded binding CO_2 emission limits without adequate monitoring and safety standards. Greece also violated Article 11(3) of the Charter, requiring to 'prevent as far as possible epidemic, endemic and other diseases, as well as accidents', because of the absence of regular health assessments. Furthermore, Greece was considered in breach of Article 11(2) of the Charter, requesting 'advisory and educational facilities for the promotion of health', for not including affected population in environmental impact assessments or providing people with access to relevant information.[104] Finally, Greece was held in violation of Articles 2 and 3 ESC on just working conditions and a safe working environment.[105] The Committee nonetheless did not award any specific remedy.

Along the same lines, in 2012 the Committee on the Environment, Agriculture, and Local and Regional Affairs of the Parliamentary Assembly of the Council of Europe adopted a Declaration on Climate Change and Human Rights, highlighting the impact of global warming on the rights to life, access to water, food, good health, decent housing and security.[106]

4. Climate change in light of the EU accession to the ECHR: from first and second generation human rights to the fundamental right to (a sustainable) environment?

Under Article 6(2) TEU, the EU 'shall accede' to the ECHR. This might have important implications for the compatibility between EU climate change policies and human rights protection provided by the ECtHR.[107] For instance, according to EU ETS Directive 2003/87/EC, from 2008 to 2012 at least 90 per cent of GHG allowances were allocated by governments for free, while after 2012 the allocation mechanism changed substantially, so that 50 per cent of all available allowances are to be auctioned off by 2020. This could entail an increase in energy costs, making households' access to essential supplies more difficult. Such an issue could be adjudicated upon by the ECtHR under Article 8 ECHR on the right to private and family life.[108] Furthermore, discriminatory issues raised by the EU climate policy might fall within the current jurisdiction of the ECtHR. For instance, the 2003 ETS Directive makes a distinction between the chemical and non-ferrous metal sectors on the one hand, and the steel sector on the other, for the purpose of including or excluding them from the scope of the EU ETS. In the *Arcelor* cases, the Grand Chamber of the ECJ held that this differential approach is justified by the dissimilar quantity of GHGs emitted in the sectors in issue as well as management questions, and therefore does not constitute a violation

104 *Marangopoulos Foundation* (2006) [194] ff.

105 Ibid, [222] ff.

106 *As the World's Warmest Year Ends, Time for Climate Change to be Seen as a Human Rights Issue* (2012).

107 See Armelle Gouritin, 'L'impact de l'adhésion de l'UE à la CEDH sur le lien entre droits de l'homme et politiques climatiques de l'UE', in Cournil and Tabau (2013) 243 ff.

108 Cournil *et al* (2012) 73.

of the general principle of equal treatment.[109] After accession of the EU to the ECHR, the ECtHR might adjudicate on similar cases, based on Article 14 ECHR (Prohibition of Discrimination). Procedurally, any individuals, non-governmental organisations and groups claiming to be victims of violations of the rights set forth in the ECHR and its Protocols because of EU-based climate policies will have the right to lodge an application with the ECtHR under Article 34 ECHR (Individual Applications).[110] Therefore, the right of access to justice will be enhanced, in line with the principles of the Aarhus Convention.[111]

Most importantly, after accession of the EU to the ECHR, the imbalance between the EU system, recognising 'environmental protection' under Article 37 CFREU, and the ECHR, which, instead, only includes first and second generation human rights, will systemically emerge. Indeed, if Article 37 CFREU is considered to set out a right, accession of the EU to the ECHR will generate a paradoxical situation, whereby the ECJ affords direct protection to the fundamental claim to a sustainable environment under EU law, whereas the Strasbourg Court only provides indirect protection via first and second generation human rights embedded in the ECHR. Even assuming that Article 37 CFREU does not establish a justiciable claim, the ECJ might adjudicate upon the right to environment based on principles inferred from domestic constitutions, most of which already recognise such a claim.[112] This inconsistency might prompt an evolution of the case law of the ECtHR, so as to afford direct protection to the right to environment, including climate sustainability. Thus, along the lines of *Tatar v Romania*, the scope of Articles 2, 3 and 8 ECHR could be further extended, so as to definitely embrace a human right to environment.[113] In this sense, some initiatives have been undertaken,

109 *Arcelor v Parliament and Council*, Case T–16/04 (ECJ, 2 March 2010); *Société Arcelor Atlantique et Lorraine and Others v Premier ministre, Ministre de l'écologie et du développement durable and Ministre de l'économie, des finances et de l'industrie*, Case C-127/07 (16 December 2008) [69] ff.

110 On the division of responsibility between the EU and its Member States under the ECHR see Quirico (2010) 40 ff.

111 This is all the more important in light of the limited access of non-State entities to justice in the EU system under Articles 256 and 263 TFEU, in spite of Article 47 CFREU (Right to an Effective Remedy and to a Fair Trial). See also Cournil *et al* (2012) 75.

112 See above section 2.1.

113 See *Hatton* (2003) Joint Dissenting Opinion of Judges Costa, Ress, Türmen, Zupančič and Steiner:

> We regret that we cannot adhere to the majority's view that there has been no violation of Article 8 of the European Convention on Human Rights in this case . . . the close connection between human rights protection and the urgent need for a decontamination of the environment leads us to perceive health as the most basic human need and as pre-eminent . . . It is true that the original text of the Convention does not yet disclose an awareness of the need for the protection of environmental human rights. In the 1950s, the universal need for environmental protection was not yet apparent. Historically, however, environmental considerations are by no means unknown to our unbroken and common legal tradition whilst, thirty-one years ago, the Declaration of the United Nations Conference on the Human Environment stated as its first principle: 'Man has the fundamental right to freedom, equality and adequate conditions of life, in an environment of quality that permits a life of dignity and well-being.' The European Union's Charter of Fundamental Rights (even though it does not

highlighting the relationship between climate change and human rights within the Council of Europe system: suggestions go as far as to require the inclusion of a right to environment in the ECHR.[114]

In Recommendation 1431 (1999),[115] the Parliamentary Assembly of the Council of Europe proposed to supplement the ECHR with a claim to a 'healthy and viable environment as a basic human right'.[116] Similarly, in Recommendation 1614 (2003) on 'Environment and Human Rights', the Parliamentary Assembly advocated the necessity of recognising a 'human right to a healthy, viable and decent environment which includes the objective obligation for States to protect the environment, in national laws, preferably at constitutional level'.[117] In Recommendation 1614, the Parliamentary Assembly also proposed the drafting of a protocol to the ECHR on environmental protection, including procedural rights grounded in the Aarhus Convention.[118] These proposals were nevertheless rejected by the Committee of Ministers of the Council of Europe, in light of the fact that the ECHR already provides environmental protection via first and second generation human rights.[119]

Recommendation 1883 (2009) on Climate Change Challenges triggered again the debate on the drafting of an additional protocol to the ECHR, concerning the Right to a Healthy Environment. Along these lines, Recommendation 1885 (2009) affirmed not only the 'fundamental right of citizens to live in a healthy environment', but also the 'duty of society as a whole and each individual in particular to pass on a healthy and viable environment to future generations'.[120] Consistently, Recommendation 1885 called for the drawing up of 'an additional protocol to the European Convention on Human Rights, recognising the right to

at present have binding legal force) provides an interesting illustration of the point . . . These recommendations show clearly that the member States of the European Union want a high level of protection and better protection, and expect the Union to develop policies aimed at those objectives. On a broader plane the Kyoto Protocol makes it patent that the question of environmental pollution is a supra-national one, as it knows no respect for the boundaries of national sovereignty. This makes it an issue par excellence for international law – and a fortiori for international jurisdiction. In the meanwhile, many supreme and constitutional courts have invoked constitutional vindication of various aspects of environmental protection – on these precise grounds. We believe that this concern for environmental protection shares common ground with the general concern for human rights.

114 See Council of Europe, *Environment: Climate Change, a Threat to Human Rights*, <http://hub.coe.int/what-we-do/culture-and-nature/climate-change>.

115 Council of Europe, Parliamentary Assembly, *Future Action to Be Taken by the Council of Europe in the Field of Environment Protection*, Recommendation 1431 (1999).

116 Ibid, [8].

117 Council of Europe, Parliamentary Assembly, *Environment and Human Rights*, Recommendation 1614 (2003) 9(ii).

118 Ibid, 10(i) and (ii).

119 See, in particular, Council of Europe, Committee of Ministers, *Environment and Human Rights*, Doc 10041 (21 January 2004).

120 Council of Europe, Parliamentary Assembly, *Drafting an Additional Protocol to the European Convention on Human Rights Concerning the Right to a Healthy Environment*, Recommendation 1885 (2009) [1].

a healthy and viable environment'.[121] In this respect, the Committee of Ministers of the Council of Europe noted the impact of climate change on fundamental rights and the importance of a 'healthy, viable and decent environment' as 'relevant to the protection of human rights'.[122] However, finally the Committee only referred to the Manual on Human Rights and the Environment,[123] an instrument drafted in 2006 by the Steering Committee for Human Rights, which is confined to environmental protection provided by classical first and second generation human rights emerging from the case law of the ECtHR.[124] Rejection of a protocol on the right to environment was nonetheless controversial, within the context of a debate exploring different facets of the issue, particularly, the possibility of grounding causes of action in damage to the environment as 'a whole' rather than in 'individual damage'.[125]

In light of the threat posed by climate change to the environment and human rights, some scholars have recently re-proposed the idea of a protocol on the right to environment additional to the ECHR.[126] Even more significantly, a Conference of international NGOs of the Council of Europe called for the recognition of a 'right to a healthy environment as a human right' and the integration of human rights into a universal climate agreement, to be adopted in December 2015 at UNFCCC COP-21 in Paris.[127] Finally, the Fourth Meeting of the Group of Experts on Biodiversity and Climate Change to the Convention on the Conservation of European Wildlife and Natural Habitats invoked the recognition of a human right to environment in the European context.[128] This right is framed as a universal and inter-temporal one and includes a correlative State and non-State duty to protect, preserve, remediate and improve the environment, within the framework of the principle of common but differentiated responsibility. Such a claim is associated with procedural rights and linked to

121 Ibid, [10.1]. This proposal was preceded by Parliamentary Assembly Recommendations on *Environmental Accounting as a Sustainable Development Tool*, 1653 (2004); *Renewable Energies and the Environment*, 1879 (2009); *Challenges Posed by Climate Change*, 1682 (2009); and *Environmentally Induced Migration and Displacement: a 21st Century Challenge*, 1655 (2009) and 1862 (2009).

122 Council of Europe, Committee of Ministers, *Joint Reply to 'The Challenges Posed by Climate Change, Parliamentary Assembly Recommendation 1883 (2009)' and 'Drafting an Additional Protocol to the European Convention on Human Rights Concerning the Right to a Healthy Environment, Parliamentary Assembly Recommendation 1885 (2009)*', CM/AS(2010) Rec1883–1885 final (18 June 2010).

123 Ibid, [10].

124 Council of Europe, *Manual on Human Rights and the Environment – Principles Emerging from the Case Law of the European Court of Human Rights*, Council of Europe Publishing (1st edn 2006, 2nd edn 2012).

125 Council of Europe, Committee of Ministers (2010) Appendix.

126 Christel Cournil, 'La relation "droits de l'homme et changements climatiques" au sein de la communauté internationale et en Europe', in Cournil and Tabau (2013) 27, 60.

127 Conference of INGOs of the Council of Europe, Standing Committee, *Climate Change and Human Rights*, Declaration to the Warsaw Climate Change Conference (5 November 2013). See also UNFCCC COP 21 (2015) <www.diplomatie.gouv.fr>.

128 Maria Blazogiannaki, *Climate Change and Human Rights*, Document prepared for the Fourth Meeting of the Group of Experts on Biodiversity and Climate Change (2009).

culture, the needs of groups disproportionately affected by environmental harm, and displaced people.[129] A reference to the principles of international solidarity, shared responsibility, and the 'balance between the right to sustainable development and environmental protection' contextualises the right to environment within the framework of the right to development.[130] In light of this, the inclusion of the right to environment in the ECHR is envisaged.[131] Of course, such an approach could further spawn direct protection for the environment as a human right in the Council of Europe system, including the ESC.

Conclusion

In the European context, on the one hand, the EU has adopted a progressive climate change policy, with respect to both mitigation and adaptation. This policy should be, and to a certain extent already is, mindful of human rights, in particular, as established in the CFREU, which provides for 'environmental protection' under Article 37. On the other hand, the main instrument adopted by the Council of Europe, that is, the ECHR, only affords environmental protection via 'classical' first and second generation human rights, particularly the right to private and family life, encompassing a human 'right to a healthy and protected environment'. Similarly, the ECteSR has established a link between GHG emissions and the right to health under the ESC, including a human 'right to a healthy environment'.

Within this framework, the accession of the EU to the ECHR is a positive issue, which may certainly improve respect for procedural environmental claims and monitoring of EU GHG policies, based on first and second generation human rights. The further evolution of the ECtHR jurisprudence, and thus possibly that of the Council of Europe system more generally, towards the recognition of a 'human right to (a climatically sustainable) environment' is not a *fait accompli*, but cannot be excluded *de lege ferenda*. This is particularly true in light of the paradoxical imbalance accession will determine between direct environmental protection, possibly regarded as a fundamental right, afforded by the ECJ, that is, a non-human rights organ, and indirect protection granted by the ECtHR, a specialised human rights body. Such a development would adequately complement procedural environmental claims, along the lines defined by Article 1 of the Aarhus Convention.

129 Ibid, 10.
130 Ibid, 10–11.
131 Ibid, 11–12.

18 Challenging the human rights responsibility of States and private corporations for climate change in domestic jurisdictions

Tineke Lambooy and Hanneke Palm

Introduction

In 2013, a Dutch NGO, the foundation Stichting Urgenda, representing a societal interest (the name combines the words 'urgent' and 'agenda'), together with 886 individual citizens (hereinafter the claimants will be referred to collectively as 'Urgenda'), commenced a law suit against the State of the Netherlands. The goal is to encourage the Dutch State to take effective measures to achieve a reduction of GHG emissions, specifically, carbon emissions, in the Netherlands by 2020 and beyond. This case is particularly interesting, because the Dutch State's climate change policy is challenged, inter alia, on the ground that the State should act in accordance with its positive obligations and duties under the ECHR.[1]

This chapter explores the possibility of holding a State responsible for causing climate change by comparing the ongoing *Urgenda* case with two other domestic cases involving climate change and human rights: *Native Village of Kivalina v ExxonMobil et al* (hereinafter '*Kivalina*')[2] and *Gbemre v Shell Petroleum Development Co* (hereinafter '*Gbemre*').[3] These cases are different from *Urgenda* in that the claimants in *Kivalina* attempted to hold private companies liable for the harm inflicted upon their Inuit community in Alaska by climate change, whereas *Gbemre* dealt with the responsibility of private companies and the Nigerian State for solving the problem of localised negative effects of GHG emissions.

The goal of comparing the three cases is to evaluate different human rights approaches to climate change from the standpoint of substantive law, whereas procedural issues of participation in decision-making and access to information

1 *Convention for the Protection of Human Rights and Fundamental Freedoms*, Rome, 4 November 1950, implemented via the ECtHR.

2 US District Court for the Northern District of California, Oakland Division, *Native Village of Kivalina v ExxonMobil Corporation et al*, Case No C 08–1138 SB (30 September 2009).

3 Federal High Court of Nigeria in the Benin Judicial Division, *Gbemre v Shell Petroleum Development Company of Nigeria Limited, Nigerian National Petroleum Corporation and Attorney-General of the Federation*, Order No FHC/B/CS/153/2005 (14 November 2005).

and justice are considered *en passant*. Where relevant, other cases dealing with climate change are also briefly considered in this chapter.

1. *Urgenda*: State responsibility for GHG emissions on a human rights basis?

On 20 November 2013, Urgenda served the summons (*dagvaarding*; hereinafter 'Summons') on the Dutch State (the Ministry of Infrastructure and the Environment)[4] in an action before the District Court of The Hague, the Netherlands, to hold the State liable for not effectively pursuing a reduction of GHG emissions.[5] On 2 April 2014, the Dutch State replied in its 'statement of defence' (*conclusie van antwoord*; hereinafter 'Defence').[6] Urgenda then filed its 'statement of reply and amended claims' (*conclusie van repliek tevens wijziging van eis;* hereinafter 'Reply') on 10 September 2014.[7] The State filed its 'rejoinder' (*dupliek*; hereinafter: Rejoinder) on 21 January 2015 and an oral hearing took place on 14 April 2015.[8]

In this section, the claims and arguments put forward by Urgenda are elaborated on in more detail than the defence by the Dutch State, for two reasons. First, in the proceedings as they stood at the time of writing, Urgenda had submitted two large documents and the State only one. The second reason lies in the manner of analysing the selected cases in this chapter, which is based on litigation strategy. This implies, in particular, examining the claimants' choice of a specific type of procedure (civil or administrative), selecting the defendant(s) (the State or GHG emitting companies), and determining the type of claim(s). Connected issues are questions of standing, legal grounds asserted by the claimants to support the claims, and the outcome.[9]

1.1. Standing: public interest litigation

In the Netherlands, the right to start a collective action, also known as class action, was developed in Dutch case law and later enacted in Article 3:305a(1) of the Dutch

4 See <www.government.nl/ministries/ienm>.

5 Urgenda, Summons (28 November 2013) <www.urgenda.nl/documents/DagvaardingKlimaatzaak 19–11–13.pdf>. English translation by Urgenda of the Summons (Summons [English Translation]) <www.wijwillenactie.nl/wp-content/uploads/2014/06/FINAL-DRAFT-Translation-Summons-in-case-Urgenda-v-Dutch-State-v.25.06.10.pdf>.

6 Dutch State Defence (2 April 2014) <http://gallery.mailchimp.com/91ffff7bfd16e26db7bee63af/files/Conclusie_van_antwoord.pdf>.

7 Urgenda, Reply (10 September 2014) <www.wijwillenactie.nl/wp-content/uploads/2013/09/Conclusie-van-Repliek-10–09–2014.pdf>.

8 See <www.wijwillenactie.nl> for up-to-date information on the Urgenda case.

9 The authors do not aim to develop a complete discussion on the facts of each case nor on the rules of evidence under the various jurisdictions. In *Urgenda*, many defences brought up by the Dutch State concern the question of (a lack of) evidence. These are only presented here in a short form.

Civil Code (DCC).[10] This Article provides an exception to the basic rule under Dutch law that only a (legal) person who has sufficient interest in a claim can bring an action before a civil court (Article 3:303 DCC). Article 3:305a determines that a foundation or an association with full legal capacity that, according to its articles of association (*statuten*), has the objective of protecting specific interests, may bring a claim in court to protect similar interests of other persons.[11] Admissibility requires that the claimant has sufficiently attempted to consult the defendant to reach a settlement over the claim,[12] which must sufficiently benefit the interests of the persons for whom protection is sought.[13] The claim may include a request to order that the defendant disclose the judicial decision to the public. However, the objective of the claim may not be to obtain compensatory damages.[14]

In the Summons, Urgenda claims that it has standing based on Article 3:305a, acting as the legal representative of the other 886 claimants.[15] Urgenda explains that it fulfils the standing requirements of Article 3:305(a), because the litigation aims to

> protect an issue of public interest that lies at the core of its [Urgenda's] constitutional purpose, that is, to protect the interests of current and future generations in order that the ecosystems and the liveability of the planet are not severely put at risk by planetary heating and climate change caused by humans.[16]

Urgenda's Articles of Association indeed state that its purpose is 'to preserve the planet as a sustainable place for future generations to live'.[17] Urgenda refers to Dutch case law in which the District Court of The Hague, the Netherlands, ruled that Article 3:305a DCC can serve as a basis for actions for the benefit of future generations.[18] Furthermore, the claimant explains how the organisation has attempted to conduct a constructive dialogue with the Dutch State in order to achieve reduction of carbon emissions, before commencing legal action.[19] Finally, Urgenda alleges that there is a gap in the legal protection under Dutch law concerning the legal interests of the persons for which protection is sought.

10 See, for instance, Dutch Supreme Court (1 July 1983) NJ 1984, 360; *Parliamentary Papers II* 1991/1992, 22 486, No 3, Article 3:305a DCC, entered into force on 1 July 1994. See also Phon van den Biesen, 'Will Civil Society Take Climate Changers to Court? A Perspective from Dutch Law', in Michael Faure and Marjan Peeters (eds), *Climate Change Liability* (Edward Elgar, 2011) 227, 227–34.

11 Article 3:305a(1) DCC.

12 Article 3:305a(2) DCC.

13 Ibid.

14 Article 3:305a(3) DCC.

15 Summons [English translation], 22, [48].

16 Ibid.

17 Ibid.

18 Urgenda refers to ECLI:NL:RBSGR:2001:AB1369, District Court The Hague (2 May 2001) [3.5].

19 Summons, 29–31, [68]–[75].

1.2. Urgenda's claims: excessive GHG emissions

In the Summons, Urgenda requests the Court to issue:[20]

i. A declaratory ruling that the State acts unlawfully (*onrechtmatig*) with respect to the claimants if the State fails to reduce or to achieve a reduction of the volume of GHG emissions[21] in the Netherlands so that by 2020 the total volume will be reduced by 40 per cent, or at least by 25 per cent, compared to the 1990 level;
ii. An order that by 2020 the State achieve a reduction of the total volume of GHG emissions in the Netherlands by 40 per cent, or at least 25 per cent, compared to the 1990 level. The request is at least an order that, within six months from the judgment's date, the State propose to the Parliament a programme with measures and corresponding budgets that can ensure the reduction of the total volume of GHG emissions in the Netherlands by 40 per cent, or at least 25 per cent, before 2020, compared to the 1990 level. The Programme should be verified and commented on by a Dutch public agency, the *Planbureau voor de Leefomgeving* (PBL – Netherlands Environmental Assessment Agency).[22]

Amending its claims in September 2014 in its Reply to the Defence of the Dutch State, Urgenda withdrew its request that the State present a plan of action to Parliament (see above under (ii)).[23] However, it maintained its other claims[24] and added two alternative claims (*subsidiaire vorderingen*).[25] The alternative claim which was added to claim (i) requests:

> a declaratory ruling that the Dutch State acts unlawfully if it fails to reduce or to achieve at least a 40% reduction of the annual volume of Dutch GHG emissions by *ultimately 2030* compared to the 1990 level.[26]

20 Free translation by the authors of the claims stated in the Summons, [143].
21 In the Summons, Urgenda also uses the term 'carbon emissions' (the concept of ppm refers to carbon, not to the broader notion of GHG emissions); see, for instance, para 45, in which Urgenda explains its claim to order the Dutch State to reduce carbon emissions. In the formal part of the claim, that is, after para 433, Urgenda uses the term 'GHG emissions'. See also paras [2.3]–[2.5] of the State's Defence and [87] of Urgenda's Reply.
22 The PBL is the Dutch National Institute for Strategic Policy Analysis in the Area of Environment, Nature and Spatial Planning. See <www.pbl.nl/en> [accessed 27 September 2014].
23 Reply, 192–4, [566]–[570].
24 The wording was slightly changed.
25 Reply, 182–3, [530]–[534] and 188–9, [553].
26 Free translation by the authors of the alternative claims, 211, [6], emphasis added. See also Urgenda, Reply, 182–3, [530]–[534].

Another alternative claim was added to the claim presented in the Summons under section (ii), first part, requesting:

> an order that the State diminish the annual volume of Dutch GHG emissions, so as to achieve a 40% reduction by *ultimately 2030* compared to the 1990 level.[27]

Third, Urgenda inserted several additional claims requiring the Court to issue a declaratory ruling affirming that:

1. Owing to the large worldwide volume of GHG emitted into the atmosphere, the temperature levels on Earth rise and, according to the best scientific judgments, will cause a dangerous change in the climate if emissions are not soon substantially reduced;
2. Climate change endangers many people and human rights;
3. Per capita emissions in the Netherlands are the highest among the group of countries which substantially emit GHGs;
4. The collective volume of the present Dutch GHG emissions is unlawful (*onrechtmatig*); and
5. The State is responsible for the collective volume of Dutch GHG emissions.

Fourth, Urgenda requested the Court to order that the State publish on its home web page and in six daily newspapers a one-page text prepared by Urgenda, informing Dutch citizens about various aspects of climate change, its dangers and harm, and explaining that the Netherlands needs to contribute to mitigating climate change by reducing Dutch emissions in a substantial way.[28]

1.3. Urgenda's statement of facts: lack of an effective GHG mitigation policy

In the Summons, Urgenda first sets out the factual context by discussing at length the scientific and political consensus on climate change and how mankind contributes to this phenomenon because of GHG emissions. It is stated that there is international consensus that breaching the two degrees Celsius threshold is dangerous for mankind and ecological systems.[29] On the basis of scientific models, it is shown that 'once carbon hits a concentration level of 450 ppm in the atmosphere,[30] there is only a 50 per cent chance that global warming can still be

27 Ibid, 212, [7], 188–9, [553] and 182–3, [530]–[534], emphasis added.

28 Ibid, 212, [8]–[9], 194–5, [571]–[576].

29 See, for instance, the 1992 UN Climate Convention (UNFCCC) and the 2010 Cancun Agreements.

30 A carbon concentration level of 450 ppm entails that out of every one million molecules, 450 molecules are carbon molecules.

kept below two degrees Celsius'.[31] Therefore, Urgenda's benchmark is that we must not exceed a carbon concentration level of 450 ppm. Furthermore, Urgenda expands on the temporal implications of emitting GHGs into the atmosphere: the consequences will only be noticeable many years later, that is, once GHGs have been emitted and are already in the atmosphere.[32]

Second, Urgenda refers to the reports from the IPCC[33] and the UNFCCC[34] and advocates that, in order to prevent the two degree Celsius rise and to stay below the 450 ppm level, industrialised countries (including the Netherlands), and the EU as a whole must reduce their GHG emissions by up to 25–40 per cent by 2020 compared to the 1990 levels. Urgenda points out that the UNFCCC considers per capita emissions the relevant standard of comparison[35] and makes clear that the Netherlands is among the five countries in the world that emit the highest volumes of GHGs in the atmosphere per capita.[36] Allegedly, this creates an even tighter responsibility for the Dutch State to effectively address climate change,[37] and reducing Dutch emissions will really matter in limiting the total volume of emissions worldwide.

Third, Urgenda explains how the Dutch government fails to adopt effective policies to meet internationally agreed targets and legal obligations and responsibilities. This is illustrated by pointing out how the Dutch State – through subsidies and tax reductions – sustains GHG-emitting industries in continuing to

31 Roger Cox, 'The Liability of European States for Climate Change' (2014) 30(78) *Utrecht Journal of International and European Law* 128, <DOI: http://dx.doi.org/10.5334/ujiel.ci>.

32 Summons [English Translation], 80, [255], referring to ECtHR, *Taşkin et al v Turkey*, Appl No 46117/99 (10 November 2004).

33 IPCC, 'Summary for Policymakers', in Susan Solomon *et al* (eds), *Climate Change 2007: the Physical Science Basis. Contribution of Working Group I to the Fourth Assessment Report of the Intergovernmental Panel on Climate Change* (Cambridge University Press, 2007) [IPCC Fourth Assessment Report, AR4]. See also the Fifth Assessment Report (AR5) (2014) <www.ipcc.ch/report/ar5>.

34 United Nations Framework Convention on Climate Change, New York, 9 May 1992.

35 'Per capita' emissions refer to the emissions per person of a population (see Summons, [170], [171], [178], [282], [320]).

36 Summons [English Translation], [348]–[354] based on World Bank data: <https://docs.google.com/spreadsheet/ccc?key=0AonYZs4MzlZbdFF1QW00ckYzOG0yWkZqcUhnNDVlSWc&hl=en#gid=0>, Exhibit 39:

> Of the 25 countries which in an absolute sense belonged to the largest CO_2 emitters in the world (with the Netherlands being the 25th largest), a ranking of the 10 largest CO_2 emitters per capita in the world would look as follows (brackets show the ranking based on absolute numbers): 1. Australia (15): 19.64 tonnes of CO_2 per capita; 2. Saudi Arabia (11): 18.65 tonnes of CO_2 per capita; 3. US (2): 17.76 tonnes of CO_2 per capita; 4. Canada (7): 16.15 tonnes of CO_2 per capita; 5. Netherlands (25): 14.89 tonnes of CO_2 per capita . . . World: 4.49 tonnes of CO_2 per capita.

According to Urgenda, similar data are provided by the PBL about Dutch emissions in 2010 and 2011.

37 Pursuant to the generally accepted principles of international law, among which are the no-harm principle and the UNFCCC approach to common but differentiated responsibilities. Summons [English Translation], [170], [178].

emit GHGs rather than developing alternatives.[38] It is subsequently explained that the manner in which the Dutch State complies with the EU ETS[39] does not and will not achieve the desired level of change in the total volume of GHGs: (i) the ceiling of the total volume of GHG emission rights available for the Dutch market participants is too high (421 million tonnes of emission credits for the period 2008–12), even higher than the total GHGs in the Netherlands (406 million tonnes emitted between 2008 and 2012);[40] and (ii) the Dutch State purchases additional GHG emission rights from other countries in order to meet its obligations. This obstructs the transition and innovation that is needed for achieving the 25–40 per cent reduction by 2020.[41]

1.4. Merits: the crucial role of the duty to protect human rights

Urgenda expounds on the obligations of the Dutch State in respect of preventing climate change and mitigating Dutch GHG emissions pursuant to the no-harm principle under international law, the UNFCCC, the Kyoto Protocol,[42] the decisions of the Conference of the Parties (COP-decisions),[43] Articles 2, 8 and 13 of the ECHR,[44] and the EU ETS and EU policy documents.[45] Urgenda also sets out how its claims are based on Dutch tort law, in light of the criteria developed by the Dutch courts on '*hinder*' (nuisance) and '*gevaarzetting*' (creating a dangerous situation, endangerment).

The *no-harm principle* is the first legal ground on which Urgenda argues that the Dutch State is acting unlawfully, that is, it is allegedly failing to take measures necessary to prevent causing 'significant damages' outside its territory through

38 For instance, under the Dutch Energy Tax Act a lower tax rate is applied for a higher level of energy use.

39 *Directive 2009/29/EC of the European Parliament and of the Council Amending Directive 2003/87/EC so as to Improve and Extend the Greenhouse Gas Emission Allowance Trading Scheme of the Community*.

40 Summons [English Translation], 92 [304]: ' [. . .] these companies received 15 million tonnes more emission credits then they needed to cover their own emissions.' See also Dutch Emission Authority, Publication 2008–2012 *Report* '*Voortgang Emissiehandel*' (1 September 2014) <https://www.emissieautoriteit.nl> .

41 Summons [English translation], 92 [306]. See also Stephen Humphreys, 'Conceiving Justice', in Stephen Humphreys (ed), *Human Rights and Climate Change* (Cambridge University Press, 2011) 305, where Humphreys explains the shortcomings of an emission trading system in light of development mechanisms. He argues:

> For many countries, especially those with urgent or emergency needs, the immediate benefits of *selling on* carbon capacity may outweigh the more substantial benefits of utilising them, particularly if the technology to do so is inefficient or lacking. Poor countries might then be stuck selling off their development potential to rich countries. It is against this background that questions must be raised about the longer-term human right simplifications of the emissions trading regime.

42 See Summons [English translation], 60–71.

43 Article 7 of the UNFCCC establishes a 'Conference of Parties' (COP) overlooking the implementation of the Convention. For that purpose, the COP gathers in principle annually.

44 Summons [English translation], 72–80, 88–9.

45 Ibid, 11, [18], 47–8.

GHGs emitted from the Netherlands. Referring to the IPCC reports, Urgenda stresses that the rising level of carbon concentration in the atmosphere causes climate change, with severe damaging consequences. Contributing to the (further) rise of GHG levels by not reducing territorial emissions implies that the State is violating the no-harm principle. To support this argument, Urgenda points to the international arbitral award in the *Trail Smelter* case (*US v Canada*).[46] In this case, an Arbitration Tribunal held Canada responsible for damage in the US: the polluting activities of a private company fell under Canada's control and therefore Canada was held responsible for transboundary damage. Urgenda also refers to the landmark decision of the ICJ in the *Corfu Channel Case (United Kingdom v Albania)*, to substantiate its argument that GHGs emitted by various actors in the Dutch territory can be attributed to the Dutch State.[47] The argument is that the Dutch State has control over such emissions and can influence the level thereof, similar to the Albanian State in the *Corfu Channel* case. According to the ICJ, Albania was aware of mine-laying operations in its territorial waters, that is, in the Corfu Channel, and thus had a responsibility to warn other States about this danger.

The second ground for arguing that the Dutch volume of carbon emissions is unlawful and that the State must take action, lies in the Dutch involvement with the UNFCCC, the Kyoto Protocol and the COP decisions.[48] State parties to this treaty, the Protocol and the international decisions explicitly acknowledge the need to prevent an increase in the average temperature of more than two degrees Celsius and, consequently, also the need for reducing GHG levels. This approach in the UNFCCC and the Protocol is based on the *precautionary principle* under international law. The Netherlands is a State party to the UNFCCC and the Kyoto Protocol. It has thus assumed responsibility to comply with the actions provided for in these international treaties and soft law documents. Moreover, the EU also has ratified both treaties. The EU committed itself under the Kyoto Protocol – together with the 15 Member States of which it was constituted at the time of ratification – to reduce its emissions of six GHGs by 8 per cent over the period 2008–12, compared to 1990. The Netherlands was and is one of those Member States.[49] Consequently, the Dutch State is also obliged to undertake urgent action as an EU Member State.

The third ground on which Urgenda's claims are based concerns human rights under Article 2 (Right to Life)[50] and Article 8 (Right to Private and Family

46 *Trail Smelter Arbitration (United States v Canada)* (1941) III *Reports of International Arbitral Awards* 1905.

47 The ICJ held that international responsibility arises when the State has knowledge about damaging activities by third parties (mine-laying in its territorial waters), and has control of a territory (*Corfu Channel Case (United Kingdom v Albania), Merits* [1949] ICJ Rep 4).

48 For instance, *Bali Action Plan* (COP13, 2007), *Copenhagen Accord* (COP15, 2009), *Cancun Agreements* (COP16, 2010) and *Durban Decision* (COP17, 2011).

49 Javier de Cendra de Larragán (2011) 59–63. It is pointed out that the common and individual commitments of the EU and its Member States are implemented in EC Council Decision 2002/358/EC (Burden-sharing Agreement).

50 Article 2(1): 'Everyone's right to life shall be protected by law. No one shall be deprived of his life intentionally save in the execution of a sentence of a court following his conviction of a crime for which this penalty is provided by law.' The exemption in the second sentence has been restricted to the death penalty in wartime by Protocol 6 to the ECHR.

Life)[51] of the ECHR. The Netherlands is a State party to the ECHR and has committed itself to securing the human rights included in this treaty.[52] It has also accepted the competence of the ECtHR to deal with any alleged violations by the Dutch State. Urgenda notes that States have negative and positive obligations in order to prevent violations and to secure the enjoyment of human rights protected under the ECHR. The claimant argues that the Dutch State must fulfil positive obligations to prevent real and direct threats to life caused by climate change. Urgenda refers to the ECtHR's case law,[53] and claims that the Dutch State must take adequate and effective measures with actual and significant results, which in fact means that the Dutch State must reduce GHG emissions in order to prevent violations of Articles 2 and 8 ECHR.[54] A 25–40 per cent reduction in the emission levels is needed, in order to maintain a 50 per cent chance of not breaching the critical two degree Celsius temperature rise, which if exceeded will undoubtedly cause harm to people, companies and others.[55]

51 Article 8(1): 'Everyone has the right to respect for his private and family life, his home and his correspondence.'

52 Article 1 ECHR: 'The High Contracting Parties shall secure to everyone within their jurisdiction the rights and freedoms defined in Section I of this Convention.'

53 The landmark case on environmental harm as a violation of Article 8 ECHR is ECtHR, *López Ostra v Spain*, 16798/90, (9 December 1994). ECtHR, *Guerra v Italy*, 14967/89 (19 February 1998) confirms that Article 8 imposes positive duties upon States to ensure respect for privacy or family life. See also ECtHR, *Tatar v Romania*, 67021/01 (27 January 2009) (precautionary principle and Article 8) and ECtHR, *Di Sarno and Others v Italy*, 30765/08 (10 January 2012) (explicit recognition of a right to a clean and healthy environment). Regarding Article 2, see ECtHR, *Öneryildiz v Turkey*, 48939/99 (30 November 2004) [71], concerning the positive obligation on States to safeguard the lives of people within their jurisdiction, applying to 'industrial activities, which by their very nature are dangerous [. . .]'.

54 Other cases include ECtHR, *Budayeva and Others v Russia*, 15339/02, 21166/02, 20058/02, 11673/02, and 15343/02 (20 March 2008): pursuant to Article 2, States are to put in place legislation and policies to combat threats to the right to life. Applying this ruling to the dangerous effects of global warming, it can be argued that the State has to adopt and implement an effective action plan to combat climate change and environmental degradation. Urgenda discusses this case in the context of Article 8 ECHR. Furthermore, see *Kyrtatos v Greece*, 41666/98 (22 May 2003), where the ECtHR set a limit to complaints about environmental damage under Article 8 ECHR. The ECHR, however, does not provide general protection for the environment as such. Only individuals whose own well-being is directly affected can force the State to protect the environment (see Council of Europe, *Manual on Human Rights and the Environment: Principles Emerging from the Case-Law of the European Court of Human Rights* (Strasbourg Council of Europe Publishing, 2012); *Grimkovskaya v Ukraine*, 38182/03 (ECtHR, 21 July 2011) (procedural right to environment); *Dubetska and Others v Ukraine*, 30499/03 (ECtHR, 10 February 2011) (coal mine causing damages to the applicant's house, health and living environment); *Bacila v Romania*, 19234/04 (ECtHR, 30 March 2010).

55 Pursuant to the ECHR, a State can be held responsible for harm caused to everyone (citizens, companies, NGOs) who falls within its jurisdiction. The extraterritorial responsibility of a State is limited. See Summons [English translation], [227]: citizens of other State Parties could file a claim against the Dutch State in case of cross-border harm violating the ECHR as long as the Dutch State has a certain degree of control.

Urgenda also refers to the citizen's right to an effective remedy, which is enshrined in Article 13 ECHR. According to this provision, remedies at the national level should exist to repair or compensate (alleged) violations of the ECHR.[56] Arguably, if there was no domestic possibility of holding the Dutch State liable for (future) harm caused by Dutch carbon emissions, an effective remedy against (possible) violations of Articles 2 and 8 ECHR resulting from climate change would be absent, and this would result in a violation of Article 13 ECHR.[57] In this regard, Urgenda maintains that the Dutch State has the competence and power to influence the volume of GHGs emitted in its territory.[58]

The fourth ground concerns tort action (nuisance and endangerment) under Dutch civil law (Article 6:162 DCC). Urgenda contends that the (un)lawfulness of the offending emissions is to be tested against the criteria of 'nuisance'[59] and/or 'endangerment'.[60] As regards the question whether the State is the correct party to sue, which in common law systems is often indicated as 'attribution', Urgenda points out that the Dutch State itself acknowledges its responsibility for the protection of the environment under Article 21 of the Constitution (*Grondwet*). This responsibility has been developed in the Environmental Management Act (*Wet milieubeheer*) and in the Environmental Permitting (General Provisions) Act (*Wet algemene bepalingen omgevingsrecht*).[61] The provisions of the EU ETS Directive have also been implemented in the Environmental Management Act.[62] Furthermore, Urgenda refers to the Dutch Supreme Court decision in the *Kalimijnen case* (*Potash Mines*)[63] to support its argument that although the Dutch State is not the only polluter (carbon/GHG emitter) and even though the damages occur and can occur both in the Netherlands and elsewhere (cross-border damages), the Dutch State can be held liable for contributing a substantial part to the aggregate world volume of carbon emissions (causation).

56 Summons [English translation], 85, [225].

57 Summons [English translation], 104, [292].

58 Pursuant to Article 2.1(1)(e) of the Environmental Permitting (General Provisions) Act, the State is the authority that issues licences, and hence in control. Furthermore, the State can reduce the level of carbon emission rights under the EU ETS allocated to the industry, and can employ financial instruments, for instance, taxes and subsidies, to reduce the use of carbon.

59 Summons [English translation], 81–3, [258]–[264].

60 Summons [English translation], 83–5, [265]–[269].

61 Wet Milieubeheer, <www.government.nl/issues/environment/roles-and-responsibilities-of-central-government/environmental-management-act>. Both Acts prescribe that companies obtain a licence before commencing activities that potentially produce health threats, hinder or nuisance to people or cause harm to the environment, including emitting gases in the air, and executing an EIA beforehand.

62 Chapter 16.

63 *Potash Mines* (Dutch Supreme Court, 23 September 1988) NJ 1989, 743. See also *Handelskwekerij GJ Bier BV v Mines de potasse d'Alsace SA*, C-21/76 (30 November 1976). The ECJ found in this preliminary ruling that 'the defendant may be sued, at the option of the claimant, either in the courts for the place where the damage occurred or in the courts for the place of the event which gives rise to and is at the origin of that damage'.

Urgenda concludes that, pursuant to Articles 2 and 8 ECHR and the no-harm principle under international law, the State has a responsibility to prevent environmental damage towards its own citizens (among which are Urgenda and the other claimants) as well as towards others. GHG emissions from Dutch territory contribute substantially (*per capita* and in an absolute sense) to the rise of GHG concentration in the global atmosphere. In order to maintain a 50 per cent chance of not breaching the two degree Celsius rise of Earth's average temperature, linked to a carbon concentration level of 450 ppm, the Netherlands is to reduce its emission levels as soon as possible and at least on a basis of 25–40 per cent by 2020, compared to the 1990 levels. By not effectively pursuing this goal, the Dutch State breaches its duty of care towards the claimants and commits a tortious act. The State is thus to be held *pro rata*[64] responsible for the threatening consequences of climate change and the claimant offers to submit proof of all statements.

1.5. The Dutch State's statement of facts: 'no regret' measures

In the Defence, the Dutch State agrees with Urgenda that climate change is taking place and that it is very likely that mankind has been the main contributor since the beginning of the 20th century. However, the State points to the uncertainties indicated in the IPCC reports.[65] It argues that policymakers (and those politically responsible) have to deal with climate change and that – in designing their policies – they should take account of these uncertainties in light of feasibility and cost effectiveness ('no regret' measures').[66] According to the Dutch State, the relationship between global warming up to a two degree Celsius rise and carbon concentration levels of 450 ppm (or other GHGs) is not as strong as Urgenda claims.[67] The Dutch State intends to await the outcome of the discussion at the European and international level, because it considers climate change a global problem, not a national or domestic problem.

Therefore, although Urgenda and the Dutch State agree to a certain extent about the effects of climate change and the need to reduce GHG emissions, the parties come to different conclusions regarding how to shape climate policies.

1.6. The statement of defence by the Netherlands: balancing different rights

The Dutch State argues first that Urgenda's claim is not admissible.[68] Next, it contests that the current level of Dutch carbon emissions is excessive and unlawful. The main defence that the Dutch State puts forward against Urgenda's claims

64 Compare with *Potash Mines* (1988).
65 Defence, 3, [1.3].
66 Ibid, 15, [5.1].
67 Ibid, 15, [5.2].
68 Ibid, 10–3, [3.1]–[3.11].

based on tort is that Urgenda does not fulfil the burden of proof to establish liability.[69] According to the Dutch State, Urgenda does not specify the harm allegedly caused by Dutch carbon emissions[70] and the *probability* of harm or damage is in itself not enough to establish a tort under Dutch law.[71] The State also reasons that Urgenda does not ascertain causation.[72] Since the State fulfils its obligations under the Kyoto Protocol and European legislation and policies,[73] the requirements for finding endangerment and nuisance are not met,[74] and hence its conduct cannot be considered tortious.[75]

With regard to the alleged violations of its obligations under the ECHR, the ECtHR gives States a 'margin of appreciation' in environmental cases.[76] Therefore, the Dutch State argues that its duty is to find a 'fair balance' in respect of environmental issues, since it also has to take into account economic interests and the uncertainties that still exist with regard to the exact consequences of climate change.[77] Within the margin of appreciation, States can prioritise economic concerns over environmental considerations as long as they do not exceed the margin afforded to them. Thus, a deviation in Dutch policy from the previously set emission targets would not be sufficient to conclude that the State exceeds its margin of appreciation and violates rights protected under the ECHR.[78] Once again, the State points out that the alleged harm and damage is not specific enough, certainly not to trigger the protection of the ECHR, and that there is no positive obligation to reduce the Dutch level of carbon emissions on the basis of Articles 2 and/or 8 ECHR.[79]

69 Article 150 of the Dutch Code of Civil Procedure. Defence, 13–4, [4.1]–[4.2], 43, [8.6].
70 Defence, 50, [8.30], 52–4, [8.38]–[8.46].
71 Ibid, 52, [8.36].
72 Indeed, Urgenda does not discuss every element of Article 6:162 DCC in detail. However, from Urgenda's perspective, the IPCC reports offer an adequate level of clarity regarding the consequences of climate change, that is, the harm, and the rise of the level of GHG in the atmosphere, that is, causation. Once Urgenda 'establishes' a link between two degrees Celsius and 450 ppm carbon level, the elements that need further substantiation are whether emissions from Dutch territory are unlawful and whether the Dutch State can be held liable.
73 Defence, 55, [8.52].
74 Ibid, 45–6, [8.11]–[8.16] (nuisance); 51–2, [8.33]–[8.36] (endangerment).
75 Ibid, 58, [8.66].
76 States are afforded a wide margin of appreciation in environment-related cases, which allows them to balance economic development against environmental protection. However, this margin shrinks when the severity of environmental pollution and its consequences for the lives and health of individuals increase. See, for instance, DIG San José, *Environmental Protection and the European Convention on Human Rights* (Strasbourg Council of Europe, 2005) 115.
77 Contrary, see Simon Caney, 'Climate Change, Human Rights and Moral Thresholds', in Humphreys (2011) 69. Caney advocates a human rights approach particularly because of the problem of 'risk and uncertainty' of climate change threats: 'it matters a great deal whether those who are taking risks are exposing just themselves to serious risks or whether they are exposing others to serious risks' (85).
78 Defence, 70–1, [10.11]–[10.12], 74, [10.19].
79 Ibid, 72–4, [10.13]–[10.19].

Another line of defence is that shaping climate policies is up to the legislator and the judge should not interfere (political question).[80] The Dutch State asserts that, since it does not violate any international or European legal obligations, a court cannot compel it to take measures to achieve a 25–40 per cent reduction of carbon emission levels by 2020, as this would interfere with the *Trias Politica* and the legislator's prerogatives. The State has 'policy discretion' and the measures it takes should be accepted by society.

1.7. Urgenda's reply: unjustifiable human rights derogations

In its Reply, Urgenda concludes that there is an agreement on the facts, as both parties refer to the IPCC reports and explicitly acknowledge their conclusions: global warming is caused by human activities, such as GHG emissions, and leads to climate change. This becomes a threatening development if global GHG emissions are not substantially and speedily mitigated. Urgenda points out that the Dutch State recognises that global warming above two degrees Celsius will result in threatening climate change and should therefore be prevented by achieving a 25–40 per cent reduction of Dutch emissions by 2020, but does not attach any legal significance to this.[81]

Urgenda summarises the State's Defence as follows: (i) Urgenda has no standing; (ii) the court may not decide on the claims (political question); (iii) the criteria to establish a tort are not met (no causation). In addition to these legal defences, the Dutch State claims that it meets its international legal obligations and that the Dutch government takes several measures to combat climate change (iv).

The authors note that these types of defences are also brought forward by defendants in other climate change cases.[82] It is therefore useful to briefly address how Urgenda responds to each of the State's defences.

Standing – Urgenda explains that Article 3:305a DCC allows litigation aimed at protecting the environment as a collective interest.[83]

Primacy of politicians – Urgenda emphasises that it claims legal protection, but does not intend to practise politics,[84] and that the *Trias Politica* does not prohibit a judge from offering legal protection.[85] Urgenda restates that its claims serve the purpose of preventing the pending threat of climate change, which will have severe consequences, including violations of rights, and therefore the requested declaratory rulings and orders are in the interest of the claimants. Urgenda reasons that

80 Ibid, 83–6, [12.10]–[12.21].

81 Reply, 2, [3].

82 See, in particular, *Kivalina* (2009).

83 Reply, 71, [200].

84 Femke Wijdekop, 'Democratie, rechtstaat en de rechten van toekomstige generaties' (2014) 89(23) *Nederlands Juristenblad* 1552–1560.

85 Reply, 74, [210].

if the court rejected the claim – for instance, based on the State's argument that climate change risks are in the hands of the legislative branch – then the consequence would be an absence of legal protection in the Netherlands against climate change, in violation of Article 13 ECHR.[86] Reference is made to *Clara Wichmann v the Dutch State*,[87] in which the Court did not oblige the State to adopt specific legislation, but 'encouraged' it to take legislative measures to end a situation which was considered unlawful.[88]

In responding to the State's defence concerning the distinction between a legal case and a political question,[89] Urgenda also refers to the judgment of the US Court of Appeals for the Second Circuit in *Connecticut v AEP*.[90] Only a political question stands in the way of a court deciding a case. However, asking for legal protection is not a political question: it is not a legislative prerogative to decide *when* the 25–40 per cent reduction in GHG emission levels should be reached.[91] Urgenda thus holds that the State has not convincingly argued the lawfulness of postponing the realisation of a 25–40 per cent reduction targets (after 2020).[92]

Requirements to establish a tort – Urgenda points out that the State did not grasp that Urgenda does not seek monetary damages, but instead seeks to *prevent damage*. Dutch case law on tort claims has increasingly developed towards preventing damage and stimulating desirable behaviour. This means that for

86 Ibid, 74, [210].

87 ECLI:NL:HR:2010:BK4549 (Dutch Supreme Court, 9 April 2010) NJ 2010/399. Reply, 187–8, [547]–[551].

88 In Reply, 188, [550], Urgenda refers to Geerten Boogaard, *Het wetgevingsbevel, over constitutionele verhoudingen en manieren om een wetgever tot regelgeving aan te zetten* (Wolf Legal Publishers, 2013), 129. Urgenda comments that Boogaard's research results do not demonstrate that claiming a declaratory ruling in fact represents an order to adopt certain legislation. See also ECLI:NL:HR:2014:523, AB 2014/230 (Dutch Supreme Court, 7 March 2014) *Staat/Norma* [4.6.2] and ECtHR, *Taşkin et al v Turkey*, Appl No 46117/99 (10 November 2004) [119], stating that the individuals concerned must be able to appeal to the courts against any decision, act or omission where they consider that their interests or their comments have not had sufficient weight in the decision-making process.

89 Reply, 202, [597].

90 US Supreme Court, *American Electric Power Co, Inc et al v Connecticut et al*, No 10-174 (20 June 2011); US Court of Appeal Second Circuit, *State of Connecticut et al v American Electric Power Co, Inc et al*, Case No 05-5104-cv; 05-5119-cv (21 September 2009) section III: *AEP*. In this case, several States, the city of New York and three private land trusts requested the Court to set carbon caps for each defendant (several power companies), to be further reduced annually. The issue was the separation of powers. The Court objected that Congress had entrusted this complex balancing to the US Environmental Protection Agency. The claims were more specific than Urgenda's claims, although also in *AEP* the claimants themselves had not suffered damages. Urgenda grounds its claims in tort, Connecticut based its claims on federal common law of interstate nuisance. For a critique on the inconsistent application of the political question doctrine, see further Elena Kosolapova, 'Liability for Climate Change-related Damage in Domestic Courts: Claims for Compensation in the USA', in Faure and Peeters (2011) 195, and Amelia Thorpe, 'Tort-Based Climate Change Litigation and the Political Question Doctrine' (2008) 24 *Journal of Land Use* 79.

91 Ibid, 206, [615].

92 Ibid, 207, [622]. On the political question issue see Chris van Dijk, 'Civil Liability for Global Warming in the Netherlands', in Faure and Peeters (2011) 212–13.

awarding an injunction, it is not required that harm has already been inflicted on the claimant(s).[93] The four requirements to grant such an order are: (a) (imminent) conduct, (b) that is unlawful, (c) in respect of the claimant, who (d) has a sufficient interest in the order. The requirements under (a), (c) and (d) have been considered in the above subsection concerning the Summons. Urgenda re-establishes them in the Reply.

As to the element of unlawfulness, Urgenda argues that the Dutch State exercises a 'lack of due care' (*maatschappelijke zorgvuldigheid*), because it does not effectively reduce the Dutch GHG emissions that contribute to aggravating climate change, endangering the enjoyment of human rights by the claimants. Urgenda then states that a violation of a provision of the ECHR suffices for establishing the requirement of unlawfulness under Article 6:162 DCC (tort). Where the State justifies its conduct by asserting that it has a margin of appreciation in deciding how to deal with the imminent threat of climate change, Urgenda clarifies that Article 2 ECHR does not allow a margin of appreciation. Referring to *Osman v United Kingdom*,[94] Urgenda insists that the right to life is an *absolute right* and that there are no grounds to justify any interference: 'It is sufficient for an applicant to show that the authorities did not do all that could be reasonably expected of them to avoid a real and immediate risk to life of which they have or ought to have knowledge.'[95]

Furthermore, again in respect to the margin of appreciation, Urgenda asserts that the State fails to substantiate its defence under Article 8(2) ECHR;[96] indeed, the State does not convincingly argue that the interference with the rights protected by Article 8 ECHR is 'necessary, effective and proportionate'. It is pointed out that 'when faced with an issue such as that raised in the instant case, the authorities cannot legitimately rely on their margin of appreciation, which in no way dispenses them from their duty to act in good time, in an appropriate and, above all, consistent manner'.[97] Urgenda contends that the burden is on the State to prove that its decision to only achieve a 16 per cent reduction of Dutch GHG emissions from 1990 levels by 2020 is a necessary, effective and proportionate measure that

93 Reply, 70, [198], 72, [202]. Urgenda cites the Dutch Handbook on Tort, that is, Titia Elizabeth Deurvorst, *Groene serie privaatrecht. Onrechtmatige daad II*, aantekening 96, (online via Kluwer Navigator).

94 ECtHR, *Osman v United Kingdom*, Appl No 23452/94 (28 October 1998) [116].

95 Reply, 120, [359].

96 Article 8(2):

> There shall be no interference by a public authority with the exercise of this right except such as is in accordance with the law and is necessary in a democratic society in the interests of national security, public safety or the economic well-being of the country, for the prevention of disorder or crime, for the protection of health or morals, or for the protection of the rights and freedoms of others.

97 *Öneryildiz* (ECtHR, 18 June 2002) [128]. Reply, 123, [366] and Grand Chamber Ruling (30 November 2004).

serves the public interest and justifies the impending interference with Urgenda's rights and the rights of future generations.[98] Indeed, under the 2010 Cancun Agreements, it has been internationally accepted that the two degrees threshold should not be exceeded because of its severe dangers and consequences.[99]

In order to further substantiate its claim that human rights will be violated as a consequence of climate change, Urgenda stresses factors such as floods, droughts, collapse of ecosystems and loss of biodiversity, food shortages, loss of human lives and an increase in human suffering in the Netherlands and elsewhere.[100] The interference with the enjoyment of family life (Article 8 ECHR) is illustrated based on a consideration of the IPCC:

> Climate change will have profound impacts on a broad spectrum of infrastructure systems (water and energy supply, sanitation, drainage, transport and telecommunication), services (including health care and emergency services), the built environment and ecosystem services. These interact with other social, economic and environmental stressors exacerbating and compounding risks to individual household well-being.[101]

Urgenda also states that even effects of climate change that will materialise outside Dutch territory can have a significant impact on the Netherlands, for instance, conflicts abroad and food insecurity. In the one-sided approach of current Dutch mitigation measures (as put forward by the State in its Defence), economic arguments prevail without sufficient justification for these imminent violations of human rights. Hence, the requirements of 'necessity, effectiveness and proportionality' (Article 8(2) ECHR) are not fulfilled, which undermines the State's argument on the margin of appreciation.[102]

Based on a recent report from the International Energy Agency (IEA),[103] the Netherlands remains one of the most fossil fuel and CO_2 intensive economies among IEA member countries and its GHG emissions have been growing.[104] Urgenda thus urges the State to take measures, referring to the analysis of the Dutch Supreme Court Advocate-General, Professor Spier:

> I have little doubt that the law as it stands, requires that it must be avoided, to the extent possible, that the fatal tipping point of an increase in global temperature with more than 2°C is passed . . . Given that the stakes are tremendously high for society at large, it seems morally and legally imperative

98 Reply, 120, [358].
99 Ibid, 40, [126].
100 Ibid, 24, [74].
101 Ibid, 49, [145], citing IPCC WGII AR5 Chapter 8, 3.
102 Ibid, 36, [113].
103 Energy Policies of IEA countries: the Netherlands, 2014 Review.
104 Executive Summary, 10, from the Report on Energy Policies of IEA countries: the Netherlands, 2014 Review.

> to arrive at the reductions needed to avoid that we pass the fatal tipping point, even if that would mean that some enterprises cannot survive and that States will have to force people and enterprises in their jurisdictions to reduce GHG emissions to the extent legally needed; even if that would necessitate painful measures. This is not to say that judges can or have to prescribe States (and arguably even enterprises) *how* they should achieve the reductions needed. States may choose the means, as long as they achieve the result.[105]

Urgenda discusses the measures taken by Denmark and Germany to illustrate that those countries did set emission targets envisaging a 40 per cent reduction by 2020, arguing that this limits the margin of appreciation: if these countries can achieve such a reduction level, why not the Netherlands? Moreover, the State cannot hide behind the margin of appreciation if it fails to meet its own standards or the precautionary principle.[106] Therefore, the State acts unlawfully by not reducing GHG emission levels as is scientifically needed to prevent exceeding the two degree threshold.[107]

Another way in which Urgenda argues the unlawfulness of the State's actions is by pointing to the criteria developed in Dutch case law concerning *endangerment.*[108] Exceeding the two degree threshold of global warming creates a dangerous situation, both within and outside the Dutch territory;[109] consequently the State's conduct endangers the lives of people and should be qualified as unlawful.[110]

Finally, Urgenda provides five reasons why the Dutch GHG emissions level is attributable to the Dutch State, basically assuming that the sovereign Dutch State can decide whether and to what extent GHG emissions are allowed in its territory, via the principle of 'system responsibility'.[111] The Dutch GHG contribution to the atmosphere is in an absolute sense and per capita significant,[112] and therefore the State should be pro-rata liable.[113]

International GHG reduction obligations – Regarding the State's remark that it meets its international reduction obligations as to GHGs, Urgenda's claims do not include an allegation that the State does not meet these duties. Rather, such obligations are insufficient to prevent dangerous climate change.[114] In other words,

105 Jaap Spier and Ulrich Magnus, *Climate Change Remedies, Injunctive Relief and Criminal Law Responses* (Eleven International Publishing, 2014) 83, 86, cited in Reply, 123, [365]. Italics in original.
106 See Reply, 142, [418], for Urgenda's conclusion on its human rights claim; 105, [319]; 106–15, [319], [322]–[345], for the supporting arguments.
107 Reply, 161, [480].
108 Criteria developed in the *Cellar Hatch* decision [*Kelderluik*-criteria] (Dutch Supreme Court, 5 November 1965) *LJN AB7079*, NJ 1966/136; Reply, 78–83, [228]–[248].
109 Reply, 84–5, [249]–[254].
110 Ibid, 91, [276].
111 Ibid, 102–4, [309]–[313].
112 Ibid, 98, [298].
113 Ibid, 99, [301].
114 Ibid, 128, [380].

the international community is acting inadequately to realise sufficient reduction levels and Urgenda rejects the argument that the State cannot be held liable for climate change as this is a subject for agreements at the international level.

In response to the State's comment that Urgenda fails to recognise adaptation measures taken by the Dutch State, the claimant explains that the purpose of the procedure is to prevent climate change rather than adapting to this danger,[115] based on Article 2 UNFCCC, which prescribes mitigation measures.[116]

Finally, in the Reply Urgenda amended its claims. It did so reluctantly, as it considers a 40 per cent GHGs reduction by 2030 insufficient in light of the two degree target. It also expressed the concern that mentioning a 40 per cent reduction by 2030 as an alternative claim may give the impression that a reduction of 40 per cent, or at least 25 per cent, by 2020 is not strictly necessary.[117] Apparently, Urgenda expects a more successful outcome for the case if it does not only claim that the State must realise the required reduction by 2020.[118]

2. *Kivalina:* challenging the liability of private corporations

Kivalia is a pioneering case that was commenced in 2007 in the US against energy and oil companies in connection with climate change,[119] inter alia to collect damages for the harm (loss of house and property) caused by Hurricane Katrina.[120] The main argument raised in this case was that the defendants had contributed to climate change, which had worsened the consequences of the hurricane. The focus of *Kivalina* is on recovering damages resulting from harm caused by climate change, also based on human rights infringements.

In other US cases, claimants requested that courts or, as the case may be, the Environmental Protection Agency (EPA), set carbon emission ceilings for the

115 Ibid, 21, [63].

116 Ibid, 22, [69].

117 Ibid, 182–83, [530]—[534].

118 Ibid, 183, [534].

119 *Kivalina* (2009). See also Kosolapova (2011) 195.

120 See also *Ned Comer et al v Murphy Oil USA, Inc et al*, Case No 12–60291 (US Court of Appeals for the Fifth Circuit, 14 May 2013) 3–9; *Ned Comer et al v Murphy Oil USA, Inc et al*, Case No 1:11CV220-LG-RHW (US District Court for the Southern District of Mississippi, Southern Division, 20 March 2012); *Comer v Murphy Oil USA, Inc*, CV 05–0436US (Supreme Court Order, 10 January 2011) 26; *Comer v Murphy Oil USA*, No 07–60756 (US Court of Appeals for the Fifth Circuit, 28 May 2010); *Comer v Murphy Oil USA*, 585 F.3d 855, 879 (US Court of Appeals for the Fifth Circuit (second panel), 2009); and *Ned Comer et al v Murphy Oil USA, Inc, et al*, Case No 1:05-cv-00436-LG-RHW (US District Court for the Southern District of Mississippi Southern Division, 30 August 2007). For a scholarly opinion, see Patrick Woods, *Reversal by Recusal? Comer v Murphy Oil USA, Inc and Mandatory Judicial Recusal Statements* (The Center for Judicial Process, 2011); Kosolapova (2011) 191–2.

industry. This happened, for instance, in *American Electric Power Co, Inc, et al v Connecticut et al (AEP)*[121] and *Massachusetts et al v Environmental Protection Agency et al.*[122] Although innovative legal strategies were attempted in these cases aimed at combating climate change, discussing them falls beyond the scope of this chapter, because the claimants did not rely upon human rights.

2.1. Facts and claims: the right to property

In *Kivalina*, the claimants were the Native Village of Kivalina, Alaska, and the City of Kivalina. Sea ice used to protect this Inuit community against winter storm waves. Because of higher temperatures, the ice now forms later and melts sooner, leaving the village unprotected and eroding the land. In 2004 and 2005, big storms hit Kivalina Island, which put people in danger and further eroded the coastline.[123] In 2007, people needed to evacuate and relocate, owing to another storm.[124] The claimants argued that Kivalina is being destroyed by flooding caused by climate change (global warming). They claimed monetary damages from the energy industry,[125] alleging that energy companies contribute to excessive carbon emissions, which in turn cause global warming.

Human rights are not as pivotal in *Kivalina* as in *Urgenda.* The claimants did not explicitly rely on fundamental rights as they sought compensation for damages in an action based on US federal common law. However, they alleged that the defendants' GHG emissions constituted an unreasonable interference with public rights, including the rights to use and enjoy public and private property: 'destruction of the entire village, is, on its face, clearly unreasonable'.[126] The claimants also stated that under federal common law the rights to the use and

121 *American Electric Power Co, Inc, and Others v Connecticut*, No 10-174, 582 F 3d 309 (US SCt, 20 June 2011).

122 US Supreme Court, *Massachusetts et al v Environmental Protection Agency et al*, Case No 05-1120 (2 April 2007). In this case, private organisations and the State of Massachusetts claimed that the EPA be ordered to regulate GHG emissions, including carbon dioxide, under the Clean Air Act. The US Supreme Court affirmed that the claimants had standing under Article III of the Clean Air Act; that Massachusetts is a sovereign State with a special position and interest as a large part of its territory is affected by GHG emissions; and that the EPA's denial to regulate GHG emissions and set standards presents a risk of harm to Massachusetts that is 'actual' and 'imminent', for instance, 'swallowing coastal land from Massachusetts, severe and irreversible changes to natural ecosystems, a significant reduction in winter snowpack with direct and important economic consequences, and increases in the spread of disease and the ferocity of weather events'.

123 See Julie Koppel Maldonado *et al*, 'The Impact of Climate Change on Tribal Communities in the US: Displacement, Relocation, and Human Rights', in Julie Koppel Maldonado, Benedict Colombi and Rajul Pandya (eds), *Climate Change and Indigenous Peoples in the US: Impacts, Experiences, and Actions* (Springer, 2014), 93.

124 Ibid, 605.

125 Defendants included ExxonMobil Corporation, BP, Chevron and Shell.

126 *Kivalina*, Petition for a Writ of Certiorari (24 February 2013) 5.

enjoyment of water not polluted by petroleum are public rights.[127] Kivalina's *displacement* is the injury ultimately sustained by the claimants.[128] The authors agree with Julie Koppel Maldonado[129] that various human rights are at stake here: the right to life, the right to health, the right to property, and the right to subsistence of the Kivalina citizens are jeopardised by climate change.[130]

2.2. Standing and the political question doctrine

In order to fulfil the criteria of causation required for standing under Article III of the US Constitution, in *Kivalina* the claimants relied upon the principles developed in the Clean Water Act cases for contributing to alleged injuries. In these cases, the courts accepted a substantial likelihood that companies' conduct caused harm once the prescribed federal limit on discharges was exceeded.

The defendants argued that deciding on global warming claims intrudes upon the political branches.[131] According to them, successfully combating global warming requires an international approach, and it would harm the bargaining position of the US in its foreign policy if courts were to establish caps on GHG emissions.[132]

The District Court in *Kivalina* noted that the need to see the environmental problem in an international context does not make it a non-justiciable issue. It pointed out that no provision in the Constitution justifies the conclusion that only the executive or legislative branch has 'the power to make the final determination regarding air pollution or global warming'.[133]

However, the Court assumed that the Kivalina claimants had not convincingly argued that the question as to whether the oil companies should bear the costs of contributing to global warming is a *legal issue*. The Court held that it cannot answer the question of what is unreasonable in the context of carbon emissions, because this requires balancing the social utility of the defendants' conduct with the harm

127 *Kivalina*, US District Court, Plaintiffs' Corrected Consolidated Memorandum of Points and Authorities in Opposition to Defendants' Motions to Dismiss Pursuant to Fed.r.civp 12(b)(1) and 12(b)(5), Case No 4:08-cv-01138-SBA (9 December 2008) 56.

128 See US Court of Appeals for the Ninth Circuit, *Native Village of Kivalina et al v ExxonMobil Corp et al*, Case No 09-17490 (21 September 2012) 11657. The Court acknowledges displacement because of the rising sea.

129 Julie Koppel Maldonado *et al* (2014) 93.

130 See also Caney (2011); Jon Barnett, 'Human Rights and Vulnerability to Climate Change', in Humphreys (2011) 257, 258–9, 267.

131 See Ashley Breakfield, 'Political Cases or Political Questions: the Justiciability of Public Nuisance Climate Change Litigation and the Impact on Native Village of Kivalina v ExxonMobil' (2011) 17(30) *Hastings W-Nw Journal of Environmental Law and Policy* 39.

132 *Kivalina* (2009) 8.

133 Ibid, 9.

it inflicts,[134] and it is up to the legislator to determine where the fault and costs of global warming are to be allocated.[135]

The District Court in *Kivalina* finally pointed out that there are no federal standards limiting GHG emissions. Therefore, no presumption arose concerning the likelihood that the oil companies' conduct causes harm to the Kivalina population.[136] Furthermore, the Court noted that there are a multitude of alternative culprits allegedly responsible and considered that the chain of causation between the alleged conduct (GHG emissions) and the harm sustained (erosion of Kivalina) was too weak.[137] The Court dismissed the claims thereby referring to the 'Political Question Doctrine'.[138]

2.3. Appeal to the Ninth Circuit and the separation of powers

The first instance decision delivered in *Kivalina* was appealed to the Ninth Circuit Court of Appeals, invoking the federal common law of public nuisance. The Ninth Circuit Court explained that federal common law develops when courts must consider federal questions that are not answered by statutes,[139] for instance, regarding transboundary pollution suits. A successful claim needs to prove 'unreasonable interference' with the use of a public right, causing 'substantial and widespread harm'. Nonetheless, referring to *AEP*,[140] the Ninth Circuit Court determined that Congress has addressed the issue of GHG emissions, leaving no room to argue on the basis of federal common law. It is noteworthy that the Kivalina-issue differs slightly from *AEP*, that is, the Kivalina claimants sought damage for the harm caused by past emissions rather than a reduction of future emissions.[141] However, in both cases separation of powers between the judiciary and legislative branches is the main issue and Congress 'spoke directly' by empowering the EPA to regulate GHG emissions based on the Clean Air Act. The Ninth Circuit Court therefore confirmed the District Court's decision, holding that 'the solution to Kivalina's dire circumstance must rest in the hands

134 Ibid, 14.

135 Ibid, 15. Similarly, in *State of California v General Motors Corporation et al*, Case No C06–05755 MJJ (17 September 2007) 15, referring to *Massachusetts v Environmental Protection Agency* (2007), the Court for the Northern District of California held that determining carbon emission *standards* lies with the political branches and not with courts. Compare also *Ned Comer* (2013).

136 *Kivalina* (2009) 19.

137 Ibid, 22. Compare with *Potash Mines* (1976); *Ned Comer* (2007).

138 The US Supreme Court has held that it is not up to federal courts to deal with disputes involving political questions (Article III of the Constitution). The case law explains that 'a non-justiciable political question exists when, to resolve the dispute, the court must make a policy judgment of a legislative nature, rather than resolving the dispute through legal and factual analysis' (US Court of Appeals for the Ninth Circuit, *Equal Employment Opportunity Commission v Peabody W Coal Co*, 400 F.3d 774).

139 *Kivalina* (2012) 11652, 11654.

140 *AEP* (2011).

141 *Kivalina* (2012) 11655.

of the legislative and executive branches of our government, not the federal common law'.[142]

On 20 May 2013, the Supreme Court declined to hear the claimants' petition for a writ of certiorari.[143]

3. *Gbemre*: acting against the State and private corporations for localised damage

In 2005, Jonah Gbemre submitted a claim against the Shell Petroleum Development Company of Nigeria Limited (Shell), the Nigerian National Petroleum Corporation (NNPC) and the Attorney-General of the Federation of Nigeria (Attorney-General). He acted on his own behalf and on behalf of the Iwherekan Community in the Delta State of Nigeria.[144]

Gbemre asserted that the gas flaring activities by the oil companies were in violation of his and the community's fundamental rights and pursued their ending. The claims were based on the Nigerian Constitution and the African Charter on Human and Peoples' Rights (African Charter),[145] specifically the 'right to respect for their lives and dignity of their persons and to enjoy the best attainable state of physical and mental health as well as a right to a general satisfactory environment favourable to their development secured by Articles 4, 16 and 24 [of the African Charter]'.[146] The case was tried before the Federal High Court of Nigeria in the Benin Judicial Division (the Court), which issued a final order on 14 November 2005 (the Order).

The analysis of this case is mainly based upon the Court Order as well as various documents submitted by the claimant, which have been made publicly accessible.[147] In contrast, the documents submitted by the defendants are not

142 Ibid, 11657.

143 US Supreme Court, *Native Village of Kivalina v ExxonMobil Corp et al*, 12–1072, Order list: 569 US (20 May 2013).

144 *Gbemre*, Order (2005). See also Abdulkadir Bolaji Abdulkadir, 'Gas Flaring in the Niger Delta of Nigeria: a Violation of the Right to Life and Comment on the Case of Johnah Gbemre v Shell Petroleum Development Company of Nigeria Limited' (2014) 22(1) *IIUM Law Journal* 75; Eferiekose Ukala, 'Gas Flaring in Nigeria's Niger Delta: Failed Promises and Reviving Community Voices' (2011) 2(1) *Journal of Energy, Climate and Environment* 97.

145 *African Charter on Human and Peoples' Rights (Ratification and Enforcement) Act*, Cap A9, Volume 1, Laws of the Federation of Nigeria, 2004. See Organisation of African Unity (OAU), *African Charter on Human and Peoples' Rights ('Banjul Charter')*, CAB/LEG/67/3 rev 5 (27 June 1981) 21 ILM 58 (1982).

146 *Gbemre*, Verifying Affidavit pursuant to Order 1 rule 2(3) of the Fundamental Rights (Enforcement Procedure) Rules (11 July 2005) [5] (hereafter: Verifying Affidavit).

147 Besides the Order and the Verifying Affidavit, the following Court documents were submitted in the *Gbemre* Suit No FHC/CS/B/153/2005 and are publicly accessible: Motion ex parte under section 46(1) of the Constitution of the Federal Republic of Nigeria, 1999; Order 1 rule 2(1)(2)(3) of the Fundamental Rights (Enforcement Procedure) Rules dated 11 July 2005 (hereinafter: Motion); Statement pursuant to Order 1 rule 2(3) of the Fundamental Rights (Enforcement Procedure) Rules dated 11 July 2005 (hereinafter: Statement); Notice of Motion for an Order Enforcing Fundamental Right under Order 2 Rule 1(1) of the Fundamental Rights (Enforcement Procedure) Rules dated 25 July 2005 (hereinafter: Notice).

available for inspection; consequently, no comments can be made regarding the argumentation in the defence statements. However, the purpose of examining *Gbemre* is to compare the litigation strategy with *Urgenda* and *Kivalina*. To that end, available information is useful.

3.1. Standing: collective action

Standing issues played no particular role in *Gbemre*. The Court decided that the claimant was 'properly granted leave to institute these proceedings in a representative capacity for himself and for each and every member of the Iweherekan Community in Delta State in Nigeria'. The Court declared that it had inherent jurisdiction to grant leave to the claimants 'who are bona fide citizens and residents of the Federal Republic of Nigeria, to apply for the enforcement of their fundamental rights to life and dignity of the human person as guaranteed by Sections 33 and 34 of the Constitution of the Federal Republic of Nigeria, 1999.'[148]

3.2. Gbemre's claims: the rights to life, health and environment

In *Gbemre*, the claimant requested the Court to issue a declaration stating that:

i. The constitutionally guaranteed fundamental rights to life and dignity of human persons[149] inevitably include the right to a clean, poison-free, pollution-free and healthy environment;
ii. The actions of the oil companies in continuing to flare gas are in violation of the claimants' fundamental rights to life (including a healthy environment) and dignity; and
iii. The failure of the oil companies to carry out an EIA in the claimants' community concerning the impact of their gas flaring activities is a violation of the Environmental Impact Assessment Act (EIA Act)[150] and contributes to the violation of the said fundamental rights.[151]

Furthermore, by virtue of section 1(3) of the Constitution the claimant requested a declaration that the Associated Gas Re-Injection Act and Regulations,[152] under which gas flaring activities may be allowed in Nigeria, are inconsistent with the

148 *Gbemre*, Order, 1, [2].
149 As provided for in sections 33(1) and 34(1) of the 1999 Constitution of the Federal Republic of Nigeria and reinforced by Articles 4, 16 and 24 of the African Charter on Human and Peoples' Rights (Ratification and Enforcement) Act, Cap A9, Vol 1, Laws of the Federation of Nigeria (2004).
150 Section 2(2) of the Environmental Impact Assessment Act, Cap. E12, Vol 6, Laws of the Federation of Nigeria (2004).
151 *Gbemre*, Statement, [B.1–3].
152 Section 3(2)(a)(b) of the Associated Gas Re-Injection Act, Cap. A25, Vol 1, Laws of the Federation of Nigeria (2004) and section 1 of the Associated Gas Re-Injection (Continued Flaring of Gas) Regulations S.1. 43 of 1984.

claimant's fundamental rights to life and dignity as enshrined in the Constitution and the African Charter, and are therefore unconstitutional, null and void.[153]

Finally, the claimant requested an 'order of perpetual injunction' to restrain the defendants Shell and NNPC themselves, their agents or workers or whosoever from further flaring gas in the claimants' community.[154]

3.3. Gbemre's statement of facts: respecting the environment

In the Verifying Affidavit,[155] Gbemre explained that gas flaring is a practice by oil companies that, on finding oil mixed with gas, burn the latter, which leaves them with the more lucrative component, the oil. Gbemre sketched the facts and alleged that gas flaring impacts the fundamental rights of the community's citizens:

> 1. [It] Poisons and pollutes the environment in our said community as it leads to the emission of carbon dioxide, the main greenhouse gas; the flares contain a cocktail of toxins that affect our health, lives and livelihood;
> 2. Exposes us to an increased risk of premature death, respiratory illnesses, asthma and cancer;
> 3. *Contributes to adverse climate change* in our community as it emits carbon dioxide and methane which causes warming of our environment;
> 4. Pollutes our food and water;
> 5. Causes painful breathing, chronic bronchitis, decreased lung function and death in our community;
> 6. Reduces our crop production and adversely impacts on our food security;
> 7. Causes acid rain . . . [which] acidifies our lakes and streams and damages our vegetation.[156]

3.4. Merits: environmental protection under constitutional and international human rights norms

The first legal basis for the claims in *Gbemre* was provided by human rights, including the right to a healthy, pollution-free air and environment.[157] Gbemre asserted that the damaging impacts gas flaring causes in his community are in contravention of these rights. The claimant pointed out that under the Nigerian Constitution he and the people he represents have fundamental rights to life and dignity of the human person. These constitutional claims are supported by the rights protected under the African Charter to life and dignity (Article 4), the right to enjoy the best attainable state of physical and mental health (Article 16), and the right to a general satisfactory environment favourable to peoples' development

153 *Gbemre*, Statement, [B.4].
154 Ibid, [B.5].
155 *Gbemre*, Verifying Affidavit.
156 *Gbemre*, Notice, 5–6, [11], emphasis added.
157 Verifying Affidavit, [19].

(Article 24), which were enacted in Nigerian law under the African Charter on Human and Peoples' Rights (Ratification and Enforcement) Act (2004).

Other grounds for seeking relief were based on the Environmental Impact Assessment Act, the Associated Gas Re-Injection Act, and the Federal Environmental Protection Agency Act. Indeed, the claimant referred to the lack of an EIA, required by law, but never undertaken by the defendants. Gbemre contended that the defendants had no valid Ministerial Certificates providing authorisation for flaring gas in the applicants' community, as they never ascertained the damaging consequences of gas flaring, but only took account of their economic interests.[158] The Federal Environmental Protection Agency Act makes gas flaring a crime 'the continuation of which should be discouraged and restrained by the court',[159] and the claimant requested the court to intervene.[160]

3.5. Court ruling: a positive outcome, but a difficult enforcement

In its Order in *Gbemre*, the High Court of Nigeria declared that the fundamental rights to life and dignity of the human person 'inevitably include the rights to a clean, poison-free, pollution-free and healthy environment'.[161] The Court held that the actions of the oil companies in continuing to flare gas in the course of their oil exploration and production activities in the claimant's community are 'a violation of the fundamental rights to life (including a healthy environment) and dignity of human persons' as enshrined in the Constitution.[162]

The Court concluded that the failure of the oil companies to carry out an EIA in the claimant's community concerning the effects of their gas flaring activities was 'a violation of Section 2(2) of the Environmental Impact Assessment Act . . . and contributed to the violation of . . . said fundamental rights'.[163] Hence, the Court ordered that the oil companies be restrained from further flaring gas in the Iwherekan Community and take immediate steps to stop gas flaring.[164]

158 Environmental Impact Assessment Act, Cap E12, Vol 6, Laws of the Federation of Nigeria (2004). A ministerial gas-flaring certificate can be obtained on the basis of the Associated Gas Re-Injection Act, Cap. A25, Vol 1, Laws of the Federation of Nigeria, 2004.

159 Statement, [7].

160 Verifying Affidavit, [20].

161 Order, 2, [3] and Dictum, 3, [1].

162 Ibid, Order 2, [4]; Dictum, 3, [1].

163 Ibid, Order, Dictum, 3, [3].

164 Ibid, Order, Dictum, 4, [5]. A consistent procedural approach to climate change in domestic litigation has been taken in three Australian cases: *Gray v Minister for Planning, Director-General of the Department of Planning and Centennial Hunter Pty Ltd* (Land and Environment Court of New South Wales, Australia, 27 November 2006) 152 LGERA 258; *Australian Conservation Foundation v Latrobe City Council, Minister for Planning* (Victorian Civil and Administrative Tribunal, 29 October 2004 VCAT 2029); *Wildlife Preservation Society of Queensland Proserphine/Whitsunday Branch Inc v Minister for the Environment and Heritage, Bowen Central Coal Management and QCoal Pty Ltd* (Federal Court of Australia, 15 June 2006) QUD 216 OF 2005. In these cases, based on environmental law, the claimants challenged with a positive outcome the failure of public decision-makers to require detailed information concerning the global warming impact of proposed new coal mines as part of EIA processes. See also Anna Rose, 'Gray v Minister for Planning: the Rising Tide of Climate Change Litigation in Australia' (2007) 29 *Sydney Law Review* 725.

Furthermore, the Court declared that certain provisions of the Associated Gas Re-Injection Act and Regulations are inconsistent with the fundamental rights of the claimant and his community, and considered them unconstitutional, null and void.[165] The third defendant, that is, the Attorney-General, was ordered to speedily amend the Act and the Regulations,[166] especially in view of the fact that the 'Act even by itself also makes the said continuous gas flaring a crime having prescribed penalties in respect thereof'.[167]

However, despite the positive outcome of the case for Gbemre and his community, the Court Order has not succeeded in guaranteeing the protection of a pollution-free environment.[168] Apparently, the defendants have not complied with the Court's order. Further, in April 2006, after the filing of contempt of court proceedings against Shell, the company was granted a 'conditional stay of execution', waiving the duty to comply with the Court order to cease flaring.'[169]

4. From *Gbemre* to *Urgenda*: State responsibility to take action?

The analysis of *Urgenda*, *Kivalina* and *Gbemre* demonstrates that action based on human rights for anthropogenic GHG emissions is possible against private companies and States where environmental consequences are localised (*Gbemre*). As Gbemre could demonstrate specific harm and establish causation, the case had a successful outcome for him and the community he represented. The Court followed the human rights approach argued by Gbemre and ordered that the oil companies cease their gas flaring activities. In *Urgenda* the claims are also based on human rights. In both cases, the right to life and the right to a healthy environment are invoked. The difference is that the right to a healthy environment is explicitly recognised in the ACHPR, whereas it is only implicitly recognised in the case law of the ECtHR concerning Article 8 ECHR (Right to Private and

165 Provision 3(2)(a)(b) (Verifying Affidavit, [20]); Order, Dictum, 4, [4].

166 In this part of the Court Order, the Act is confusingly referred to as 'Associated Gas Regulation Act'.

167 Order, Dictum, 5, [6].

168 Frank Maes *et al*, *Biodiversity and Climate Change: Linkages at International, National and Local Levels* (Edward Elgar Publishing, 2013) 17–20. These scholars provide some critical remarks regarding this far-reaching judgment. They also provide information on the subsequent proceedings and discuss how Shell and the Court staff obstructed access to the Court. See also Kaniye Ebeku, 'Constitutional Right to Healthy Environment and Human Rights Approaches to Environmental Protection in Nigeria: Gbemre v Shell Revisited' (2007) 16(3) *RECIEL* 312 ff.

169 Friends of the Earth International, Press release, *Shell Fails to Obey Court Order to Stop Nigeria Flaring, Again* (2 May 2007). According to a press release from Environmental Rights Action/Friends of the Earth, Gbemre was arrested by soldiers guarding gas flaring sites (Environmental Rights Action/Friends of the Earth, Nigeria, *ERA Flays Arrest of Activists, Journalists* (2 September 2008)). This arrest took place in 2008, after the judge of the Federal High Court in Benin, which issued the 2005 Order, had been transferred and the case file was reported as lost. See also Eferiekose Ukala (2011) 111; Abdulkadir Bolaji Abdulkadir (2014).

Family Life). Furthermore, the harm suffered by Gbemre and his community was more tangible and individualised than the harm stated by Urgenda. It is difficult for Urgenda to itemise the harm caused by Dutch GHG emissions, because for a large part the damage will materialise in the future. The question in *Urgenda* is thus whether it is necessary to substantiate actual harm at an individual level for establishing liability under Dutch tort law. Indeed, the claimant only requests the Court to issue certain declaratory statements and an injunctive order aimed at preventing rights violations and harm in the future. As in *Gbemre*, in *Kivalina* the damage was also tangible and localised, that is, the Court acknowledged that the community was displaced because of the rising sea, but it still proved difficult for the claimants to obtain a positive result in their litigation against the oil, coal and power companies. The causal relationship between the damage and the defendants was less obvious and direct than in *Gbemre*.

When the impact of GHG emissions is difficult to ascertain, as in *Urgenda* and *Kivalina*, a human rights-based action in a domestic court is challenging. Standing can be a first obstacle. As Dutch law offers a special approach for collective action, Urgenda bases its claims thereon, and additionally relies on the independent standing of the co-claimants, that is, 886 individual citizens. The claimants seem to fulfil the technical requirements for collective action, but the question of submitting a claim on behalf of future generations requires the Court's consideration. Although this question has been positively answered in Dutch case law, the facts presented by Urgenda slightly differ from previous cases. In *Gbemre*, the standing of the claimant was accepted, as well as on behalf of his community. In *Kivalina*, the claimants had to present sufficient facts and grounds to prove standing. This constituted an extra hurdle for the Kivalina community to be heard. The Ninth Circuit Court concluded that the claimants could not base their claims on federal common law. The Court bypassed the fact that the Kivalina community had already suffered damages.

Second, an action against private companies is complicated by the principle of causation. In *Kivalina*, it turned out to be difficult to clarify the exact contribution of the oil and energy sector to the aggravation of climate change, specifically the contributions of individual companies within these sectors, as many parties worldwide contribute thereto. However, an action might be less problematic at the aggregate State level based on pro rata responsibility. This is the argument Urgenda makes when it refers to *Potash Mines* (Dutch Supreme Court). Related is Urgenda's argument on attribution of responsibility to the Dutch State, emphasising that the State is the sovereign decision-maker, which has control over the level of emissions originating from its territory, along the lines of the decision delivered by the ICJ in *Corfu Channel*.

Third, the political question doctrine can be an obstacle in a domestic climate change case. The question to be answered in *Urgenda* is to what extent the legislative and executive branches have discretion in (not) adopting climate change mitigation policies, and in changing them yearly, in view of the imminent impacts of climate change on the enjoyment of human rights. In *Gbemre*, this doctrine did not prevent a decision on the merits. On the contrary, the Court ordered, inter alia, that the Advocate-General start a process to amend certain legal provisions, so as to bring them in line with the fundamental rights provided for in the Nigerian Constitution. In *Urgenda*, the claims do not include a request to review certain acts or regulations. This is inherent to differences in domestic legal systems, as Dutch courts are not allowed to conduct a constitutional review. Nonetheless, they can declare that a certain act or provision of an act is in violation of human rights protected by international human rights treaties, such as the ECHR, the ICCPR, and the ICESCR. In *Kivalina*, although it stated that the US Constitution does not support the conclusion that only the executive or legislative branch of the government has 'the power to make the final determination regarding air pollution or global warming', the Court finally considered it to be the legislator's task to determine where the fault and costs of global warming are to be allocated. Consequently, the political question doctrine impeded a decision on the merits. In *Urgenda*, the claimants emphasise that they invoke protection of human rights against imminent danger rather than practising politics, and that Dutch law on this point is fairly different from the US political question doctrine. Dutch precedents show that, as long as the claimants do not specify the exact actions expected from the executive or the legislature, courts seem to be more willing to award claims against the government than in the US.[170]

Finally, regarding the type of claims made by the claimants, it is interesting that in *Gbemre* the request was made for a declaratory ruling that the rights to life and dignity include 'the right to a clean, poison-free, pollution-free and healthy environment'. Such a declaratory order has not been requested in *Urgenda*, but might have stimulated a discussion on the adoption of this most fundamental right in the Netherlands.

Conclusion

The 'database of cases' thus far available is too limited to draw far-reaching conclusions on critical success factors or best practices in seeking protection for human rights in view of global warming.

Within this context, *Urgenda* proves particularly relevant, because there are no other cases where State responsibility for human rights breaches arising from GHG emissions has been discussed so openly. At the international level, two Inuit petitions to the IAComHR are so far the only international cases where State responsibility for climate change has been in issue. However, the first petition was dismissed without being fully considered and the status of the second is uncertain.[171]

170 Chris van Dijk (2011) 212–13.

171 IAComHR, *Inuit Petition v United States of America*, No 1413–05 (16 November 2006) and *Petition to the Inter-American Commission on Human Rights Seeking Relief from Violations of the Rights of Arctic Athabaskan Peoples Resulting from Rapid Arctic Warming and Melting Caused by Emissions of Black Carbon by Canada* (23 April 2013).

The allegations and motivations in *Urgenda* assert that problems of imputation and causation should be overcome. The case is nevertheless still pending and only the final decision will shed light on these crucial issues. It is clear that the claimants have a strong feeling of urgency with regard to the reduction of GHG emissions in the Netherlands, whereas the Dutch State interprets its legal obligations in a restrictive manner.[172]

172 Research for this chapter was concluded at the beginning of 2015. On 24 June 2015, the District Court of The Hague delivered its decision in *Urgenda*. The Court ruled that the Dutch State is obliged to cut back GHG emissions in the Netherlands by at least 25 per cent by 2020, compared to emission levels in 1990, so as to minimise global warming and protect its people. An English version of the decision is available at <http://uitspraken.rechtspraak.nl/inziendocument?id=ECLI:NL:RBDHA:2015:71968&keyword=Urgenda>.

Conclusion

A positivist-oriented legal analysis of climate change and human rights shows that relevant theoretical and practical obstacles currently determine a divide between the two poles of the relationship. To a certain extent, climate change policies may already be interpreted as human rights measures, but only marginally can it be said that they are informed by fundamental claims. Key substantive and institutional issues hamper the possibility of tearing down the 'wall'.

Substantively, the individually oriented nature of first and second generation human rights makes it difficult to identify specific 'victims' of anthropogenic GHG emissions. Objectively, the causal link between GHG emissions and human rights breaches is blurred by a set of circumstances that lead to violations. Subjectively, imputing GHG emissions in breach of human rights to a given legal person is not simple, in light of the number of physical and legal entities acting globally. Such problems are perfectly exemplified by the right to life: is it possible to assume that the death of a person caused by Typhoon Haiyan is ultimately imputable to GHGs emitted in specific countries? However, this is less true of other fundamental rights. For instance, establishing a link between climate change, drought and loss of biodiversity is more straightforward, which entails relevant implications for the rights to water, food, health and culture. In particular, because of their welfare nature and lesser resistance to extraterritoriality second generation human rights emerge as precious tools in the struggle against climate change.[1] This also applies to the collective right to self-determination, since it is not excessively complex to prove an objective link between anthropogenic GHG emissions, rising sea level and loss of land, notably in low-lying island States. Within this framework, vulnerable groups, specifically indigenous peoples, are the main victims of climate change. Besides imputation, and thus responsibility, restitution and compensation, human rights can be exploited as interpretive and regulatory instruments, taking into account the needs of specific groups and competing rights; think, for instance, about the constraints implied by the obligation to fulfil human rights in terms of adaptation and the fact that the right to self-determination could be relied upon to preserve existing claims over the territorial sea and the EEZ, beyond Statehood.

1 OHCHR, *Report on the Relationship between Climate Change and Human Rights*, UN Doc A/HRC/10/61 (15 January 2009) 25, [75]–[77].

Institutionally, with regard to law-making, at the international level UNFCCC organs have not yet concretely taken human rights into account, despite some explorative studies and resolutions on the issue by bodies such as the HRC, CESCR and States, especially low-lying island countries. Investment and trade regimes, particularly the WTO, seem relatively more advanced as to consideration of fundamental rights. Regionally, action can be hampered by the lack of a coherent legal framework, namely in the Asia-Pacific. When consistent regulation exists, its practical implementation can prove difficult, as in Africa and America. The latter case is particularly interesting because the American continent is partially included in the Arctic Circle, which is inhabited by indigenous peoples and severely affected by climate change. Paradoxically, Europe is the least affected region, but has taken the relationship between climate change and fundamental rights most seriously, despite some inconsistencies arising from the separation of human rights protection between the EU and the Council of Europe system. From the standpoint of adjudication and enforcement, the case law of human rights jurisdictions in Europe, Africa and America provides useful tools to address climate change. However, between 2005 and 2015 no clear answer has yet been given to the only cases where the impact of global warming on fundamental rights has been raised, that is, the two Inuit petitions to the IAComHR. Maybe, the long-term effects of climate change necessitate long-term procedures. Domestically, in addition to the above-mentioned substantial problems, the political question doctrine seems to be a major obstacle to crossing the barrier between climate change and human rights; these issues are nonetheless challenged in pending disputes.

Within such a context, collective third generation human rights emerge as important means to overcome the divide between climate change and human rights. In particular, despite its substantively uncertain legal status the procedural components of the right to development should facilitate participation in decision-making and access to justice. Furthermore, a human rights-oriented interpretation of the no-harm rule, that is, the recognition of a human right to (a climatically sustainable) environment, would permit a simplification of the causal link between anthropogenic GHG emissions and fundamental rights' violations. Whereas imputation may prove difficult when faced with innumerable private corporations, proportionate attribution of responsibility does not seem impossible with regard to around 200 States. Unsustainable emissions in breach of the limits established under the UNFCCC and the Kyoto Protocol, or a new climate agreement, possibly assessed in light of fundamental rights, would thus trigger human rights protection mechanisms. Along these lines, for instance, the justiciability of Article 11 of the San Salvador Protocol to the ACHR could provide a new lens for considering the Inuit petitions. Given the limited efficacy of human rights protection systems, this would basically result in giving voice to vulnerable people.

Is the tide turning?

Ottavio Quirico and
Mouloud Boumghar

Appendix

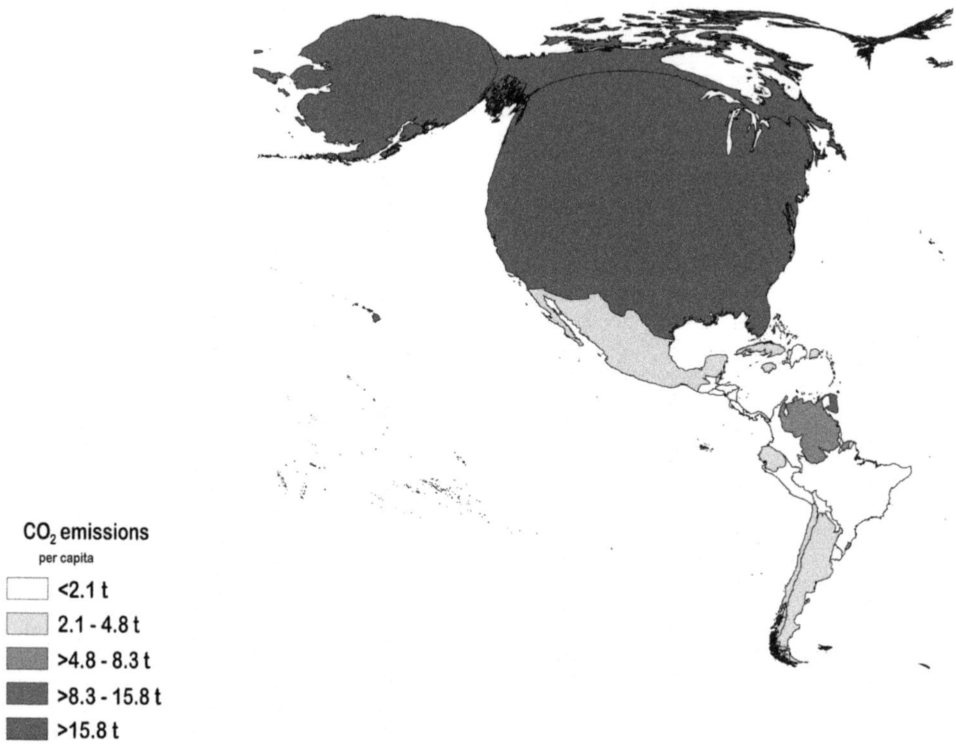
CO_2 emissions
per capita
<2.1 t
2.1 - 4.8 t
>4.8 - 8.3 t
>8.3 - 15.8 t
>15.8 t

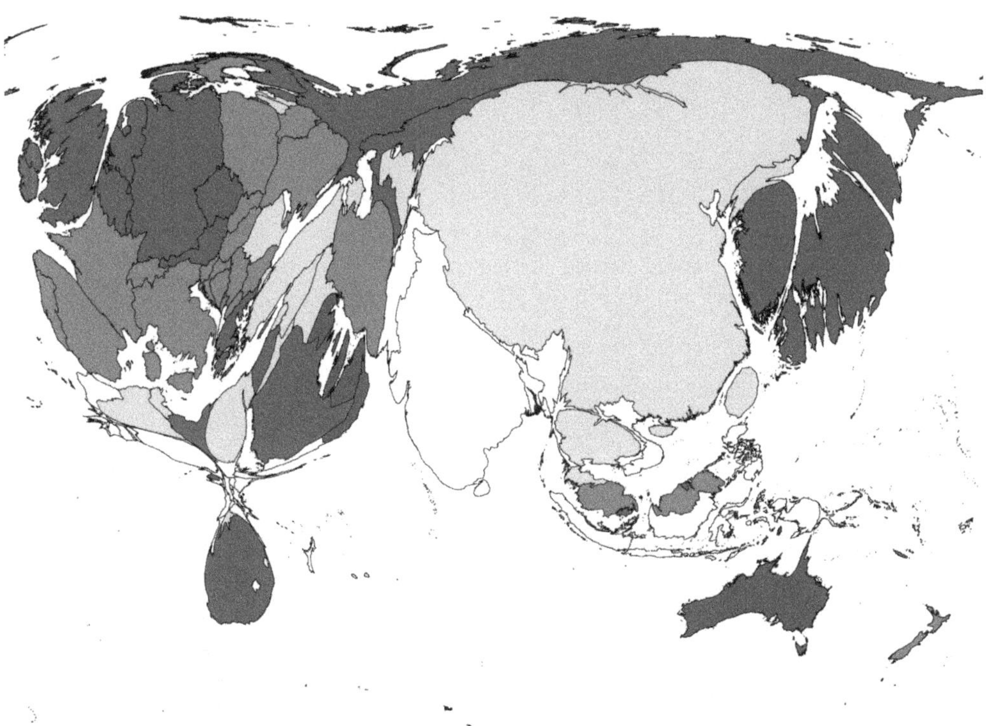

Documents

Cases

Bibliography

Adams, Barbara and Luchsinger, Gretchen, *Climate Justice for a Changing Planet: a Primer for Policy Makers and NGOs* (UN Non-Governmental Liaison Service, 2009)

Aerni, Philipp *et al*, 'Climate Change and International Law: Exploring the Linkages between Human Rights, Environment, Trade and Investment' (2010) 53 *Ger Yb Int'l L* 139

Ahlgren, Ingrid, Yamada, Seiji and Wong, Allen, 'Rising Oceans, Climate Change, Food Aid, and Human Rights in the Marshall Islands' (2014) 1 *Health and Human Rights Journal* 69

Ailincai, Mihaela and Lavorel, Sabine, *Exploitation des ressources naturelles et protection des droits de l'homme* (Pedone, 2013)

Alland, Denis *et al* (eds), *Unité et diversité du droit international – Mélanges en l'honneur du Professeur Pierre Marie Dupuy* (Brill, 2013)

Alston, Philip, 'Conjuring up New Human Rights: a Proposal for Quality Control' (1984) 78(3) *AJIL* 607

Alston, Philip, 'Making Space for New Human Rights: the Case of the Right to Development' (1988) 1 *Harvard Human Rights Year Book* 3

Alvarez, José E, 'The Return of the State' (2011) 20 *Minn J Int'l L* 223

Anton, Donald K and Shelton, Dinah, *Environmental Protection and Human Rights* (CUP, 2011) 863

Atapattu, Sumudu, *Climate Change in South Asia: towards an Equitable Legal Response within a Framework of Sustainable Development and Human Security* (IDLO, 2011)

Baddeley, Shane, Cheng, Peter and Wolfe, Robert, 'Trade Policy Implications of Carbon Labels on Food' (2012) 13(1) *The Estey Centre Journal of International Law and Trade Policy* 59

Baetens, Freya, *Foreign Investment Law and Climate Change* (International Development Law Organisation, Sustainable Development Law on Climate Change Legal Working Paper No 1, 2010)

Baetens, Freya (ed), *Investment Law within International Law — Integrationist Perspectives* (Cambridge University Press, 2013)

Bakker, Christine and Francioni, Francesco (eds), *The EU, the US and Global Climate Governance* (Ashgate, 2014)

Barnett, Jon and Adger, Neil, 'Climate Change, Human Security and Violent Conflict' (2007) 26 *Political Geography* 639

Beck, Ulrich, *Risikogesellschaft, Auf dem Weg in eine andere Moderne* (Suhrkamp, 1986)

Bell, Derek, 'Climate Change and Human Rights' (2013) 4 *WIREs Climate Change* 159

Bellamy, Gretchen, Blecher, Lara and Kaymar-Stafford, Nancy (eds), *Corporate Responsibility for Human Rights Impacts* (American Bar Association, 2014)

Betts, Alexander, *Survival Migration: Failed Governance and the Crisis of Displacement* (Cornell University Press, 2013)

Biermann, Frank and Boas, Ingrid, 'Preparing for a Warmer World: towards a Global Governance System to Protect Climate Refugees' (2010) 10(1) *Global Environmental Politics* 60

Biondi, Andrea and Eeckhout, Piet, *EU Law after Lisbon* (Oxford University Press, 2012)

Blake, Janet and Boer, Ben, 'Human Rights, the Environment and the Tehran Declaration' (2009) 39/6 *Environmental Policy and Law* 302

Bobbio, Norberto, *Teoria Generale del Diritto* (Giappichelli, 1993)

Boer, Ben (ed), *Environmental Law Dimensions of Human Rights* (OUP, 2015)

Boer, Ben and Clarke, Pepe, *Ecosystem-Based Adaptation to Climate Change in the Pacific Islands*, (SPREP, 2011)

Bolaji Abdulkadir, Abdulkadir, 'Gas Flaring in the Niger Delta of Nigeria: a Violation of the Right to Life and Comment on the Case of Johnah Gbemre v Shell Petroleum Development Company of Nigeria Limited' (2014) 22(1) *IIUM Law Journal* 75

Bonnitcha, Jonathan and McCorquodale, Robert, *Is the Concept of 'Due Diligence' in the Guiding Principles Coherent?* (SSRN Paper, 2013)

Boogaard, Geerten, *Het wetgevingsbevel, over constitutionele verhoudingen en manieren om een wetgever tot regelgeving aan te zetten* (Wolf Legal Publishers, 2013)

Bosselmann, Klaus, *The Principle of Sustainability: Transforming Law and Governance* (Ashgate, 2008)

Boute, Anatole, 'The Potential Contribution of International Investment Protection Law to Combat Climate Change' (2009) 27 *J En & Nat Res L* 333

Boyd, David R, *The Environmental Rights Revolution: a Global Study of Constitutions, Human Rights, and the Environment* (University of British Columbia Press, 2012)

Boyd, David R, *The Constitutional Right to a Healthy Environment* (online, 2013)

Boyle, Alan, 'Environment and Development: Accountability through International Law' (1993) 12 *Third World Legal Studies* 95

Boyle, Alan, 'Human Rights and the Environment: Where Next?' (2012) 23 *EJIL* 613

Boyle, Alan and Anderson, Michael R (eds), *Human Rights Approaches to Environmental Protection* (OUP, 1996)

Braimoh, Ademola K and Qing Huang, He, *Vulnerability of Land Systems in Asia* (Wiley, 2014)

Breakfield, Ashley, 'Political Cases or Political Questions: the Justiciability of Public Nuisance Climate Change Litigation and the Impact on Native Village of Kivalina v ExxonMobil' (2011) 17(30) *Hastings W-Nw Journal of Environmental Law and Policy* 39

Brenton, Paul, Edwards-Jones, Gareth and Friis Jensen, Michael, 'Carbon Labelling and Low-income Country Exports: a Review of the Development Issues' (2009) 27–3 *Development Policy Review* 243

Brown, Abbe (ed), *Environmental Technologies, Intellectual Property and Climate Change: Accessing, Obtaining and Protecting* (Edward Elgar, 2013)

Brown, Chester and Miles, Kate (eds), *Evolution in Investment Treaty Law and Arbitration* (Cambridge University Press, 2011)

Brown Weiss, Edith, 'Our Rights and Obligations to Future Generations for the Environment' (1990) 84 *Am J Int'l Law* 198

Brown Weiss, Edith (ed), *Environmental Change and International Law: New Challenges and Dimensions* (United Nations University Press, 1992)

Bruce, Stuart A, *An International Court for the Environment and Climate Change Law Compliance and Enforcement* (AID, 2012)

Bruch, Carel, Cocker, Wole and VanArsdale, Chris, 'Constitutional Environmental Law: Giving Force to Fundamental Principles in Africa' (2001) 26 *Colum J Entl L* 131

Burgorgue-Larsen, Laurence and Ubeda de Torres, Amaya (eds), *The Inter-American Court of Human Rights: Case Law and Commentary* (OUP, 2011)

Burns, William CG and Osofsky, Hari M (eds), *Adjudicating Climate Change: State, National, and International Approaches* (CUP, 2009)

Butt, Peter *et al*, *Legal Dictionary* (LexisNexis, 4th edn, 2014)

Cameron, Edward, 'Human Rights and Climate Change: Moving from an Intrinsic to an Instrumental Approach' (2010) 38(3) *Georgia Journal of International and Comparative Law* 673

Cardona, Natalia, *Climate Justice for a Changing Planet: Placing the Right to Development and Justice at the Heart of the Response to Climate Change* (UN NGLS, 2009)

Carlarne, Cinnamon, 'The Kyoto Protocol and the WTO: Reconciling Tensions between Free Trade and Environmental Objectives' (2006) 17 *Colorado J International Environmental Law & Policy* 45

Cassese, Antonio, *Self-Determination of Peoples: a Legal Reappraisal* (CUP, 1995)

Chalmers, Damien, Davies, Gareth and Monti, Giorgio, *European Union Law* (CUP, 3rd edn, 2014)

Chapman, Megan, 'Climate Change and the Regional Human Rights Systems' (2010) 10(2) *Sustainable Development Law and Policy* 37

Charnovitz, Steve (ed), *Trade Law and Global Governance* (Cameron May, 2002)

Cheyne, Ilona, 'Proportionality, Proximity and Environmental Labelling in WTO Law' (2009) 12(4) *Journal of International Economic Law* 927

Chinkin, Christine M, 'The Challenge of Soft Law: Development and Change in International Law' (1989) 38 *International and Comparative Law Quarterly* 850

Clapham, Andrew, *Human Rights: a Very Short Introduction* (OUP, 2007)

Cleveland, Sarah, 'Human Rights Sanctions and International Trade: a Theory of Compatibility' (2002) 5–1 *Journal of International Economic Law* 133

Cole, Matthew and Elliott, Robert, 'Determining the Trade-Environment Composition Effect: the Role of Capital, Labor and Environmental Regulations' (2003) 46–3 *Journal of Environmental Economics and Management* 363

Condon, Bradley and Sinha, Tapen, *The Role of Climate Change in Global Economic Governance* (Oxford University Press, 2013)

Cook, John *et al*, 'Quantifying the Consensus on Anthropogenic Global Warming in the Scientific Literature' (2013) 8(2) *Environmental Research Letters* 1

Cordes-Holland, Owen, 'The Sinking of the Strait: the Implications of Climate Change for Torres Strait Islanders' Human Rights Protected by the ICCPR' (2008) 9 *Melbourne Journal of International Law* 405

Cotula, Lorenzo, 'Reconciling Regulatory Stability and Evolution of Environmental Standards in Investment Contracts: towards a Rethink of Stabilization Clauses' (2008) 1 *J W En L Bus* 158

Cournil, Christel and Tabau, Anne-Sophie, *Politiques climatiques de l'UE et droits de l'homme* (Bruylant, 2013)

Cox, Roger, 'The Liability of European States for Climate Change' (2014) 30(78) *Utrecht Journal of International and European Law* 128

Crawford, James (ed), *The Rights of Peoples* (Clarendon Press, 1988)

Crawford, James, *The Creation of States in International Law* (Oxford University Press, 2nd edn, 2006)

Dabrowska, Anna Olga, *Legal Status of the Right to Development* (Unversitas Islandiae, 2010)

D'Amato, Anthony *et al*, 'Agora: What Obligation Does Our Generation Owe to the Next? An Approach to Global Environmental Responsibility' (1990) 84 *Am J Int'l Law* 190

David, Steven J and Caldeira, Ken, 'Consumption-based Accounting of CO_2 Emissions' (2010) 107(12) *Proceedings of the National Academy of Sciences* 5687

De Brabandere, Eric and Gazzini, Tarcisio (eds), *Foreign Investment in the Energy Sector: Balancing Private and Public Interests* (Brill, 2014)

De Cendra de Larragán, Javier, 'Tying the Knot of Energy Security and Climate Change Mitigation: a Tale of Solidarity?' (2011) 22 *Yb Int'l Env L* 76

De Liva, Charles and Morita, Sachiko, *Maritime Rights of Coastal States and Climate Change: Should States Adapt to Submerged Boundaries?* (World Bank, 2008)

De Sadeleer, Nicolas, 'Enforcing EUCHR Principles and Fundamental Rights in Environmental Cases' (2012) 81 *Nordic Journal of International Law* 39

De Schutter, Olivier, *Competing for Standards in Global Food Supply Chains* (Cridho Working Paper, 2012)

De Schutter, Olivier, *International Human Rights Law* (CUP, 2nd edn, 2014)

De Vries, Sybe A, 'Balancing Fundamental Rights with Economic Freedoms According to the European Court of Justice' (2013) 9 *Utrecht Law Review* 169

Deva, Surya and Bilchitz, David, *Human Rights Obligations of Business: beyond the Corporate Responsibility to Respect?* (CUP, 2013)

Divac Oberg, Marko, 'The Legal Effects of United Nations Resolutions in the *Kosovo* Advisory Opinion' (2011) 105(1) *AJIL* 81

Docherty, Bonnie and Giannini, Tyler, 'Confronting a Rising Tide: a Proposal for a Convention on Climate Change Refugees' (2009) 33 *Harvard Environmental Law Review* 349

Dominicé, Christian, 'The International Responsibility of States for Breach of Multilateral Obligations' (1999) 10 *EJIL* 353

Dommen, Caroline, 'Claiming Environmental Rights: Some Possibilities Offered by the United Nations' Human Rights Mechanisms' (1998) 11 *Geo Int'l Envtl L Rev* 1

Donnelly, Jack, 'In Search of the Unicorn: the Jurisprudence and Politics of the Right to Development' (1985) 15 *Cal W Int'l LJ* 473

Dörr, Oliver, Grote, Rainer and Marauhn, Thilo, *EMRK/GG Konkordanzkommentar zum europäischen und deutschen Grundrechtsschutz* (Mohr Siebeck, 2nd edn, 2013)

Dudai, Ron, 'Climate Change and Human Rights Practice: Observations on and around the Report of the OHCHR on the Relationship between Climate Change and Human Rights' (2009) 1 *Journal of Human Rights Practice* 294

Dupuy, Pierre-Marie, *Le droit à l'eau, un droit international?* (EUI Working Paper, 2006)

Ebbeson, Johnas, *Participatory and Procedural Rights in Environmental Matters: State of Play*, Unpublished Draft Paper (2009)

Ebeku, Kaniye, 'Constitutional Right to Healthy Environment and Human Rights Approaches to Environmental Protection in Nigeria: Gbemre v Shell revisited' (2007) 16(3) *RECIEL* 312

Falk, Richard, 'The *Kosovo* Advisory Opinion: Conflict Resolution and Precedent' (2011) 105(1) *AJIL* 50

Farber, Daniel A, 'Apportioning Climate Change Costs' (2008) 26 *UCLA J Envtl L and Pol'y* 21, 39–40

Faure, Michael and Nollkaemper, André, 'International Liability as an Instrument to Prevent and Compensate for Climate Change' (2007) 26A *Stan Envtl LJ* 123

Faure, Michael and Peeters, Marjan (eds), *Climate Change Liability* (Edward Elgar, 2011)

Feliziani, Chiara, 'The Duty of Member States to Guarantee the Right to a Healthy Environment: a Consideration of *European Commission v Italy*' (C-297/08)' (2012) 24 *JEL* 535

Fitzmaurice, Malgosia (ed), *Contemporary Issues in International Environmental Law* (Edward Elgar, 2009)

Flückiger, Alexandre, *Keeping Domestic Soft Law Accountable: towards a Gradual Formalization* (OUP, 2012)

Francioni, Francesco, 'International Human Rights in an Environmental Horizon' (2010) 21 *EJIL* 41

Francioni, Francesco, *Access to Justice as a Human Right* (OUP, 2010)

French, Duncan, 'Developing States and International Environmental Law: the Importance of Differentiated Responsibilities' (2000) 49 *International & Comparative Law Quarterly* 35

Gagain, Michael, 'Climate Change, Sea Level Rise, and Artificial Islands: Saving the Maldives' Statehood and Maritime Claims through the "Constitution of the Oceans"' (2012) 23 *Colo J Int'l Envtl L & Pol'y* 77, 81

Gai, Yashi, *Whose Human Right to Development?* (Commonwealth Secretariat Series of Occasional Papers on the Right to Development, 1989)

Gaja, Giorgio, 'The Protection of General Interests in the International Community' (2013) 364 *Hague Academy Collected Courses* 9

Garcia, Kelly K and Gostin, Lawrence O, 'One Health, One World – The Intersecting Legal Regimes of Trade, Climate Change, Food Security, Humanitarian Crisis and Migration' (2012) 1 *Laws* 4

Genevois, Bruno, 'La Convention européenne des droits de l'homme et la Charte des droits fondamentaux de l'Union européenne: complémentarité ou concurrence?' (2010) 26 *Revue française de droit administratif* 437

Gerrard, Michael and Wannier, Gregory (eds), *Threatened Island Nations: Legal Implications of Rising Seas and a Changing Climate* (Cambridge University Press, 2013)

Giddens, Anthony, 'Risk and Responsibility' (1999) 62(1) *Modern Law Review* 1

Gleick, Peter H, 'Basic Water Requirements for Human Activities: Meeting Basic Needs' (1996) 21 *Water International* 83

Gómez, Manuel A, 'The Global Chase: Seeking the Recognition and Enforcement of the Lago Agrio Judgment outside Ecuador' (2013) 1 *Stanford Journal of Complex Litigation* 429

Gordon, Jessica, 'Inter-American Commission on Human rights to Hold Hearing after Rejecting Inuit Climate Change Petition' (2006–7) 7(2) *Sustainable Development Law and Policy* 55

Greiber, Thomas (ed), *Conservation with Justice: a Rights-based Approach* (International Union for Conservation of Nature, 2009)

Gromilova, Mariya, 'Revisiting Planned Relocation as a Climate Change Adaptation Strategy: the Added Value of a Human Rights-Based Approach' (2014) 10 *Utrecht L Rev* 76

Gupta, Joyeeta and Lebel, Louis, 'Access and Allocation in Earth System Governance: Water and Climate Change Compared' (2010) 10 *Int Environ Agreements* 377

Habermas, Jürgen, *Die postnationale Konstellation: Politische Essays* (Suhrkamp, 1998)

Haines, Andy *et al*, 'Climate Change and Human Health: Impacts, Vulnerability and Public Health' (2006) 120 *Public Health* 585

Hall, Margaux J, 'Advancing Climate Justice and the Right to Health through Procedural Rights' (2014) 1 *Health and Human Rights Journal* 8

Hall, Margaux J and Weiss, David C, 'Avoiding Adaptation Apartheid: Climate Change Adaptation and Human Rights Law' (2012) 39 *Yale J Int'l L* 309

Hannum, Hurst, 'Rethinking Self-Determination' (1993) 34 *Virginia Journal of International Law* 1

Hanski, Raija and Scheinin, Martin, *Leading Cases of the Human Rights Committee* (Åbo Akademi University, 2003)

Harrington, Alexandra R, *Life as We Know It: the Expansion of the Right to Life under the Jurisprudence of the Inter-American Court of Human Rights* (Albany Law School, 2012–13)

Hassan, Parvez and Azfar, Azim, 'Human Rights and the Environment: a South Asian Perspective' (2014) 5(2) *Journal of Human Rights and the Environment* 192

Hathaway, James, 'Are International Institutions Doing their Job? Discussion' (1996) 90 *Proceedings of the Annual Meeting of the American Society of International Law* 558

Heil, Mark and Selden, Thomas, 'International Trade Intensity and Carbon Emissions: a Cross-Country Econometric Analysis' (2010) 10(1) *Journal of Environment and Development* 35–49

Henkin, Louis, 'The Universal Declaration at 50 and the Challenge of Global Markets' (1999) 25 *Brooklyn Journal of International Law* 17

Higgins, Rosalyn, *Problems and Process: International Law and How We Use It* (OUP, 1999) 50)

Hohfeld, Wesley N, *Fundamental Legal Conceptions as Applied in Judicial Reasoning* (Yale UP, 1964)

Hollo, Erkki J, Kulovesi, Kati and Mehling, Michael (eds), *Climate Change and the Law* (Springer, 2013)

Holzer, Kateryna, 'Proposals on Carbon-related Border Adjustments: Prospects for WTO Compliance' (2010) 1 *Carbon and Climate Law Review* 51

Hossain, Kamrul, 'The Realisation of the Right to Environment and the Right to Development in respect to the Arctic Indigenous Peoples' (2011) 3 *The Yearbook of Polar Law* 129

Hossain, Moazzem and Selvanatha, Anthony (eds), *Climate Change and Growth in Asia* (Edward Elgar, 2011)

Howse, Robert, 'The World Trade Organization and the Protection of Workers' Rights' (1999) 3 *Journal of Small and Emerging Small Business Law* 131

Howse, Robert, 'Back to Court After Shrimp? Almost but Not Quite Yet: India's Short-lived Challenge to Labour and Environmental Exceptions in the European Union's Generalized System of Preferences' (2003) 18 *Amer Univ International Law Review* 1333

Howse, Robert and Regan, Donald, 'The Product/Process Distinction – An Illusory Basis for Disciplining "Unilateralism" in Trade Policy' (2000) 11 *EJIL* 249

Humphreys, Stephen (ed), *Human Rights and Climate Change* (Cambridge University Press, 2010)

Hunter, David, 'Human Rights Implications for Climate Change Negotiations' (2009) 11 *Or Rev Int'l L* 331

Hunter, David, Salzman, James and Zaelke, Durwood, *International Environmental Law and Policy* (Foundation Press, 2nd edn, 2002)

Iancu, Lidia and Ghilain, Timothy, *The Role of the Court of Justice in Environmental Justice as a Third Generation Human Right* (Effectius, 2011)

Jacobs, Rebecca Elizabeth, 'Treading Deep Waters: Substantive Law Issues in Tuvalu's Threat to Sue the United States in the International Court of Justice' (2005) 14 *Pac Rim L & Pol'y J* 103.

Jing, Fang and Leduc, Brigitte, *Potential Threats from Climate Change to Human Wellbeing in the Eastern Himalayan Region* (MacArthur Foundation, 2010)

Jodoin, Sébastien and Lofts, Katherine (eds), *Economic, Social, and Cultural Rights and Climate Change: a Legal Reference Guide* (Centre for International Sustainable Development Law *et al*, 2013)

Johl, Alyssa and Duyck, Sébastien, 'Promoting Human Rights in the Future Climate Change Policy' (2012) 15 *Ethics Policy and Environment* 298

Jones, Rhys, Bennett, Hayley, Keating, Gay and Blaiklock, Alison, 'Climate Change and the Right to Health for Māori in Aotearoa/New Zealand' (2014) 16 *Health and Human Rights Journal* 54

Joseph, Sarah, *Blame it on the WTO? A Human Rights Critique* (Oxford University Press, 2011)

Joseph, Sarah and Castan, Melissa, *The International Covenant on Civil and Political Rights* (OUP, 3rd edn, 2013)

Joshi, Manoj, 'Are Eco-Labels Consistent with WTO Agreements?' (2004) 38(1) *Journal of World Trade* 69

Kälin, Walter, 'From the Nansen Principles to the Nansen Initiative' (2012) 41 *Forced Migration* 48

Kant, Immanuel, *Zum ewigen Frieden. Ein philosophischer Entwurf* (1795)

Kelsen, Hans, *Pure Theory of Law* (Peter Smith, 1989)

Kirchmeier, Felix, *The Right to Development – Where Do We Stand?* (Friedrich Ebert Stiftung, 2006)

Knox, John, 'Climate Change and Human Rights Law' (2009) 50 *Va J Int'l L* 163

Knox, John, 'Linking Climate Change and Human Rights at the UN' (2009) 33 *Harvard Environmental Law Review* 477

Koivurova, Timo, 'International Legal Avenues to Address the Plight of Victims of Climate Change: Problems and Prospects' (2007) 22(2) *J Envtl Law and Litigation* 267

Koppel Maldonado, Julie, Colombi, Benedict and Pandya, Rajul (eds), *Climate Change and Indigenous Peoples in the US: Impacts, Experiences, and Actions* (Springer, 2014)

Kornicker Uhlmann, Eva M, 'State Community Interest, Jus Cogens and Protection of the Global Environment: Developing Criteria for Peremptory Norms' (1998–9) 11 *Geo Int'l Envtl L Rev* 101

Kravchenko, Svitlana, 'Procedural Rights as a Crucial Tool to Combat Climate Change' (2010) 38 *Georgia J Int'l & Comp L* 613

Kumssa, Asaf and Jones, John F, 'Climate Change and Human Security in Africa' (2010) 17 *International Journal of Sustainable Development & World Ecology* 453

Lawrence, Peter, *Justice for Future Generations: Climate Change and International Law* (Edward Elgar, 2014)

Lenzerini, Federico, *The Culturalisation of Human Rights Law* (OUP, 2014)

Lenzerini, Federico and Vrdoljak, Ana (eds), *International Law for Global Common Goods: Normative Perspectives on Nature, Culture and Human Rights* (Hart Publishing, 2014)

Limon, Marc, 'Human Rights and Climate Change: Constructing a Case for Political Action' (2009) 33 *Harv Envtl L Rev* 439

Limon, Marc, 'Human Rights Obligations and Accountability in the Face of Climate Change' (2009–10) 38 *Ga J Int'l & Comp L* 543

Lixinski, Lucas, 'Treaty Interpretation by the Inter-American Court of Human Rights: Expansion at the Service of the Unity of International Law' (2010) 21 *EJIL* 585

Lord, Richard, Goldberg, Silke, Rajamani, Lavanya and Brunnée, Jutta, *Climate Change Liability – Transnational Law and Practice* (CUP, 2012)

Lourme, Louis, *Le nouvel âge de la citoyenneté mondiale* (PUF, 2014)

Lowe, Vaughan *et al* (eds), *Fifty Years of the International Court of Justice* (CUP, 1996)

Lutz, Robert E *et al*, 'Environment, Economic Development and Human Rights: a Triangular Relationship? Remarks' (1988) 82 *Am Soc'y Int'l L Proc* 40

McAdam, Jane, *Complementary Protection in International Refugee Law* (OUP, 2007)

McAdam, Jane, *Climate Change and Displacement: Multidisciplinary Perspectives* (Hart, 2010)

McAdam, Jane, *Climate Change, Forced Migration, and International Law* (OUP, 2012)

McAdam, Jane and Saul, Ben, *An Insecure Climate for Human Security? Climate-Induced Displacement and International Law* (Sydney Centre for International Law, online)

McBeth, Adam, *International Economic Actors and Human Rights* (Routledge, 2010)

McCorquodale, Robert and Simons, Penelope, 'Responsibility beyond Borders: State Responsibility for Extraterritorial Violations by Corporations of International Human Rights Law' (2007) 70(4) *Modern Law Review* 598, 602–3

MacDonald, Karen E, 'A Right to a Healthful Environment – Human Rights Habitats: Re-thinking Rights in an Age of Climate Change' (2008) 17 *EEELR* 213

McInerney-Lankford, Siobhan, 'Climate Change and Human Rights: an Introduction to Legal Issues' (2012) 33 *Harv Envtl L Rev* 431

M'Baye, Keba, 'Le droit au développement comme un droit de l'homme' (1972) 5 *Revue des Droits de L'Homme* 503

Maes, Frank, Cliquet, An, Du Plessis, Willemien and McLeod-Kilmurray, Heather (eds), *Biodiversity and Climate Change. Linkages at International, National and Local Levels* (Edward Elgar, 2013)

Maguire, Rowena, 'Designing REDD+ to Be Just: Considerations for a Legally Binding Instrument' (2014) 4(1) *Asian Journal of International Law* 169.

Maljean-Dubois, Sandrine (ed), *Changements climatiques. Les enjeux du contrôle international* (Université Paul Cézanne Aix-Marseille, 2007)

Marks, Stephen, 'Emerging Human Rights: a New Generation for the 1980's' (1981) 33 *Rutgers Law Review* 435

Markulev, Ana and Long, Anthea, *On Sustainability: an Economic Approach* (Commonwealth of Australia, 2013)

Martin, Susan F, Weerasinghe, Sanjula and Taylor, Abbie, *Humanitarian Crises and Migration: Causes, Consequences and Responses* (Routledge, 2014)

Mason Meier, Benjamin and Mori, Larisa M, 'The Highest Attainable Standard: Advancing a Collective Human Right to Public Health' (2005) 35 *Columbia Human Rights Law Review* 101

Mayer, Benoît, 'Pour en finir avec la notion de "réfugiés environnementaux": critique d'une approche individualiste et universaliste des déplacements causés par des changements environnementaux' (2011) 7(1) *McGill International Journal for Sustainable Development Law and Policy* 33

Mégret, Frédéric, 'The Disabilities Convention: Human Rights of Persons with Disabilities or Disability Rights?' (2008) 30(2) *Human Rights Quarterly* 494

Messerlin, Patrick, 'Climate and Trade Policies: from Mutual Destruction to Mutual Support' (2012) 11 *World Trade Review* 53

Miles, Kate, *The Origins of International Investment Law* (Cambridge University Press, 2013)

Moloo, Rahim and Jacinto, Justin, 'Environmental and Health Regulation: Assessing Liability under Investment Treaties' (2011) 29 *Berkeley J Int'l Inv L* 1

Morgera, Elisa and Savaresi, Annalisa, 'A Conceptual and Legal Perspective on the Green Economy' (2013) 22 *RECIEL* 14

Morhead, John and Nixon, Tim, *Global 500 GHGs Performance 2012–13: 2014 Report on Trends* (Thomson Reuters, 2014)

Morten Haugen, Hans, 'Food Sovereignty: an Appropriate Approach to Ensure the Right to Food?' (2009) 78 *Nordic Journal of International Law* 263

Mwiturubani, Donald Anthony and Van Wyk, Sanita Jo-Ansie (eds), *Climate Change and Natural Resources Conflicts in Africa* (Institute of Security Studies, 2010)

Narula, Smita, 'The Right to Food: Holding Global Actors Accountable under International Law' (2006) 44 *Columbia Journal of Transnational Law* 691

Nuffer, Sarah, 'Human Rights Violations and Climate Change: the Last Days of the Inuit People?' (2010) 37 *Rutgers Law Record* 182

Orellana, Marcos A, Kothari, Miloon and Chaudry, Shivani, *Climate Change in the Work of the Committee on Economic, Social and Cultural Rights* (CIEL, 2010)

Osofsky, Hari M, 'The Inuit Petition as a Bridge? Beyond Dialectics of Climate Change and Indigenous Peoples' Rights' (2006–7) 31 *American Indian Law Review* 675

Osofsky, Hari M, 'Is Climate Change "International"? Litigation's Diagonal Regulatory Role' (2009) 49 *Va J Int'l L* 585

Overmars, Marc and Beth Gottlieb, Sasha, *Adapting to Climate Change in Water Resources and Water Services in Caribbean and Pacific Small Island Countries* (World Water Forum, 2009)

Pavoni, Riccardo, *Interesse pubblico e diritti individuali nella giurisprudenza ambientale della Corte europea dei diritti umani* (Editoriale scientifica, 2013)

Pavoni, Riccardo, *Climate Change and Human Rights: a Viable and Unavoidable Relationship* (University of Delhi, 2014)

Payne, Rachel, 'Climate Change, Basic Rights and International Obligations' (2012) 2 *Yale Rev Int'l Stud* 64

Pearson, Charles, *An Environmental Code of Conduct for Multinational Companies?* (Duke University Press, 1987)

Pedersen, Ole W, 'The Janus-Head of Human Rights and Climate Change: Adaptation and Mitigation' (2011) 80 *Nordic Journal of International Law* 403

Pedersen, Ole W, 'An International Environmental Court and International Legalism' (2012) 24 *Journal of Environmental Law* 547

Peel, Jacqueline, Godden, Lee and Keenan, Rodney, 'Climate Change Law in an Era of Multi-Level Governance' (2012) *Transnational Environmental Law* 245

Peers, Steve, Hervey, Tamara, Kenner, Jeff and Ward, Angela (eds), *Commentary on the EU Charter of Fundamental Rights* (Hart, 2013)

Picone, Paolo, *Comunità internazionale e obblighi 'erga omnes'* (Jovene, 2006)

Postiglione, Amedeo, 'A More Efficient International Law on the Environment and Setting up an International Court for the Environment within the United Nations' (1990) 20 *Envtl L* 321

Quirico, Ottavio, 'Substantive and Procedural Issues Raised by the Accession of the European Union to the European Convention on Human Rights' (2010–11) 20 *Italian Yearbook of International Law* 31

Quirico, Ottavio, 'Amidst Fragmentation and Coherence: a Systemic Interpretation of the World Heritage Convention and the UNFCCC Regime' (2012) 10 *New Zealand Yearbook of International Law* 33

Rayfuse, Rosemary and Crawford, Emily, *Climate Change, Sovereignty and Statehood* (University of Sydney, 2011)

Rayfuse, Rosemary and Scott, Shirley V (eds), *International Law in the Era of Climate Change* (Edward Elgar, 2012) 243

Richardson, Benjamin J, Le Bouthillier, Yves, McLeod-Kilmurray, Heather and Wood, Stepan (eds), *Climate Law and Developing Countries: Legal and Policy Challenges for the World Economy* (Edward Elgar, 2009)

Risbey, James S *et al*, 'Well-estimated Global Surface Warming in Climate Projections Selected for ENSO Phase' (2014) *Nature Climate Change* 835

Roberts, Anthea, 'Power and Persuasion in Investment Treaty Interpretation: the Dual Role of States' (2010) 104 *Am J Int'l L* 179

Robinson, Mary, 'Climate Change and Justice' (2007) 36 *Scottish Human Rights Journal* (online)

Roht-Arriaza, Naomi, '"First, Do No Harm": Human Rights and Efforts to Combat Climate Change' (2010) 38 *Georgia J Int'l & Comp L* 593

Roht-Arriaza, Naomi, 'Human Rights in the Climate Change Regime' (2010) 1(2) *Journal of Human Rights and the Environment* 211

Rose, Anna, 'Gray v Minister for Planning: the Rising Tide of Climate Change Litigation in Australia' (2007) 29 *Sydney Law Review* 725

Rosin, Christopher, Stock, Paul and Campbell, Hugh (eds), *Food Systems Failure: the Global Food Crisis and the Future of Agriculture* (Routledge, Earthscan, 2012)

Ruppel, Oliver C, Roschmann, Christian and Ruppel-Sehlichting, Katharine (eds), *Climate Change: International Law and Global Governance* (Nomos, 2013)

Sang, Joseph, *The Ogiek in Mau Forest* (online, 2000)

Sauvant, Karl P (ed), *The Rise of Transnational Corporations from Emerging Markets: Threat or Opportunity?* (Edward Elgar, 2008)

Scelle, Georges, *Précis de droit des gens – Principes et systématique* (Dalloz, 11th edn, 2008)

Schemmel, Mathew L and De Regt, Bas, 'The European Court of Justice and the Environmental Protection Policy of the European Community' (1994) 17 *Boston College Int'l & Comp L Rev* 53

Schill, Stephan, 'Do Investment Treaties Chill Unilateral State Regulation to Mitigate Climate Change?' (2007) 24(5) *Journal Int'l Arb* 469

Schill, Stephan, *The Multilateralisation of International Investment Law* (Cambridge University Press, 2009)

Schmidt, Hans-Jürgen, 'Carbon Footprinting, Labelling and Life Cycle Assessment' (2009) 14 *The International Journal of Life Cycle Assessment* 6

Schreuer, Christoph, 'Derogation of Human Rights in Situations of Public Emergency' (1982) 9 *Yale Journal of World Public Order* 113, 122

Seymour, Frances, *Forests, Climate Change, and Human Rights: Managing Risk and Trade-offs* (CIFOR, 2008)

Shelton, Dinah, 'Human Rights, Environmental Rights and the Right to Environment' (1992) 28 *Stanford Journal of International Law* 103

Shelton, Dinah, *Human Rights and Climate Change* (Northwestern University, 2009)

Shelton, Dinah, *Litigating a Rights-Based Approach to Climate Change* (GWU Legal Studies Research Paper No 2013–57, 2009)

Shelton, Dinah (ed), *The Oxford Handbook of International Human Rights Law* (OUP, 2013)

Shue, Henry, *Basic Rights: Subsistence, Affluence and US Foreign Policy* (Princeton University Press, 1980)

Simma, Bruno, 'Self-contained Regimes' (1985) 16 *Netherlands Yearbook of International Law* 111

Sinden, Amy, 'Climate Change and Human Rights' (2007) 27 *Journal of Land Resources and Environmental Law* 255

Sohn, Louis B, 'The Stockholm Declaration on the Human Environment' (1973) 14 *Harvard Int'l LJ* 451

Song, Yann-huei, 'The Application of Article 121 of the Law of the Sea Convention to the Selected Geographical Features Situated in the Pacific Ocean' (2010) 9 *Chinese Journal of International Law* 66

Spieler, Paula, 'The La Oroya Case: the Relationship between Environmental Degradation and Human Rights Violations' (2010) 18(1) *Human Rights Brief* 19

Spier, Jaap and Magnus, Ulrich, *Climate Change Remedies, Injunctive Relief and Criminal Law Responses* (Eleven International Publishing, 2014)

Steiner, Henry J, Alston, Philip and Goodman, Ryan, *International Human Rights in Context: Law, Politics, Morals* (OUP, 2007)

Stone, Christopher D, 'Common but Differentiated Responsibilities in International Law' (2004) 98 *AJIL* 276

Sussman, Edna, 'The Energy Charter Treaty's Investor Protection Provisions: Potential to Foster Solutions to Global Warming and Promote Sustainable Development' (2008) 14(2) *ILSA Journal of International & Comparative Law* 391

Swinarski, Christophe (ed), *Studies and Essays on International Humanitarian Law and Red Cross Principles* (Martinus Nijhoff, 1984)

Tarlock, Dan, 'How Well Can International Water Allocation Regimes Adapt to Global Climate Change?' (2000) 15 *J Land Use & Transnational Environmental Law* 423

Thorpe, Amelia, 'Tort-Based Climate Change Litigation and the Political Question Doctrine' (2008) 24 *Journal of Land Use* 79

Tignino, Mara, *International Law and Freshwater* (Edward Elgar Publishing, 2013)

Tirado, Joseph and Bloom, Jerry R, 'Renewable Energy Reforms in Europe: Growing Threats to International Investors' (2014) *Lexology* (online)

Tol, Richard SJ and Verheyen, Roda, 'State Responsibility for Climate Change Damages – A Legal and Economic Assessment' (2004) 32 *Energy Pol'y* 1109

Tomuschat, Christian, Lagrange, Evelyne and Oeter, Stefan (eds), *The Right to Life* (Leiden, Koninklijke Brill, 2010)

Trachtman, Joel, 'First Tuna/Dolphin dispute' (1992) 86(1) *AJIL* 142

Tsaltas, Grigoris, Bourtzis, Tilemachos and Rodotheatos, Gerasimos, *Artificial Islands and Structures as a Means of Safeguarding State Sovereignty against Sea Level Rise: a Law of the Sea Perspective* (UNSW, online)

Tully, Stephen, 'The Contribution of Human Rights as an Additional Perspective on Climate Change: Impacts within the Pacific' (2007) 5 *New Zealand Journal of Public International Law* 175

Ukala, Eferiekose, 'Gas Flaring in Nigeria's Niger Delta: Failed Promises and Reviving Community Voices' (2011) 2 *Washington and Lee Journal of Energy, Climate, and the Environment* 97

Vadi, Valentina, *Public Health in International Investment Law and Arbitration* (Routledge, 2012)

Van der Have, Nienke, *The Right to Development and State Responsibility: Can States Be Held to Account?* (SHARES Research Paper 23, 2013)

Vandenbergh, Michael P, Ackerly, Brooke A and Forster, Fred E, 'Micro-Offsets and Macro-Transformation: an Inconvenient View of Climate Change Justice' (2012) 33 *Harv Envtl L Rev* 303

Vandenbogaerde, Arne, 'The Right to Development in International Human Rights Law: a Call for Its Dissolution' (2013) 31 *Netherlands Quarterly of Human Rights* 187, 208

Vasak, Karel, 'A Thirty Year Struggle: the Sustained Efforts to give Force of Law to the Universal Declaration of Human Rights' (1977) *UNESCO Courier*

Verheyen, Roda, *Climate Change Damage and International Law* (Martinus Nijhoff, 2005)

Vidas, Davor, 'Sea-Level Rise and International Law: at the Convergence of Two Epochs' (2014) 4 *Climate Law* 70

Villaroman, Noel G, 'The Right to Development: Exploring the Legal Basis of a Supernorm' (2010) 22 *Florida Journal of International Law* 299

Villaroman, Noel G, 'Rescuing a Troubled Concept: an Alternative View of the Right to Development' (2011) 29 *Netherlands Quarterly of Human Rights* 13

Viñuales, Jorge E (ed), *The Rio Declaration on Environment and Development: a Commentary* (CUP, 2014)

Voigt, Christina, 'State Responsibility for Climate Change Damage' (2008) 77 *Nordic J Int'l L* 9

Von Doussa, John, 'Human Rights and Climate Change: a Tragedy in the Making' (2008) 31 *UNSW LJ* 953

Von Doussa, John, Corkery, Allison and Chartres, Renée, 'Human Rights and Climate Change' (2007) 14 *Austl Int'l LJ* 161

Wapner, Paul, 'The Challenges of Planetary Bandwagoning' (2011) 1(3) *Global Environmental Politics* 137

Ward, Tara, 'The Right to Free, Prior and Informed Consent: Indigenous Peoples' Participation Rights within International Law' (2011) 10(2) *Northwestern Journal of International Human Rights* 52

Warner, Robin and Schofield, Clive (eds), *Climate Change and the Oceans: Gauging the Legal and Policy Currents in the Asia Pacific and Beyond* (Edward Elgar, 2013)

Warnock, Alisa Ceri, 'Small Island Developing States of the Pacific and Climate Change: Adaptation and Alternatives' (2007) 4 *NZYIL* 247

Weil, Prosper, 'Towards Relative Normativity in International Law' (1983) 77 *AJIL* 413

Weissbrodt, David and Kruger, Muria, 'Norms on the Responsibilities of Transnational Corporations and Other Business Enterprises with Regard to Human Rights' (2003) 97 *AJIL* 901

Wellman, Carl, 'Solidarity, the Individual, and Human Rights' (2000) 22 *Human Rights Quarterly* 639

Weston, Burns H and Bollier, David, *Green Governance: Ecological Survival, Human Rights, and the Law of the Commons* (CUP, 2013)

Wewerinke, Margaretha, *Climate Change, Human Rights and the International Legal Order: the Role of the UN Human Rights Council* (University of Cambridge WP, 2013)

Whelan, Daniel J, *Indivisible Human Rights: a History* (University of Pennsilvanya Press, 2010)

Wijdekop, Femke, 'Democratie, rechtstaat en de rechten van toekomstige generaties' (2014) 89(23) *Nederlands Juristenblad* 1552

Willcox, Susanna, 'A Rising Tide: the Implications of Climate Change Inundation for Human Rights and State Sovereignty' (2012) 9 *Essex Human Rights Review* 1

Williams, Jay, 'The Impact of Climate Change on Indigenous People – The Implications for the Cultural, Spiritual, Economic and Legal Rights of Indigenous People' (2012) 14 *Int'l J L Hum Rts* 648

Wolfrum, Rüdiger (ed), *Max Planck Encyclopaedia of Public International Law* (OUP, 2009)

Woods, Patrick, *Reversal by Recusal? Comer v Murphy Oil USA, Inc and Mandatory Judicial Recusal Statements* (The Center for Judicial Process, 2011)

Wright, Jane, 'Minority Groups, Autonomy, and Self-Determination' (1999) 19(4) *Oxford Journal of Legal Studies* 605

Wuerth, Ingrid B, '*Kiobel v Royal Dutch Petroleum Co*: the Supreme Court and the Alien Tort Statute' (2013) 107 *AJIL* 601

Zengerling, Cathrin, *Greening International Jurisprudence: Environmental NGOs before International Courts, Tribunals, and Compliance Committees* (Martinus Nijhoff Publishers, 2013)

Zolo, Danilo, 'The Water Right as a Social and Collective Right' (2005) 1 *Jura Gentium. Journal of Philosophy of International Law and Global Politics* (online)

Index

Abbreviations

AAC	Arctic Athabaskan Council
AB	Appellate Body
ACHPR	African Charter on Human and Peoples' Rights
ACHR	American Convention on Human Rights
AComHPR	African Commission on Human and Peoples' Rights
ACRWC	African Charter on the Rights and Welfare of the Child
ACtHPR	African Court on Human and Peoples' Rights
ADRDM	American Declaration of the Rights and Duties of Man
AMCEN	African Ministerial Conference on Environment
AR	Assessment Report
ArCHR	Arab Charter on Human Rights
ASEAN	Association of Southeast Asian Nations
AU	African Union
BIT	bilateral investment treaty
BTA	border tax adjustment
CA	Court of Appeal
CAHOSCC	Committee of African Heads of State and Government on Climate Change
CBDR	common but differentiated responsibility
CDM	clean development mechanism
CER	certified emission reduction
CESCR	Committee on Economic, Social and Cultural Rights
CFREU	Charter of Fundamental Rights of the European Union
CIDI	Inter-American Council for Integral Development
CISDL	Centre for International Sustainable Development Law
CO_2	carbon dioxide
COP	Conference of the Parties
CRC	Convention on the Rights of the Child
CSR	corporate social responsibility
CteRC	Committee on the Rights of the Child
DCC	Dutch Civil Code
DICED	Draft International Covenant on Environment and Development

EC	European Community
ECHR	European Convention on Human Rights
ECJ	European Court of Justice
ECOSOC	Economic and Social Council
ECT	Energy Charter Treaty
ECteSR	European Committee of Social Rights
ECtHR	European Court of Human Rights
EEC	European Economic Community
EEZ	exclusive economic zone
EIA	environmental impact assessment
EPA	Environmental Protection Agency
ESC	European Social Charter
ESCAP	Economic and Social Commission for Asia and the Pacific
ETS	emission trading system
EU	European Union
FAO	Food and Agriculture Organisation
FDI	foreign direct investment
FIT	feed-in tariff
FTA	free trade agreement
GA	General Assembly
GATS	General Agreement on Trade in Services
GATT	General Agreement on Tariffs and Trade
GDP	gross domestic product
GEN	Global Eco-labelling Network
GHG	greenhouse gas
GP	guiding principle
HC	High Court
HCHR	High Commissioner for Human Rights
HCR	High Commissioner for Refugees
HRC	Human Rights Council
HRCte	Human Rights Committee
HRD	Human Rights Declaration
HREOC	Human Rights and Equal Opportunity Commission
IACERDI	Inter-American Convention on the Elimination of Racism, Racial Discrimination and Related Forms of Intolerance
IAComHR	Inter-American Commission on Human Rights
IACtHR	Inter-American Court of Human Rights
ICAO	International Civil Aviation Organisation
ICC	International Criminal Court
ICCPR	International Covenant on Civil and Political Rights
ICE	International Court for the Environment
ICESCR	International Covenant on Economic, Social and Cultural Rights
ICHR	Intergovernmental Commission on Human Rights
ICJ	International Court of Justice

ICpC	Inuit Circumpolar Council
ICSID	International Centre for Settlement of Investment Disputes
ICTY	International Criminal Tribunal for the Former Yugoslavia
IDLO	International Development Law Organisation
IDP	internally displaced person
IEA	International Energy Agency
IFAD	International Fund for Agricultural Development
IISD	International Institute for Sustainable Development
ILA	International Law Association
ILC	International Law Commission
ILO	International Labour Organisation
IMERS	International Maritime Emission Reduction Scheme
IMF	International Monetary Fund
IPCC	Intergovernmental Panel on Climate Change
ISEAL	international social and environmental accreditation and labelling
ISO	International Organization for Standardization
IUCN	International Union for Conservation of Nature
LAWASIA	Law Association of Asia and the Pacific
LCA	life-cycle assessment
LOSC	Law of the Sea Convention
MDGs	Millennium Development Goals
MEA	multilateral environmental agreement
NAFTA	North America Free Trade Agreement
NCP	National Contact Point
NEASPEC	Northeast Asian Subregional Programme for Environmental Cooperation
NEPAD	New Partnership for Africa's Development
NGO	non-governmental organisation
NIEO	New International Economic Order
NNPC	Nigerian National Petroleum Corporation
NSW	New South Wales
NZ	New Zealand
OAS	Organisation of American States
OAU	Organisation of African Unity
OECD	Organisation for Economic Co-operation and Development
OHCHR	Office of the High Commissioner for Human Rights
PACC	Pacific Adaptation to Climate Change
PCA	Permanent Court of Arbitration
PPM	production or process method
PRC	People's Republic of China
REDD	reduction of emissions from deforestation and forest degradation
RRT	Refugee Review Tribunal
SAARC	South Asian Association for Regional Cooperation

SACEP	South Asia Co-operative Environment Programme
SC	Security Council
SCt	Supreme Court
SCC	Stockholm Chamber of Commerce
SG	Secretary General
SPC	Secretariat of the Pacific Community
SPREP	Secretariat of the Pacific Regional Environment Programme
SPS	sanitary and phytosanitary
TBT	technical barriers to trade
TEC	Treaty Establishing the European Community
TEU	Treaty on European Union
TFEU	Treaty on the Functioning of the European Union
TRIPS	Trade-related Aspects of Intellectual Property Rights
UDHR	Universal Declaration of Human Rights
UK	United Kingdom
UN	United Nations
UNCED	United Nations Conference on Environment and Development
UNCITRAL	United Nations Commission on International Trade Law
UNDP	United Nations Development Programme
UNDRD	United Nations Declaration on the Right to Development
UNDRIP	United Nations Declaration on the Rights of Indigenous Peoples
UNECE	United Nations Economic Commission for Europe
UNEO	United Nations Environment Organisation
UNEP	United Nations Environment Programme
UNESCO	United Nations Educational, Scientific and Cultural Organisation
UNFCCC	United Nations Framework Convention on Climate Change
UNICEF	United Nations Children's Fund
UNWFP	United Nations World Food Programme
US	United States
USSR	Union of Soviet Socialist Republics
VCAT	Victorian Civil and Administrative Tribunal
WCED	World Commission on Environment and Development
WG	working group
WHC	World Heritage Committee
WHO	World Health Organisation
WRI	World Resources Institute
WTO	World Trade Organisation

Notes on contributors

Roland Adjovi
adjovir@arcadia.edu
Roland Adjovi is Assistant Professor of International Law and Human Rights at Arcadia University, Glenside, US. He has also taught in the Ivory Coast (1998–99) and France (1999–2002). He worked at the International Criminal Court and International Criminal Tribunal for Rwanda, and was Lead Counsel before the African Court on Human and Peoples' Rights. He is a member of the Editorial Committee of the African Yearbook of International Law and of the Editorial Advisory Committee of International Legal Materials (ASIL).

Spyridon Aktypis
spyridon.aktypis@ouc.ac.cy; aktypis@mfhr.gr
Spyridon-Heracles Aktypis is Assistant Professor of International Law and Human Rights at the Open University of Cyprus and Director of the Marangopoulos Foundation of Human Rights in Athens, Greece. He has also been Assistant Professor of Law at Sciences Po, Paris, and at the University of Rouen, France, Intern on Human Rights of Persons with Disabilities at the UN, and Reporter on international law in domestic courts at Oxford University Press.

Christine Bakker
christine.bakker@eui.eu
Christine Bakker, PhD, is a Research Fellow at the European University Institute, Academy of European Law, Florence, Italy, specialising in human rights law, including children's rights and international environmental law. She also carried out research for the UNICEF Innocenti Research Centre in Florence, and previously worked at the European Commission (DG Development).

Ben Boer
ben.boer@sydney.edu.au
Ben Boer has taught and researched in environmental law since 1979. In 1992 he was appointed as Professor in Environmental Law at Sydney Law School, where he became Emeritus Professor in 2008. He is also Distinguished Professor at the Research Institute of Environmental Law at Wuhan University, PRC, and has served as the Deputy Chair of the World Commission on Environmental Law of the International Union for Conservation of Nature since 2012.

Mouloud Boumghar
mouloud.boumghar@u-picardie.fr
Mouloud Boumghar has been Agrégé des facultés de droit since 2008 and is currently Professor of Public Law at Université de Picardie Jules Verne, Amiens, France, where he teaches international and human rights law. He was previously a Lecturer at Universités Paris XIII and Évry-Val d'Essonne, and a Max Weber Fellow at the European University Institute.

Jürgen Bröhmer
j.brohmer@murdoch.edu.au
Jürgen Bröhmer is Professor and Dean of the Law School at Murdoch University, Perth, Australia. Prior to joining Murdoch University, he held positions at Europa Institute (Saarland University), Saarbücken, Germany, and at the University of New England, Australia. His areas of expertise include international and human rights law.

Christel Cournil
christelcournil@yahoo.fr
Christel Cournil is Associate Professor of Public Law at University Paris 13 – PRES, Sorbonne Paris Cité, where she is a member of the Institute for Interdisciplinary Research on Social Issues and an Associate Researcher at the Centre for Administrative and Political Research and Studies. Her research concerns environmental migration and the link between global environmental change and human rights. In 2010, she was given the Award of Scientific Excellence by University Paris 13.

Olivier De Schutter
olivier.deschutter@uclouvain.be
Olivier De Schutter is Professor of Law at the University of Louvain (UCL), Belgium, and at Sciences Po, Paris, France. He is also a member of the Global Law School Faculty at New York University and of the UN Committee on Economic, Social and Cultural Rights. Between 2008 and 2014 he was the UN Special Rapporteur on the Right to Food.

Emmanuel Decaux
emm.decaux@wanadoo.fr
Emmanuel Decaux is Professor of Public Law at Université Panthéon-Assas (Paris 2), France. He is currently President of the UN Committee on Enforced Disappearances and has held several other human rights positions at the UN. He chairs the Sub-Commission on International Issues of the French National Advisory Commission on Human Rights.

Pierre-Marie Dupuy
pierre-marie.dupuy@graduateinstitute.ch
Pierre-Marie Dupuy is Emeritus Professor of Law at Université Panthéon-Assas, Paris, France, and at the Graduate Institute for International Studies, Geneva,

Switzerland. He was also Professor at the European University Institute, Florence, Italy, and Director of the Paris Institute for Advanced International Studies. He often acts as a PCA and ICSID Arbitrator, Counsel for States before the ICJ, and worked as a consultant for UNEP, OECD and UNESCO.

Alessandra Franca
alessandra@ccj.ufpb.br
Alessandra Franca is Professor of International and Environmental Law at the Federal University of Paraiba (UFPB), João Pessoa, Brazil. She has a Master of Laws from UFPB and a PhD from the University of Geneva. She is specialised in the field of environmental law, with particular regard to the right to water.

Francesco Francioni
francesco.francioni@eui.eu
Francesco Francioni is Professor Emeritus of International Law and Co-Director of the Law Academy at the European University Institute (EUI), Florence, Italy. He is also part-time Professor of International Law at LUISS University, Rome, and Member of the Institut de droit international. Prior to joining the EUI, he was Professor of International and European Law and Director of the International Peace Studies Centre at the University of Siena, Italy.

Tineke Lambooy
t.lambooy@nyenrode.nl; t.e.lambooy@uu.nl
Tineke Lambooy is Professor at Nyenrode Business Universiteit (Centre for Entrepreneurship, Governance and Stewardship) and Associate Professor at the Law School of Utrecht Universiteit (Centre for Water, Oceans and Sustainability Law). Her areas of interest include corporate law and sustainability, corporate social responsibility, and public and corporate governance.

Federico Lenzerini
federico.lenzerini@unisi.it
Federico Lenzerini is PhD and Professor of International Law at the University of Siena, Italy. He is also Professor in the LLM Programme on Intercultural Human Rights at St Thomas University School of Law, Miami, Florida, US. He is a Consultant to UNESCO and Rapporteur to the International Law Association's Committee on the Implementation of the Rights of Indigenous Peoples.

Bronwen Leroy
bjackman@une.edu.au
Bronwen Leroy is a Lecturer in Law at the University of New England, Australia. She teaches and researches in the areas of environmental law, human rights and international law.

Bridget Lewis
b.lewis@qut.edu.au
Bridget Lewis is a Lecturer in the Faculty of Law at Queensland University of Technology, Brisbane, Australia. Bridget's principal research interests relate to international human rights law, particularly the area of environmental human rights. She has a PhD from Monash University on 'The Human Right to a Good Environment in International Law and the Implications of Climate Change'.

Benoît Mayer
bmayer@u.nus.edu
Benoît Mayer, Master of Laws (McGill), MA Political Sciences (Sciences Po), is Associate Professor at the Institute of International Law of Wuhan University, PRC. He is also a Researcher at the Centre for International Sustainable Development Law at the National University of Singapore. In 2010, he received the CISDL-IDLO Award of Excellence in Legal Scholarship on Sustainable Development.

Cameron Moore
cmoore6@une.edu.au
Cameron Moore is a Senior Lecturer in Law at the University of New England, Australia. His work focuses on international law, environmental law, and the law of the sea. Prior to joining UNE, he was a Navy Legal Officer.

Faustin Ntoubandi
faustin.ntoubandi@mail.utoronto.ca
Faustin Zacharie Ntoubandi is a Visiting Scholar at the Law Faculty of the University of Toronto. Prior to that, he was Assistant Professor in Public Law, International Law, and European Law at Justus-Liebig Universität, Giessen, Germany. He also taught at the University of Wisconsin in Madison, US.

Hanneke Palm
hannekepalm@gmail.com
Hanneke Palm is a Legal Advisor in the field of human rights law at the Dutch Ministry of Security and Justice. In 2012, she completed a Legal Research Master at Utrecht Universiteit, with a thesis on 'The Right to a Clean and Healthy Environment and Corporate Activities: the State's Responsibility and Its Positive Obligations'.

Erika Piergentili
erikapiergentili@gmail.com
Erika Piergentili, JD, University of Siena, Italy, is an Intern at the Court of First Instance in Florence, Italy. She has actively participated in various research programmes in the field of international law, particularly concerning human rights and culture.

Ottavio Quirico

oquirico@une.edu.au; ottavio.quirico@eui.eu

Ottavio Quirico is a Senior Lecturer in the School of Law at the University of New England, Australia. Prior to joining UNE, he was, inter alia, a Marie Curie Fellow at Université Panthéon-Assas, Paris, France, Lecturer at Université Lille Nord de France and at the UN Academia in Turin, Max Weber Fellow at the European University Institute, Florence, Italy, and Visiting Fellow at the Lauterpacht Centre for International Law, Cambridge, UK.

Anna Riddell

a.riddell@biicl.org

Anna Riddell is Publications Editor at the British Institute of International and Comparative Law in London, UK, while finishing her PhD at the European University Institute in Florence, Italy. Her research focuses on the connection between international environmental protection and human rights.

Marcel Szabó

szabo.marcel@jak.ppke.hu; szabo.marcel@ajbh.hu

Marcel Szabó is the Chair of the European Law Department at Péter Pázmány University and Ombudsman for Future Generations in Budapest, Hungary. He appeared before the ICJ in *Gabčíkovo-Nagymaros* and was a Visiting Fellow at the Lauterpacht Research Centre for International Law and at the Centre for European Legal Studies in Cambridge, UK.

Valentina Vadi

v.vadi@lancaster.ac.uk

Valentina Vadi is Professor in International Economic Law at Lancaster University, UK, where she was formerly a Reader (2013–15). Prior to that, she was Emile Noël Fellow at New York University (2013–14) and Marie Curie Fellow at Maastricht University (2011–13). Her main areas of research are international economic law and international cultural law.

Same Varayudej

svarayud@une.edu.au

Same Varayudej is a Lecturer at the University of New England, Australia. He also taught at the University of Sydney and at the University of New South Wales, practised for several years as a barrister, and acted for the Council of State of Thailand. His research focuses on international and trade law.

Detailed contents

Acknowledgements

Our warmest thanks go to all authors for their critical and timely contribution to this volume.

We wish to express appreciation to those who provided assistance and suggestions, namely Paul Akon, Ognjen Aleksic, Jennifer Cant, Emmanuel Decaux, Pierre-Marie Dupuy, Francesco Francioni, Deborah Hey, Tamsyn Hopkins, Susan Leaper, Bronwen Leroy, John Morss, Anna Riddell, Mark Sapwell, and Amy Wheeler.

We are also grateful for the support of the European Commission, Université Panathéon-Assas (Paris II), especially the Centre de recherches sur les droits de l'homme, the European University Institute, notably the Law Department and Max Weber Programme, the Lauterpacht Centre for International Law, the Max Planck Institute for Comparative and International Law, the University of New England, Australia, particularly the Law School, Université du Littoral, Université de Picardie Jules Verne, and Routledge.

Ottavio Quirico and
Mouloud Boumghar
Sydney and Paris
8 June 2015

For Product Safety Concerns and Information please contact our EU representative GPSR@taylorandfrancis.com
Taylor & Francis Verlag GmbH, Kaufingerstraße 24, 80331 München, Germany

www.ingramcontent.com/pod-product-compliance
Ingram Content Group UK Ltd.
Pitfield, Milton Keynes, MK11 3LW, UK
UKHW022007180926
475701UK00003B/233

* 9 7 8 0 8 1 5 3 5 5 3 4 2 *